Nanomaterials in the Battle Against Pathogens and Disease Vectors

Emerging Materials and Technologies

Series Editor

Boris I. Kharissov

Biotribology: Emerging Technologies and Applications
T.V.V.L.N. Rao, Salmiah Binti Kasolang, Xie Guoxin, Jitendra Kumar Katiyar, and Ahmad Majdi Abdul Rani

Bioengineering and Biomaterials in Ventricular Assist Devices
Eduardo Guy Perpétuo Bock

Semiconducting Black Phosphorus: From 2D Nanomaterial to Emerging 3D Architecture
Han Zhang, Nasir Mahmood Abbasi, Bing Wang

Biomass for Bioenergy and Biomaterials
Nidhi Adlakha, Rakesh Bhatnagar, and Syed Shams Yazdani

Energy Storage and Conversion Devices: Supercapacitors, Batteries, and Hydroelectric Cell
Anurag Gaur, A.L. Sharma, and Anil Arya

Nanomaterials for Water Treatment and Remediation
Srabanti Ghosh, Aziz Habibi-Yangjeh, Swati Sharma, and Ashok Kumar Nadda

2D Materials for Surface Plasmon Resonance-Based Sensors
Sanjeev Kumar Raghuwanshi, Santosh Kumar, and Yadvendra Singh

Functional Nanomaterials for Regenerative Tissue Medicines
Mariappan Rajan

Uncertainty Quantification of Stochastic Defects in Materials
Liu Chu

Recycling of Plastics, Metals, and Their Composites
R.A. Ilyas, S.M. Sapuan, and Emin Bayraktar

Viral and Antiviral Nanomaterials
Synthesis, Properties, Characterization, and Application
Devarajan Thangadurai, Saher Islam, Charles Oluwaseun Adetunji

Drug Delivery using Nanomaterials
Yasser Shahzad, Syed A.A. Rizvi, Abid Mehmood Yousaf and Talib Hussain

Nanomaterials for Environmental Applications
Mohamed Abou El-Fetouh Barakat and Rajeev Kumar

Nanotechnology for Smart Concrete
Ghasan Fahim Huseien, Nur Hafizah A. Khalid, and Jahangir Mirza

Nanomaterials in the Battle Against Pathogens and Disease Vectors
Kaushik Pal and Tean Zaheer

MXene-Based Photocatalysts: Fabrication and Applications
Zuzeng Qin, Tongming Su, and Hongbing Ji

Advanced Electrochemical Materials in Energy Conversion and Storage
Junbo Hou

For more information about this series, please visit: https://www.routledge.com/Emerging
-Materials-and-Technologies/book-series/CRCEMT

Nanomaterials in the Battle Against Pathogens and Disease Vectors

Edited by
Kaushik Pal and Tean Zaheer

CRC Press
Taylor & Francis Group
Boca Raton London New York

CRC Press is an imprint of the
Taylor & Francis Group, an **informa** business

First edition published 2022
by CRC Press
6000 Broken Sound Parkway NW, Suite 300, Boca Raton, FL 33487-2742

and by CRC Press
2 Park Square, Milton Park, Abingdon, Oxon, OX14 4RN
CRC Press is an imprint of Taylor & Francis Group, LLC

Library of Congress Cataloging-in-Publication Data

Names: Pal, Kaushik, author. | Zaheer, Tean, author.
Title: Nanomaterials in the battle against pathogens and disease vectors /
Kaushik Pal, Tean Zaheer.
Description: First edition. | Boca Raton : CRC Press, 2022. | Includes
index.
Identifiers: LCCN 2021044470 (print) | LCCN 2021044471 (ebook) | ISBN
9780367647810 (hardback) | ISBN 9780367647834 (paperback) | ISBN
9781003126256 (ebook)
Subjects: LCSH: Medicine, Preventive. | Nanostructured materials.
Classification: LCC RA425 .P193 2022 (print) | LCC RA425 (ebook) | DDC
610.28--dc23/eng/20211115
LC record available at https://lccn.loc.gov/2021044470
LC ebook record available at https://lccn.loc.gov/2021044471

ISBN: 978-0-367-64781-0 (hbk)
ISBN: 978-0-367-64783-4 (pbk)
ISBN: 978-1-003-12625-6 (ebk)

DOI: 10.1201/9781003126256

Typeset in Times
by Deanta Global Publishing Services, Chennai, India

Contents

Preface..vii
Editors..ix
Contributors xi

Chapter 1   Introduction to Nanoscience.. 1

Ekta Poonia, Krishan Kumar, Jasbir Sangwan, Vijay Kiran,
and Brij Mohan

Chapter 2   Multifaced Applications of Nanoparticles in Biological Science 17

Jayamanti Pandit, Md. Sabir Alam, Jamilur R. Ansari,
Monisha Singhal, Nidhi Gupta, Aafrin Waziri, Kajal Sharma,
and Faheem Hyder Pottoo

Chapter 3   Synthesis Approaches for Higher Yields of Nanoparticles................ 51

Md. Sabir Alam, Md. Farhan Naseh, Jamilur R. Ansari, Aafrin
Waziri, Md. Noushad Javed, Amirhossein Ahmadi, Muhammad
Khalid Saifullah, and Arun Garg

Chapter 4   Methods for Characterization and Quantitation of Nanomaterials....83

Kuna Lakshun Naidu, Thirupathi Gadipelly, Rajakumar
Anbazhagan, Shweta Yadav, and Nisha Gautam

Chapter 5   Metal Nanoparticles against Bacteria 119

Amjad Islam Aqib, Iqra Muzammil, Saad Ahmad, Muhammad
Luqman Sohail, Ahmad Ali, Kashif Prince, Amna Ahmad, and
Hina Afzal Sajid

Chapter 6   Non-metallic Nanoparticles Eliminating Bacteria........................... 161

Muhammad Ijaz, Amjad Islam Aqib, Muhammad Muddassir
Ali, and Yung-Fu Chang

Chapter 7   Biological Nanomaterials for Toxicity of Bacteria........................... 187

Diptikanta Acharya, Sagarika Satapathy, Prasanna Kumar
Dixit, Gitanjali Mishra, Padmaja Mohanty, Jayashankar Das
and Sushma Dave

Chapter 8   Metallic and Non-metallic Nanoparticles against Viruses 205

*Aqsa Ahmad, Iqra Muzammil, Tariq Munir, Muhammad Aamir
Naseer, Amjad Islam Aqib, Muhammad Muddassir Ali,
Imran Khan Sohrani, Arslan Rasool, Ibrahim Hakki Cigerci and
Muhammad Imran Arshad*

Chapter 9   Nanomaterials against Parasites: The Developments
and the Way Forward 229

*Tean Zaheer, Sadia Muneer, Rao Zahid Abbas, Muhammad
Kasib Khan, Muhammad Imran, Amna Ahmed, Iqra Zaheer
and Nighat Perveen*

Chapter 10   Toxicity of Metal Oxide Nanoparticles in Freshwater Fish 257

Sana Aziz and Sajid Abdullah

Chapter 11   The Future of Nanoscience: Where To, Where From? 283

*Anum Haleem, Tean Zaheer, Rao Zahid Abbas, Sadia Muneer,
Kaushik Pal, Afrah Nawaz and Alisha Tahir*

Index 307

Preface

The book is a true reflection of its title, detailing the contemporary advancements and scope for research and development in nanotechnology. It has a beautiful consortium of research ideas on the applications of nanotechnology against pathogens, parasites, and disease-transmitting vectors. There's an ever-growing concern and demand on the ways to curtail antimicrobial and antiparasitic resistance. In this scenario, nanotechnology has devised a promising channel for the diagnosis, treatment, and prevention of diseases. The book highlights the background and escalating developments in nanotechnology for diseases of human, animal, and zoonotic concern.

This book offers broad coverage on insights from different aspects of nanotechnology and its applications. The unique combination of antibacterial, antiviral, and antiparasitic applications of nanomaterials in the book make it worth reading. Readers from academia and industry will find it a comprehensive guide to the synthesis, analysis, and applications of nanomaterials. Based on the originality of viewpoints, schematics, and future research directions, this book is also recommended for high school, graduate, and nontechnical learners of nanotechnology. The learner is expected to grasp the basic and applied concepts of in vitro and in vivo laboratory models concerning nanoparticle study.

The editors acknowledge the encouragement and motivation from academic and industry colleagues in writing this book. They also gratefully acknowledge the support of their families throughout the writing journey. The assistance of Taylor & Francis in editing this book is also well-acknowledged.

Editors

Kaushik Pal

Prof. (Dr.) Kaushik Pal is an Indian citizen. He is a former Distinguish Research Professor at Laboratório de Biopolímeros e Sensores, Instituto de Macromoléculas, Universidade do Rio de Janeiro, Brazil. Most recently, he received Honorìs Cauśa Doctor of Science (DSc) award from the Ministry of Education, Govt. of Malaysia, at Selangor Higher National Youth Skills Institute (IKTBN) Sepang, 2020. He served as a visiting professor at School of Energy Materials, Mahatma Gandhi University, Kottayam, Kerala, and as a full-time research professor in the Department of Nanotechnology at Bharath University (BIHER), Chennai. He completed his Doctorate in Philosophy (PhD) in Physics (Expt. Materials Science and Nanotechnology) from the University of Kalyani. He has received most significant prestigious awards such as Marie-Curie Experienced Researcher (Postdoctoral Fellow) offered by the European Commission in Greece and Brain Korea National Research Foundation Visiting Scientist Fellowship in South Korea during his research career. He worked as senior Postdoctoral Fellow at Wuhan University, China, and within a year achieved the prestigious position of Chief-Scientist and Faculty Fellow offered by Chinese Academy of Science. His research focuses are: nanofabrication, chemical nanoengineering, solid-state condensation chemistry, renewable materials, energy materials, functional materials, CNTs/graphene, liquid crystalline optical materials, polymeric nanocomposite, switchable device, spectroscopy and electron microscopy, bio-inspired materials for nano-biochemistry, drug delivery, tissue engineering, cell culture and integration, switchable device modulation, flexible and transparent electrodes, supercapacitor, optoelectronics, green chemistry, green chemistry, and sensor applications. He supervises a significant number of bachelors, masters, PhDs and postdoctoral scholars' thesis. His outstanding research findings 120 articles in several international top-tier journals. Prof. Pal is an expert group leader as well as an associate member in various scientific societies, organizations, and professional bodies. He was a recognized chairperson/convener of 36 national or international events, symposiums, conferences, workshops, and summer internship programs, and has contributed around 25 plenary, 30 keynote, and 32 invited lectures worldwide.

Tean Zaheer

Tean Zaheer earned her doctor of veterinary medicine (DVM) degree from the University of Agriculture Faisalabad, Pakistan. She pursued MPhil in parasitology from the same institution. She has been working on alternative control options for disease vectors and vector-borne pathogens of animal, human, and zoonotic concern. Her areas of research interest include genome engineering, nanotechnology, vaccinology, and immunology. She has been actively involved in writing, conducting, and moderating training sessions on the prior-stated technical subjects in her field of study and research. Currently, she is working as a technical editor to the largest veterinary periodical of her country: *The Veterinary News and Views*. She has presented her work on several national and international forums in the form of extension articles, presentations, book chapters, and journal articles. She is an active member of Genetics Society of America, American Society of Microbiology, World's Poultry Science Association, British Ecological Society, British Society for Parasitology, Biological Survey of Canada, and American Mosquito Control Association, among others.

Contributors

Rao Zahid Abbas
Department of Parasitology
University of Agriculture
Faisalabad, Pakistan

Sajid Abdullah
Department of Zoology, Wildlife and
    Fisheries
University of Agriculture
Faisalabad, Pakistan

Diptikanta Acharya
Department of Biotechnology
GIET University
Rayagada, Odisha, India

Amirhossein Ahmadi
Pharmaceutical Sciences Research
    Centre
Faculty of Pharmacy
Mazandaran University of Medical
    Sciences
Sari, Iran

Amna Ahmad
Department of Parasitology
Faculty of Veterinary Science
University of Agriculture
Faisalabad, Pakistan

Aqsa Ahmad
Institute of Microbiology
University of Agriculture
Faisalabad, Pakistan

Saad Ahmad
Lanzhou Institute of Husbandry
    and Pharmaceutical Sciences
Lanzhou, China

Ahmad Ali
Department of Medicine,
Cholistan University of
    Veterinary and Animal
    Sciences
Bahawalpur, Pakistan

Muhammad Muddassir Ali
Institute of Biochemistry and
    Biotechnology
University of Veterinary and Animal
    Sciences
Lahore, Pakistan

Md. Sabir Alam
SGT College of Pharmacy
SGT University
Gurugram, Haryana, India

Rajakumar Anbazhagan
Section on Molecular
    Endocrinology
Division of Developmental
    Biology
Eunice Kennedy Shriver
    National Institute of Child
    Health and Human
    Development
National Institutes of Health
Bethesda, Maryland

Jamilur R. Ansari
Department of Humanities and
    Applied Science
Dronacharya College of
    Engineering
Gurugram, Haryana, India

Muhammad Imran Arshad
Institute of Microbiology
University of Agriculture
Faisalabad, Pakistan

Amjad Islam Aqib
Department of Medicine
Cholistan University of Veterinary and
    Animal Sciences
Bahawalpur, Pakistan

Muhammad Muddassir Ali
Institute of Biochemistry and
    Biotechnology
University of Veterinary and Animal
    Sciences
Lahore, Pakistan

Sana Aziz
Department of Zoology, Wildlife and
    Fisheries
University of Agriculture
Faisalabad, Pakistan

Yung-Fu Chang
Department of Population Medicine and
    Diagnostics Sciences
Cornell University
Ithaca, New York

Ibrahim Hakki Cigerci
Department of Molecular Biology and
    Genetic, Afyon Kocatepe University
Afyonkarahisar, Turkey

Jayshankar Das
SOA University
Bhubaneswar, Odisha, India

Sushma Dave
Jodhpur Institute of Engineering and
    Technology
Jodhpur, Rajasthan, India

Prasanna Kumar Dixit
Department of Zoology
Berhampur University
Berhampur, Odisha, India

Thirupathi Gadipelly
Department of Physics, Basic Sciences
    and Humanities
Madanapalle Institute of Technology and
    Sciences
Andhra Pradesh, India

Arun Garg
Department of Pharmacy
School of Medical and Allied Sciences
K R Mangalam University
Gurgaon, Haryana, India

Nisha Gautam
School of Physics
University of Hyderabad
Hyderabad, Telangana, India

Nidhi Gupta
Department of Biotechnology
IIS (Deemed to be University)
Jaipur, Rajasthan, India

Anum Haleem
Department of Chemistry
Minhaj University
Lahore, Pakistan

Muhammad Ijaz
Department of Veterinary Medicine
University of Veterinary and Animal
    Sciences
Lahore, Pakistan

Muhammad Imran
Department of Parasitology
University of Agriculture
Faisalabad, Pakistan

Md Noushad Javed
Department of Pharmacy
School of Medical and Allied Sciences
K R Mangalam University
Gurgaon, Haryana, India

Krishan Kumar
Physical Chemistry Research Laboratory
Department of Chemistry
Deenbandhu Chhotu Ram University of
    Science & Technology
Murthal, Haryana, India

Vijay Kiran
Department of Chemistry
C.R.A. College
Sonipat, Haryana, India

Muhammad Kasib Khan
Department of Parasitology
University of Agriculture
Faisalabad, Pakistan

Gitanjali Mishra
Department of Zoology
Berhampur University
Berhampur, Odisha, India

Brij Mohan
School of Science
Harbin Institute of Technology
    (Shenzhen)
Shenzhen, China

Padmaja Mohanty
Department of Botany
Gujarat University
Ahmedabad, Gujarat, India

Iqra Muzammil
Department of Clinical Medicine and
    Surgery
University of Agriculture
Faisalabad, Pakistan

Sadia Muneer
Institute of Microbiology
University of Agriculture
Faisalabad, Pakistan

Kuna Lakshun Naidu
Department of Electronics and Physics
GITAM (Deemed to be University)
Visakhapatnam, Andhra Pradesh, India

Md. Farhan Naseh
University School of Basic & Applied
    Sciences
Guru Gobind Singh Indraprastha
    University
New Delhi, India

Muhammad Aamir Naseer
Department of Clinical Medicine and
    Surgery
University of Agriculture
Faisalabad, Pakistan

Afrah Nawaz
Department of Physical Chemistry
Quaid e Azam University
Islamabad, Pakistan

Kaushik Pal
Laboratório de Biopolímeros
    e Sensores
Instituto de Macromoléculas
Universidade do Rio de Janeiro
Rio de Janeiro, Brazil

Jayamanti Pandit
Women Scientist C-KIRAN IPR
National Research and Development
    Corporation
New Delhi, India

Nighat Perveen
United Arab Emirates University
Abu Dahbi, Dubai

Kashif Prince
Department of Medicine
Cholistan University of Veterinary and
   Animal Sciences
Bahawalpur, Pakistan

Ekta Poonia
Physical Chemistry Research
   Laboratory
Department of Chemistry
Deenbandhu Chhotu Ram University of
   Science & Technology
Murthal, Haryana, India

Faheem Hyder Pottoo
Department of Pharmacology
College of Clinical Pharmacy
Imam Abdul Rahman Bin Faisal
   University
Dammam, Saudi Arabia

Arslan Rasool
Department of Biochemistry
University of Agriculture
Faisalabad, Pakistan

Muhammad Khalid Saifullah
Faculty of Pharmacy
Department of Pharmaceutical
   Chemistry
Umm Al-Qura University
Makkah, Saudi Arabia

Jasbir Sangwan
Department of Chemistry
Tau Devi Lal Government College for
   Women
Murthal, Haryana, India

Imran Khan Sohrani
Department of Clinical Medicine and
   Surgery
University of Agriculture
Faisalabad, Pakistan

Hina Afzal Sajid
Center of Excellence in Molecular Biology
Punjab University
and
School Education Department
Government of Punjab, Punjab Pakistan

Sagarika Satapathy
Department of Biotechnology
GIET University
Rayagada, Odisha, India

Kajal Sharma
NIMS Institute of Pharmacy
NIMS University
Jaipur, Rajasthan, India

Monisha Singhal
Department of Biotechnology
IIS (Deemed to be University)
Jaipur, Rajasthan, India

Muhammad Luqman Sohail
Department of Medicine
Cholistan University of Veterinary and
   Animal Sciences
Bahawalpur, Pakistan

Alisha Tahir
Department of Physical Chemistry
Quaid e Azam University
Islamabad, Pakistan

Aafrin Waziri
University School of Biotechnology
Guru Gobind Singh Indraprastha
   University
New Delhi, India

Shweta Yadav
Department of Electronics and Physics,
   GIS
GITAM (Deemed to be University)
Visakhapatnam, Andhra Pradesh, India

Iqra Zaheer
Department of Pathology
University of Agriculture
Faisalabad, Pakistan

Tean Zaheer
Department of Parasitology
University of Agriculture
Faisalabad, Pakistan

1 Introduction to Nanoscience

*Ekta Poonia, Krishan Kumar, Jasbir Sangwan,
Vijay Kiran, and Brij Mohan*

CONTENTS

1.1 Introduction 1
1.2 Basic Terminology in Nano 2
1.3 Historical Milieu of Nanoscience 3
1.4 Significance of Nanoscale 4
1.5 Incredible Innovations in Nanoscience 5
    1.5.1 High-Resolution Transmission Electron Microscope (HRTEM) 5
    1.5.2 Field Emission Scanning Electron Microscope (FESEM) 6
1.6 Applications of Nanoscience 7
    1.6.1 Energy 7
    1.6.2 Medicine and Healthcare 9
    1.6.3 Environmental Remediation 10
    1.6.4 Nanosensors 11
1.7 Future Aspects of Nanoscience 12
    1.7.1 Nanoscience in Futuristic Gadgets 12
    1.7.2 Nanoscience in Infectious Diseases 13
1.8 Conclusion 13
References 14

1.1 INTRODUCTION

In recent years, a small word with enormous potential has been quickly forcing itself into the world's cognizance; the word is *nano*. The study of nanomaterials has made extraordinary strides in the most recent decades. The prefix *nano-* signifies one-billionth, for example, a nanometer is one-billionth of a meter. Nanoscience is the investigation of principal properties of particles and structures generally in the range of 1 to 100 nm [1]. The nanoworld plays a key role between a solid and an atom, from the enormous particle or the little strong item to the solid connection along with surface and volume. Carefully, the nanoplanet has subsisted for quite a while and it is dependent upon scientific experts to examine properties and configurations of particles. Nanoscience is the assessment and exploitation of minuscule matter and it can be employed in a wide range of scientific fields, including chemistry, materials science, biology, physics and engineering [2].

DOI: 10.1201/9781003126256-1

Nanoscience is the science in which the material science merges along with the science of physical and organic frameworks. The cosmos of atoms (where average nanogroups may have from hundreds to an immense amount of atoms) is ruled by quantum mechanics, a place where nuclear materials science merges with physical science and the science of multifaceted frameworks. The universe of the atoms ruled by quantum mechanics, and the average nanosystem may include hundreds to a vast number of molecules. Inside nanostructures, we have included along with quantum mechanics, measurable activity of a huge assortment of connecting particles. From this combination of quantum activity and factual unpredictability, numerous wonders arise [3]. They range from the array of nanoscale materials science to substance responses to biological progressions. The estimation of this important activity is upgraded when one understands that the complete number of atoms in the frameworks is still small enough that numerous issues in nanoscience are agreeable to present-day computational procedures. The practicality of the nanometer scale is wide ranging, regardless of whether it is in physical science, materials science, life science or chemistry. Similarly as significant as the innovation prospects, in my view, is the binding together of logical thoughts at the core of nanoscience. This book tries to assemble these thoughts and exhibit their application in numerous disciplines.

We will begin with innovation, that is, those applications that stem from the capacity to control materials on the nanometer scale. We will then proceed to look at the scientific phenomena that govern nanoscale frameworks. To welcome the mechanical ramifications of working at the nanoscale, one must value the extraordinary scale contrast between our standard microscopic world and the nuclear world. There are brilliant websites that permit the user to zoom in from galactic scales to subatomic scales by venturing through variables of ten in size. The quintessence of nanoscience is the formation and utilization of particles, subatomic congregations, materials and gadgets in the range of 1 nm to 100 nm, and exploitation of the novel assets and observable facts in this range [4].

1.2   BASIC TERMINOLOGY IN NANO

Nanoscience is an investigation of structures in the range of 1 nm to 100 nm in size. Nanoscience is concerned about materials and frameworks whose structure segments display novel and essentially enhanced chemical, biological and physical properties, phenomena and succession due to their nanoscale size. The structural highlights in the range of ~10^{-9} to 10^{-7} m, for example 1 nm to 100 nm, reveal significant changes when contrasted with the conduct of segregated atoms of mass materials. It is an interdisciplinary science including chemistry, biology, physics, materials science, software engineering and so forth [5–8].

Nanoscience utilizes properties of matter at a size greater than atomic size or the length of a usual chemical bond yet smaller than the size of a cell. The size of a chemical bond is short of 1 nm. Change in the surface-to-volume proportion is the reason for the change in the physical properties. The proportion of surface area to volume increases at the nano scale. Researchers are attempting to utilize this trait to manufacture applications. The term *nanotechnology* is nowadays frequently perceived to include both nanoscience and the wide scope of innovations/applications that work at the nanoscale [9–11].

Nanoscience – The science where the properties differ essentially from those in bigger ranges with analysis of materials and phenomena at nuclear, atomic and macromolecular scales.

Nanotechnology – Plan, portrayal, fabrication and use of structures, gadgets and frameworks by controlling shape and size at the nanoscale.

Nanomaterials – Particles, quantum dots, nanowires, nanotubes, fullerenes and so forth that exist in the size of 100 nm or less, or that have at any rate one dimension that influences their useful conduct at this scale.

1.3 HISTORICAL MILIEU OF NANOSCIENCE

Nano, represented by "n" as a prefix, implies 10^{-9}, signifying a feature in a decimal standard for measuring. This is a derivative of the Greek word ῖανος, which means "dwarf", and was formally settled as a norm in 1960. Gordon Moore, head of Intel Corporation, in 1965, anticipated that the number of transistors that could be fitted in a given region would be twofold in number every 1.5 years for the next decade [12–13]. This phenomenon is famously known as Moore's law. We have seen this in critical expansion of housing transistors from several thousand transistors in Intel 4004 processors in the year 1971 to more than 700,000 transistors in the Intel Core 2 Duo PC motherboard. It appears that there has been a dramatic decline in the size of some electronic devices in contemporary hardware, from millimeters to nanometers. Simultaneously, biochemistry, applied chemistry and atomic hereditary qualities networks are moving in a similar fashion [14].

At the beginning of another developmental era, a surprisingly different development came very close to the new nanoscale with the guarantee of changing scientific fields and gadgets [15]. The imminent atomic field of bionanotechnology opens new avenues from critical experiments in biology and subatomic biology to biosensing, bioinformatics, biolabeling, pharma and medicine, genomics, data retention, as well as power switch applications, and furthermore in figuring [16]. We refer to a lecture given by scientist Richard Feynman, a Nobel Prize winner, at the American Physical Society conference on December 29, 1959, titled "There's Plenty of Room at the Bottom" [17]. The term *nanotechnology* first came to light through the renowned researcher Norio Taniguchi of the University of Science, Tokyo, in a 1974 paper [18]. In the mid-1980s, experimenting with nanotechnology was simpler than deterministic. The nanoscale phenomenon was proposed by Dr. K. Eric Drexler for the direct treatment of certain particles and atoms by deception [19]. Educational information on nanotechnology in life was provided by R. Jones in a book "Soft Machines" or nanotechnology and health. The concept of DNA nanotechnology was invented by a crystallographer, Nadrian Seeman, in 1980, which showed a cross-section of 3-D DNA to identify molecules [20]. In 1991, the Seeman's laboratory published a compilation of 3-D DNA, the first 3-D nanoscale, for which he was awarded in 1995 with the Feynman Prize in Nanotechnology [21]. Even though nanotechnology has magnificent future applications, it raises numerous issues, as is normal for any innovation, including concerns about the harmful effect of nanoresidues on the climate and the worldwide monetary impacts [22]. This gave legislatures misgivings and lawmakers debated the need for guidelines for nanotechnology.

1.4 SIGNIFICANCE OF NANOSCALE

Nanoscience refers directly to any technique performed on a nanoscale that actually has applications. Nanoscience joins the formation and application of chemical, physical and natural structures on scales ranging from individual particles or atoms to submicron scales as a mixture of subsequent nanostructures into larger structures [23–24]. Nanoscience can be thought of as having a profound impact on our economy and society in the mid 21st century, such as that of information technology, new semiconductor innovations or the science of cells and subatoms. Areas of scientific research and innovation with the possibility of nanoscience advancing include building and integrating materials, medicine and medical care, energy, biotechnology, information technology and public safety [25]. It is widely reported that nanoscience will be the next technological explosion.

Nanometer-scale structures are built primarily from their natural components. Chemical fusion is unrestricted self-assembly of subatomic groups from basic reagents in solutions or organic particles used as components for the construction of three-dimensional nanostructures, including quantum dots. An assortment of vacuum affirmations and nonequilibrium plasma chemistry methods are used to deliver cooled nanotubes and nanocomposites. Cell-controlled structures are made using subatomic beam epitaxy and organometallic fume stage epitaxy. Micro- and nanosystem components are made using non-lithographic creation processes and range in size from micro- to nanometers [26]. It was preceded by the development of lithography for use in the construction of nanoparticles brought to the line width as small as 10 nm in experimental models. Nanoscience, in addition to making nanosystems, gives impetus to the continuation of experiments with computer equipment.

Micro- and nanosystems contain micro or nanomicro mechatronics, frame reconciliation, optoelectronics, microfluidics and electromechanical components. These structures can detect, control and operate on a micro-/nanoscale and operate separately or on displays to produce macroscale results [27–28]. Because of the legitimate view of these structures and the significant impact, they can have applications for business and defense, investors, industries, just as central governments have made clear consideration in supporting development in the sector. Micro-/nanosystems will likely be the next logical next step in the silicon revolution [29]. Science and innovation continue to advance in the development of micro or nano gadgets and frameworks that will enable the realization of equipment, consumer and biomedical applications. A wide range of microelectromechanical systems (MEMS) gadgets have been created, some of which are financially viable. Sensor assortments are used in various programs such as biomedical, customer and mechanical [30]. Different microstructures or microsegments are used in small tools and other modern applications, for example, micromirror displays. Combined with the capacitive type, silicon accelerometers are used in the charging of airbags in cars. Other key applications include optical switching, pressure sensors and inkjet printer heads [31]. Silicon-based pressure sensors for pressurizing high-pressure motor vehicles were shipped in 1991 by Nova-Sensor and their annual deals were 25 million units in 2002 [32]. Annual deals for inkjet printer heads with usable microscale components were 400 million units in 2002. Pressure sensors capable of measuring tire pressure were supplied by Motorola. Different uses of MEMS gadgets include gas sensors, integrated sensors, infrared detectors, intermediate aerial displays, space science applications and rocket protection, picosatellites for space applications, and many operational presses, pneumatic and other consumer items [33]. MEMS gadgets are also increasingly sought after in attractive power structures, where they are designed for the availability of large recordings and ultrahigh sizes. Integrated configuration gadgets are included for contact recording programs. For high-speed data transmission, servo-controlled microactuators are produced by ultrahigh tracking systems, which fill the control component of a two-component, fine servo frame integrated with a standard actuator. Millimeter-sized engines and actuators of tips-based account systems have also been developed [34]. Figure 1.1 shows the significance of nanoscience in different areas.

1.5 INCREDIBLE INNOVATIONS IN NANOSCIENCE

The wave properties of electrons are utilized by electron microscopy. Without resistance, as particles, they need a vacuum to move. The magnifying lens is as a metal fence enclosed in an area where the following are found, for example, an electron gun in the cathode beam tubes used in TVs.

Various electronic components, for example, electromagnetic focused lenses and control electron indices act as an aid to the object to be tested. Two types of electron microscopes are described next.

1.5.1 HIGH-RESOLUTION TRANSMISSION ELECTRON MICROSCOPE (HRTEM)

The development of the first-ever high-resolution transmission electron microscope (HRTEM) took place almost 20 years before the scanning electron microscope (SEM). In 1932, Knoll and Ruska described an instrument that was able to take photographs by focusing electrons through magnetic lenses [35]. At first, the images were of poor quality, but due to advancement in techniques, these modern-day microscopes can easily take images of every single atom present within the crystalline lattices of a cell. The productive operation of an HRTEM depends on the thickness of the sample. While preparing the sample one must take the utmost care with the thickness of the film (should not exceed a few hundred nanometers) [36]. The challenge for operating an HRTEM lies in sample preparation. If a prepared film is very thick or contains clumps, then the projected beam of electrons will not be able to pass entirely through the film section. Partially for this reason, the operating beam voltage of an HRTEM is higher than that of field emission scanning electron microscope and falls between 100 and 400 kilovolts.

Comfort and safety of life	Industry	Technological progress control
Air conditioning instrument	Diagnostic quality control	Pharmaceuticals industry
Weather forecasting	Agriculture	Cryogenic processing
Health and medical cares	Food	Electronic technology
Inflammable-gas instrument	Chemicals	Food technology
Civil engineering	Semiconductors	Chemical processing
Aeronautical safety	Petroleum products	Energy
		Photography

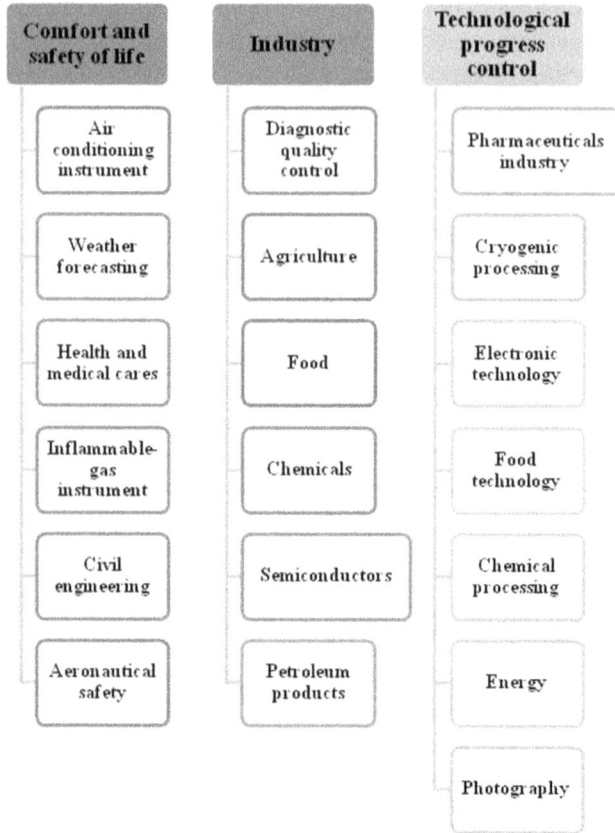

FIGURE 1.1   Significance of nanoscience in different areas.

In this type of microscope, similar to X-beams, the beam interacts with a flexible sample and creates a deviation image or image with a wide range. Investigation of the posterior figure gives us the ability to examine the atomic structure of a damaged sample. The related number of electrons in this way and their energy shows the ultimate goal. The most notable machines operate at voltages instead of a large number of volts.

1.5.2 FIELD EMISSION SCANNING ELECTRON MICROSCOPE (FESEM)

The field emission scanning electron microscope (FESEM) is a type of field emission electron magnifying lens that envisions a test surface by checking it through a large vitality emission of light in a raster examines design. Discussions about the first-ever SEM are found in the works of Manfred von Ardenne. He discussed various possibilities and theories of SEM instrument construction with different scanning modes along with techniques related to achieving highly magnified results [37]. Generally, the SEM was invented after the transmission electron microscope (TEM), but it took almost 30 years to be developed into a fully practical tool for medical as well as scientific research. It was the Cambridge Scientific Instrument Company that developed the first-ever commercial SEM in 1965. The FESEM is the most widely used among all electron microscopes because of its versatility. It is a mapping device that provides complete details about the topography of the material under study by detecting reflected electrons from the surface of a material with help of multiple detectors [38–39]. The FESEM is user-friendly, and micrographs generated by it can be easily interpreted with excellent spatial resolution.

The surface of the sample under examination is scanned with an electron shaft. The size of the scanned area depends on the level of desired magnification. The interaction between electrons and the sample provides an increase in the various signals that when collected and tested combine the image of the observed sample face without using a numerical process, unlike the TEM process [40]. The purpose of this type of metal, which is restricted by mechanical design, gives researchers the ability to detect atoms (one-tenth of a nanometer). The biggest drawback of this magnifying glass, similar to the TEM case, is that it needs a vacuum. Samples should be set up in a certain way, and at the end of the day they should be smeared, cooled and cut in pieces. Some years of SEM have hit this limit; they are known as electron scanning microscopes. These scanned microscopes give researchers the ability to see objects in their natural state. The difference between the electron magnifier and the standard amplifier, which requires a high vacuum at all degrees of the components that make up the magnification tool, is that the sample resides at a certain pressure due to the differential diaphragm siphon frame used in the viewing chamber.

1.6 APPLICATIONS OF NANOSCIENCE

Nanoscience is a rapidly evolving field with the critical potential to provide another era of both scientific and innovative approaches as well as clinical instruments and gadgets. Nanoscience engineering is the construction and use of particles, atomic alloys, building materials and gadgets at a scale of 1–100 nm. Nanoscience works to significantly improve, and even transform, many new and industrial areas: data innovation, national security, medicine, transportation, energy, sanitation and environmental science, among many others. Figure 1.2 shows the various applications of nanoscience. Following is an examination of the rapid application of nanoscience.

1.6.1 Energy

Nanoscience finds application in general energy sources, and a variety of energy sources are being developed to help integrate the world's increasing energy needs. Many researchers are exploring ways to grow more comprehensive, rational and sustainable energy sources, with the aim of reducing energy consumption and reducing climate problems.

- Nanoscience restores the product of the creation of fuel in crude oil products through better catalysis. It also enables reduced fuel consumption in cars and compels crops through the use of high-tech lighting and reduced scrubbing.

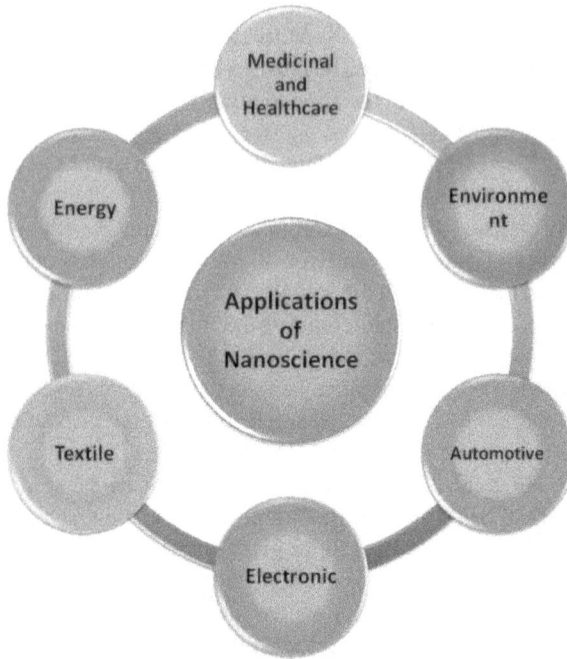

FIGURE 1.2   Applications of nanoscience.

- Nanoscience is also used in the extraction of oil and gas by, for example, the use of nanotechnology valves in marine operations or the use of nanoparticles to separate small cracks in oil pipelines.
- Analysts test carbon nanotube "scrubbers" and films to separate carbon dioxide from the power plant.
- Experts make cables containing carbon nanotubes that will have much lower resistance than high-pressure cables currently used in electrical wiring, reducing the risk of power transmission.
- Nanoscience can be integrated into solar panels to convert sunlight into completely electrical energy, promoting cheap solar energy over time. Solar-based cells are less expensive to make and easier to launch because they can use assembly methods such as printing and can be made on flexible rolls instead of separate boards. More recent research suggests that future sun powered converters may even be paintable.
- Nanoscience is currently being used to create new types of batteries that are faster, more efficient, lighter, have a much larger size and last longer.
- Epoxy-containing carbon nanotube is used to make the edges of the wind blower longer, heavier and lighter than the other cutting edges to increase the amount of energy that can be produced by the wind.
- In the field of energy collection, experts are developing solar panels with thin films that can be installed on compact PCs and piezoelectric nanowires

wicker in clothing to create instant energy from light and body heat in addition to controlling portable electronic gadgets. Also, a variety of nanoscience-based options are needed after switching to waste heaters for PCs, homes, power plants, cars, etc., in electric utilities.

- Energy proficiency and energy-saving technologies are growing in number and types of uses. In addition to the aforementioned, nanoscience empowers very efficient light structures, lightweight vehicle-based materials based on tow, low power consumption of low-quality hardware, and savvy coating that responds easily to glass.

1.6.2 MEDICINE AND HEALTHCARE

Nanoscience currently expands the clinical instruments, information and treatments currently available to medical professionals. Nanomedicine, the use of nanoscience in treatment, attracts miraculous biological sizes to deliver specific solutions against infection, determination and treatment. Following are a few examples of ongoing developments.

- Nanoscience is studied for the treatment of atherosclerosis, or plaque formation, in arteries. Alternatively, scientists are developing a nanoparticle that mimics the body's "acceptable" cholesterol, known as HDL (high density lipoprotein), which helps detect plaque.
- Industrial applications have replaced gold nanoparticles as experiments with nucleic acid-based preparations, and gold nanoparticles are also being clinically studied as a treatment for cancer and various diseases.
- Better thinking and nanoscience-enabled imaging devices that prepare for previous analysis, more individualized treatment alternatives, and better recovery rates.
- Nanoscience specialists are taking a shot with a variety of drugs where the nanoparticle can mimic or help transfer the drug directly to cancer cells and reduce the risk of solid tissue damage. This could change the way professionals treat disease and significantly reduce the toxic effects of chemotherapy.
- The aim and design of the dynamic edges of the nanopore materials can allow for the improvement of novel quality sequencing advances, which enables the detection of a single molecule easily and quickly through non-essential sample formation.
- Nanomedicine experts are taking steps on how nanoscience can improve techniques, including the distribution of antibodies without the use of needles. More experts are trying to develop a standard antibody vaccine platform that will cover a wide range of strains and require fewer building resources each year.
- Research into the application of nanoscience to regenerative medicine traverses several regions of the application, including bone and neural tissue formation. For example, novel objects may be designed to mimic the mineral composition of human bone or be used as a laughing gas for dental

applications. Scientists are exploring ways to build more complex tissue with the goal of someday developing human organs to move. Experts are also considering ways to use graphene nanoribbons to help repair spinal cord injuries; preliminary tests show that neurons grow well on the graphene surface.

- Delayed ultraviolet (UV) spotlight causes skin cancer. Sunscreen lotions with nano-TiO2 provide improved sun protection factor (SPF) while losing firmness. An additional positive area of nano skin blocks (ZnO and TiO_2) appears as it protects the skin by sitting on it instead of entering the skin. As a result, they block UV rays viably for delayed span. Also, they are straight forward; consequently hold a different skin tone while working in a way better than regular skin moisturizers.

1.6.3 ENVIRONMENTAL REMEDIATION

In addition to demonstrating the ways in which nanotechnology can help improve energy production, there are many additional ways that it can help differentiate and collect natural toxins:

- Nanotechnology can help solve the issue of clean drinking water with quick, easy access to and treatment of water pollution.
- Experts have developed a layer of thin film with nanopores to eliminate energy dissipating in water. This layer of molybdenum disulfide (MoS) filters water twice or more often than normal channels.
- Scientists have developed a nanofabric "paper towel" woven with minuscule coatings of potassium manganese oxide that can penetrate several times the weight of oil for cleaning programs. Scientists have similarly applied nanoparticles that attract water to oil reservoirs using magnets to extract oil directly from water.
- Nanoparticles are made to clean toxic equipment in groundwater using composite solutions that make toxins less harmful. This cycle was relatively inexpensive such as methods that require drainage to be treated.
- Sensors and arrangements powered by nanotechnology are currently not ready to distinguish and recognize an object or biological operator that is visible around the ground with a much higher sensitivity than at any other time. Analysts are examining particles, for example, monolayers that have accumulated on mesoporous foundations, dendrimers and carbon nanotubes to determine how to use their artificial architecture structures and actual types of harmful site layouts. Another sensor has been developed by the National Aeronautics and Space Administration (NASA) as an extension of the cell phones that firefighters can use to monitor air quality around flames.
- Most airplane cabins and various types of air channels are nanotechnologies designed for "mechanical filtration", in which fiber material creates nanoscale holes that hold particles larger than the size of the pores. Channels similarly may contain odorless charcoal.

1.6.4 Nanosensors

Nanosensors show a few particular focal points over their microscale and macroscale partners:

- Utilization of physical phenomena showing up on the nanoscale
- Reduction in cost
- Mass creation
- Reduction in the general size and weight of the related framework
- Less power utilization
- Enhanced sensitivity
- Higher degree of reconciliation
- Some programs include nanoscale frameworks for active applications, for example, the installed clinical sensors must be nanoscale

Nanoscale sensors can be programmed to rely on sensory applications such as chemical, physical or organic nanosensors [41]. Compared to sensor configurations, nanoscale sensors can also be controlled by a power transducer. Table 1.1 shows the types and measured properties of different sensors.

Despite this, nanosensors can likewise be ordered by

- Sensor element material
- Consequence or transduction phenomena
- Technological perspectives
- Measurands

Specifically, nanosensors can be identified according to the materials used, for example, nanoparticles, nanotubes, nanocomposites, nanowires, nanostructures and quantum dots. Although there are many sensors, no single sensor can adequately detect all the relevant parameters of all imaginary weather. After all, the development of

TABLE 1.1
Assorted Nanosensors

Type	Measured Property
Thermal	Entropy, flux, temperature, heat flow, specific heat
Chemical	Concentration, pH, composition, reaction rate, reduction/oxidation potential
Electrical	Resistance, current, voltage, impedance, capacitance, electric field, inductance, polarization, dipole moment, dielectric constant, frequency
Mechanical	Acceleration, pressure, mass flow, force, size, strain, velocity, stress, torque, acoustic wave, piezoelectric
Optical	Refractive index, phase, wavelength, intensity, polarization, frequency, reflectance, transmittance
Magnetic	Magnetic moment, flux density, field strength, permeability

sensory collections to integrate multiple structures in a variety of conditions is now popular. Sensor displays contain a variety of multicolored sensory combinations and many amazing simultaneous discoveries, such as the human sensory framework with the nose as a gas sensor, the eyes as the visual sensor, the ear as an acoustic sensor and the tongue as sensory fluids. This creates data acquisition and augmentation, and is used in material or chemical applications. Nanosensors and nanopowered sensors have applications in many systems, for example, transportation, correspondences, buildings and offices, medicine, welfare and public safety, including both national security and military operations. There are various types of nanowire sensors used, for example, to identify synthetics and biologics; nanosensors are placed on platelets to aid in early identification [42].

1.7 FUTURE ASPECTS OF NANOSCIENCE

Nanoscience offers a multifunctional design that will add to the design and maintenance of lighter, safer, smarter and more efficient materials, aircraft, rockets and ships. Also, nanoscience offers various purposes for improving the transportation framework.

1.7.1 NANOSCIENCE IN FUTURISTIC GADGETS

As noted earlier, engineers in automotive materials include polymer and nanocomposites components, dynamic frames, thermal control devices, advanced grounding tires, efficient/flexible sensors and gadgets, thinfilm brilliant solar panels and fuel added substances, as well as advanced smoke extraction systems and increased access. The recycling of asphalt, steel, aluminum, concrete, and various reinforcing cement materials and their recycled materials provide an unconventional possibility to enhance the display, strength and longevity of highway and transport infrastructure components while reducing their life cycle costs. New structures can incorporate creative energy into traditional products, for example, architectural structures or production or communication capacity.

Nanoscale sensors and gadgets can provide an inexpensive look at the basic reliability and design of bridges, parking structures and asphalts for long-distance travel. Nanoscale sensors, special gadgets and unique developments powered by nanoelectronics can also maintain an improved transport base that can communicate with car-based components to help drivers stay on the road, maintain a distance from accidents, change go courses to prevent blockage and improve driver-friendly connectors.

The advantage of "game modification" through the use of nanoscience-enabled light allows high-quality materials to work in almost any vehicle. For example, it has been suggested that reducing the weight of a commercial aircraft by 20% would reduce its fuel consumption by 15%. Preliminary studies conducted by NASA have shown that the flexibility of the event and the use of nanomaterials with double-quality traditional compounds could reduce the craft's heavy load by 63%. Not only will this save much of the energy expected to send a spacecraft into space, but it will also enable the development of a single platform to track launch vehicles, further reduce delivery costs, increase equipment reliability and create away for elective impetus ideas.

1.7.2 NANOSCIENCE IN INFECTIOUS DISEASES

Human respiratory infections are a foremost cause of illness and death worldwide. Among the various respiratory viruses, coronaviruses (SARS-CoV-2) have produced a major challenge and a serious health threat worldwide. Human coronaviruses tend to infect the upper respiratory tract, causing illness from common cold symptoms to severe respiratory infections. Many promising vaccines have been available since early 2021. However, successful herd protection is far from being achieved. Isolation is still the only way to fight coronavirus infection. Nanoscience has opened new horizons in many different fields of medical science, such as targeted genetic delivery, targeted drug delivery, biosensor platforms, imaging and diagnostics [43–44]. Nanomaterials have been developed to fight germs, viruses and fungal diseases [45] due to their unique environmental characteristics, such as surface area, nanoscale size, and easily accessible modification of the surface. These frameworks empower scientists to improve drug pharmacokinetics, control drug withdrawal, strengthen drug density, facilitate cell mobility and improve the availability of antiretroviral drugs, such as antibodies for herpes simplex virus and hepatitis B virus [46–47]. Nanomaterials are promising tools for the diagnosis and treatment of COVID-19. Nanoscience-based technologies offer many roles in the fight against coronavirus infection, such as nanosensors, nano-based vaccines and smart drugs.

1.8 CONCLUSION

The distinguishing feature of nanomaterials is the presence of a high concentration of imperfections. *Nano* refers to a Greek word that means something small or something very small. It is the one-billionth unit. Nanomaterials refer to the class of objects within the nanometric range. They can be polymers, composites, ceramics or metals. Nanomaterials show interestingly extraordinary chemical, physical and mechanical properties contrasted with mass materials. Various graphic design machines have been developed over the years and have helped to understand the effectiveness of nanomaterials and nanostructures. Novel objects can be made with a few geometric designs including bars, holes, cords, shells, tubes and horns. A huge number of nanoparticles and gadgets have been developed. Nanoscience will undoubtedly have a great affect because of the enormous scope of applications in industry and everyday items. Even though nanomaterials have been involved in causing well-being concerns, definitive data isn't accessible as of now. There is a need for administrative bodies to set rules and safe cutoff points for guaranteeing appropriate use of nanoscience without causing human or environmental harm. To validate their positive potential, nanotechnological techniques and items must be assessed for focal points in contrast with traditional other options. The dangers related to nanoscience must be considered. It is visualized that nanoscience will prompt minuscule automated gadgets, utilizing nanoelectronics and sensors for observing and diagnosing problems within the human body. Regardless of the numerous guarantees that nanoscience advocates have been making about the capacity of nanotechnology to take care of our health and create more feasible products, barely any projects have been conveyed to date. Many thought-provoking techniques in the area of water treatment and environmental rehabilitation/anointing treatment are in the testing stage. Globalization of this could take five to ten years after approval. Significantly, a large number of these substances or methods are performed without proper concern for environmental, social and safety issues. As the field of nanotoxicology is slowly gaining momentum, the stressful indicators of human and environmental toxicity are increasing. This gives NGOs the impetus to request a way to deal with the commercialization of these items.

REFERENCES

1. Ventra M, Evoy S, Heflin JR, editors. *Introduction to Nanoscale Science and Technology.* Springer Science & Business Media; 2006 Apr 11.
2. Brune H, Ernst H, Grunwald A, Grünwald W, Hofmann H, Krug H, Janich P, Mayor M, Rathgeber W, Schmid G, Simon U. *Nanotechnology: Assessment and Perspectives.* Springer Science & Business Media; 2006 Jun 23.
3. Afzali A, Maghsoodlou S. *Nanoelement Manufacturing: Quantum Mechanics and Thermodynamic Principles, Engineering Materials: Applied Research and Evaluation Methods.* CRC Press; 2014 Nov 24:67.
4. Schummer J, editor. Interdisciplinary issues in nanoscale research. *Discovering the Nanoscale,* Amsterdam: IOS Press; 2004:9.
5. Mamalis AG. Recent advances in nanotechnology. *Journal of Materials Processing Technology.* 2007 Jan 1;181(1–3):52–8.
6. Goyal MR. *Scientific and Technical Terms in Bioengineering and Biological Engineering.* CRC Press; 2018 Jan 3.
7. Porter AL, Youtie J. How interdisciplinary is nanotechnology? *Journal of Nanoparticle Research.* 2009 Jul;11(5):1023–41.
8. Tarafdar JC, Sharma S, Raliya R. Nanotechnology: Interdisciplinary science of applications. *African Journal of Biotechnology.* 2013;12(3):219–26.
9. Ni B, Shi Y, Wang X. The sub-nanometer scale as a new focus in nanoscience. *Advanced Materials.* 2018 Oct 1;30(43):1802031(1–24).
10. Baer DR, Engelhard MH, Johnson GE, Laskin J, Lai J, Mueller K, Munusamy P, Thevuthasan S, Wang H, Washton N, Elder A. Surface characterization of nanomaterials and nanoparticles: Important needs and challenging opportunities. *Journal of Vacuum Science & Technology A: Vacuum, Surfaces, and Films.* 2013 Sep 27;31(5):050820.
11. Goddard III WA, Brenner D, Lyshevski SE, Iafrate GJ. *Handbook of Nanoscience, Engineering, and Technology.* CRC Press; 2002 Oct 29.
12. Brock DC, Moore GE, editors. *Understanding Moore's Law: Four Decades of Innovation.* Chemical Heritage Foundation; 2006.
13. Schaller RR. Moore's law: Past, present and future. *IEEE Spectrum.* 1997 Jun;34(6):52–9.
14. Gennis RB, editor. *Biomembranes: Molecular Structure and Function.* Springer Science & Business Media; 2013 Apr 17.
15. Bayda S, Adeel M, Tuccinardi T, Cordani M, Rizzolio F. The history of nanoscience and nanotechnology: From chemical–physical applications to nanomedicine. *Molecules.* 2020 Jan;25(1):112.
16. Roco MC, Bainbridge WS, editors. *Converging Technologies for Improving Human Performance: Nanotechnology, Biotechnology, Information Technology and Cognitive Science.* Springer Science & Business Media; 2013 Apr 17.

17. Feynman RP. Plenty of room at the bottom. In *APS Annual Meeting*; 1959 Dec 29.
18. Singh A, Dubey S, Dubey HK. Nanotechnology: The future engineering. *Nanotechnology*. 2019 Apr;6(2):230–3.
19. Cerofolini G. *Nanoscale Devices: Fabrication, Functionalization, and Accessibility from the Macroscopic World*. Springer Science & Business Media; 2009 Aug 26.
20. Douglas K. *DNA Nanoscience: From Prebiotic Origins to Emerging Nanotechnology*. CRC Press; 2016 Oct 14.
21. Sadowski JP. *Design and Synthesis of Dynamically Assembling DNA Nanostructures* (Doctoral dissertation, Harvard University).
22. Taghavi SM, Momenpour M, Azarian M, Ahmadian M, Souri F, Taghavi SA, Sadeghain M, Karchani M. Effects of nanoparticles on the environment and outdoor workplaces. *Electronic Physician*. 2013 Oct;5(4):706.
23. Ratner MA, Ratner D. *Nanotechnology: A Gentle Introduction to the Next Big Idea*. Prentice Hall Professional; 2003.
24. Nasrollahzadeh M, Sajadi SM, Sajjadi M, Issaabadi Z. An introduction to nanotechnology. In*Interface Science and Technology*. 2019 Jan 1;28:1–27. Elsevier.
25. Von Schomberg R, Richard Owen, John Bessant, Maggy Heintz, editors. A vision of responsible research and innovation. In *Responsible Innovation: Managing the Responsible Emergence of Science and Innovation in Society*. Wiley Online Library; 2013 Apr 26;51–74.
26. Brodie I, Muray JJ. *The Physics of Micro/Nano-Fabrication*. Springer Science & Business Media; 2013 Jun 29.
27. Ashraf MW, Tayyaba S, Afzulpurkar N. Micro electromechanical systems (MEMS) based microfluidic devices for biomedical applications. *International Journal of Molecular Sciences*. 2011 Jun;12(6):3648–704.
28. Staufer U, Akiyama T, Gullo MR, Han A, Imer R, de Rooij NF, Aebi U, Engel A, Frederix PL, Stolz M, Friederich NF. Micro-and nanosystems for biology and medicine. *Microelectronic Engineering*. 2007 May 1;84(5–8):1681–4.
29. Mahalik NP. *Micromanufacturing and Nanotechnology*. Springer; 2006 Jan 16.
30. Baltes H, Brand O, Fedder GK, Hierold C, Korvink JG, Tabata O, editors. *Enabling Technology for MEMS and Nanodevices: Advanced Micro and Nanosystems*. John Wiley & Sons; 2013 Mar 27.
31. Allen JJ. *Introduction to MEMS (MicroElectroMechanical Systems)*. Sandia National Lab.(SNL-NM); 2007 Jul 1.
32. Song P, Ma Z, Ma J, Yang L, Wei J, Zhao Y, Zhang M, Yang F, Wang X. Recent progress of miniature MEMS pressure sensors. *Micromachines*. 2020 Jan;11(1):56.
33. Liu B, Li X, Yang J, Gao G. Recent advances in MEMS-based microthrusters. *Micromachines*. 2019 Dec;10(12):818.
34. Chollet F. Devices based on co-integrated MEMS actuators and optical waveguide: A review. *Micromachines*. 2016 Feb;7(2):18.
35. Williams DB, Carter CB, editors. The transmission electron microscope. In *Transmission Electron Microscopy*. Springer; 1996;3–17.
36. Dömer H, Bostanjoglo O. High-speed transmission electron microscope. *Review of Scientific Instruments*. 2003 Oct;74(10):4369–72.
37. Niedrig H. The early history of electron microscopy in Germany. *Advances in Imaging and Electron Physics*. 1996 Jan 1;96:131–147. Elsevier.
38. Wells OC, Joy DC. The early history and future of the SEM. *Surface and Interface Analysis*. 2006 Dec;38(12–13):1738–42.
39. Havrdova M, Polakova K, Skopalik J, Vujtek M, Mokdad A, Homolkova M, Tucek J, Nebesarova J, Zboril R. Field emission scanning electron microscopy (FE-SEM) as an approach for nanoparticle detection inside cells. *Micron*. 2014 Dec 1;67: 149–54.

40. Prior DJ, Boyle AP, Brenker F, Cheadle MC, Day A, Lopez G, Peruzzi L, Potts G, Reddy S, Spiess R, Timms NE. The application of electron backscatter diffraction and orientation contrast imaging in the SEM to textural problems in rocks. *American Mineralogist.* 1999 Dec 1;84(11–12):1741–59.

41. Bogue R. Nanosensors: A review of recent progress. *Sensor Review.* 2008 Jan 25;28(1):12–7.

42. Bogue R. Nanosensors: A review of recent research. *Sensor Review.* 2009;29(4):310–5.

43. Kargozar S, Mozafari M. Nanotechnology and nanomedicine: Start small, think big. *Materials Today: Proceedings.* 2018;5(7):15492–500.

44. Yang Y, Chawla A, Zhang J, Esa A, Jang HL, Khademhosseini A, Anthony Atala, Robert Lanza, Robert Nerem, editors. Applications of nanotechnology for regenerative medicine; healing tissues at the nanoscale. In *Principles of Regenerative Medicine.* Elsevier; 2019;485–504.

45. Lin LCW, Chattopadhyay S, Lin JC, Hu CMJ. Advances and opportunities in nanoparticle-and nanomaterial-based vaccines against bacterial infections. *Advanced Healthcare Materials.* 2018;7(13):1701395.

46. Cojocaru F-D, Botezat D, Gardikiotis I, Uritu C-M, Dodi G, Trandafir L, Rezus C, Rezus E, Tamba B-I, Mihai C-T. Nanomaterials designed for antiviral drug delivery transport across biological barriers. *Pharmaceutics.* 2020;12(2):171.

47. Nasrollahzadeh M, Sajjadi M, Soufi GJ, Iravani S, Varma RS. Nanomaterials and nanotechnology-associated innovations against viral infections with a focus on coronaviruses. *Nanomaterials.* 2020;10(6):1072.

2 Multifaced Applications of Nanoparticles in Biological Science

Jayamanti Pandit, Md. Sabir Alam, Jamilur R. Ansari,
Monisha Singhal, Nidhi Gupta, Aafrin Waziri,
Kajal Sharma, and Faheem Hyder Pottoo

CONTENTS

Abbreviations 18
2.1 Introduction 18
2.2 Metallic Nanoparticles (MNPs) 18
2.3 Biomedical Applications of Metallic Nanoparticles 19
2.4 Chemical Methods 22
    2.4.1 Chemical Reduction Method 22
    2.4.2 Microwave-Assisted Synthesis Method 22
    2.4.3 Irradiation Methods 22
    2.4.4 Photoinduced Reduction 22
    2.4.5 Microemulsion Techniques 22
2.5 Physical Methods 23
    2.5.1 Laser Ablation 23
    2.5.2 Mechanical Milling Method 23
2.6 Biological Methods 23
    2.6.1 Bacteria 23
    2.6.2 Fungi 23
    2.6.3 Algae 23
    2.6.4 Plants 23
2.7 Green synthesis of metallic nanoparticles 24
2.8 Mechanism of Antibacterial Action of Metallic Nanoparticles 25
2.9 Anticancer Mechanism of Metallic Nanoparticles 26
2.10 Catalytic Activity of Metal Nanoparticles (MNPs) 30
2.11 Catalytic Change of 4-Nitrophenol (4-NP) into 4-Aminophenol (4-AP) 30
2.12 Catalytic Activity of MNPs by Methylene Blue (MB) and Methyl
    Orange (MO) 36
2.13 Conclusion and Future Perspectives 38
Acknowledgments 38
References 38

DOI: 10.1201/9781003126256-2

ABBREVIATIONS

MB: methylene blue
MNPs: metallic nanoparticles
UV–Vis: ultraviolet visible

2.1   INTRODUCTION

Nanotechnology has emerged as a powerful tool or technique to solve any problem related to the efficacy, stability, and safety of medicine and food products. Every dimension of life has benefited from the use of nanotechnology [1]. Nanotechnology is the branch of science that deals with nano-size molecules or particles ranging from 1 to 100 nanometers (nm). The biological property of any molecule depends on the dimension, charge, and other physicochemical behavior of the molecules. Size plays a very important role from its functionalization to absorption and delivery of medicine at the target site [2, 3]. The use of nanoparticles (NPs) in different fields of chemistry, physics, molecular biology, medicine, and material science has increased [4–6]. The exclusive features of nanoparticles made them the potential choice for application in the fields of medicine and nutrition [7]. Nanoparticles can behave like solutes and have separate particle-phase properties. They have a larger surface volume ratio than microparticles and have a strong surface reactivity that is size-dependent [8]. These properties proved their application in drug delivery carriers, diagnoses, sensing, and implants. They are widely explored for their use as antimicrobials due to the availability of a larger surface area for interaction with microorganisms [9]. Nanotechnology has emerged with advanced tools in biotechnology, which has more portable, safer and cheaper nanoparticles and diverse biomedical application [10–12].

Nanoparticles can be broadly classified as organic and inorganic nanoparticles. The latter gained more focus from researchers due to their ability to withstand harsh process conditions [13]. Metals like gold, silver, and platinum have significant effects on health, and therefore metallic nanoparticles have proved their utility for catalysis, diagnoses and treatment, sensors, labeling of optoelectronic recorded media, and preparation of polymer composite [14–21].

2.2   METALLIC NANOPARTICLES (MNPS)

Metallic nanoparticles (MNPs) from different metals, such as gold, silver, copper, zinc, titanium, and magnesium, are well known for their properties [22]. The methods used for the preparation of metallic nanoparticles can be broadly categorized into two types: top-down methods and bottom-up methods. This classification is based on the starting material used in the preparation of nanoparticles. If bulk material is used as the starting material and the size reduction is achieved by employing various physicochemical methods, the process is called a top-down method, whereas small atoms or molecules are used as the starting material in a bottom-up method [23, 24]. MNPs can be synthesized either by a physical and chemical method or by biological ways. Physicochemical methods based on chemical reductions, photochemical reductions, and electrochemical changes are used for the preparation of metallic nanoparticles [25, 26]. The properties of the metallic nanoparticles such as size, morphology, physicochemical characteristics, and stability are dependent on the experimental conditions, the process of adsorption of the stabilizer, and the interaction kinetics of the metal ions with the reducing agents [27, 28].

The biological method is preferred over the chemical method due to the associated toxicity of the traces of chemicals present on the surface of nanoparticles. The eco-friendly green synthesis of nanoparticles is done by using biological materials such as enzymes [29], microorganisms [30, 31], fungus [32], and plant extracts [33, 34]. In recent years, the green synthesis of nanoparticles has evolved as a separate branch of nanotechnology especially for gold and silver nanoparticles, which has enormous applications [35–37]. Biosynthesis of metallic nanoparticles may be extracellular or intracellular according to the location of the nanoparticle synthesis in the microorganism. During intracellular synthesis, the metal ions are transported inside the microbial cells to form nanoparticles in the presence of coenzymes. Extracellular synthesis has more applications, as the synthesized nanoparticles are free from unnecessary cellular components from the microbial cell. Intracellular synthesis forms smaller-sized particles compared to extracellular synthesis [38].

2.3 BIOMEDICAL APPLICATIONS OF METALLIC NANOPARTICLES

The biomedical applications of metallic nanoparticles range from antimicrobial effects to anticancer and diagnostic application to biosensors, neurodegenerative disorders, imaging, and many more [39–41]. They are efficiently utilized as drug delivery carriers with synergistic properties and in the targeted delivery of anticancer agents.

Most metallic nanoparticles have antimicrobial, antiviral, and antifungal properties. MNPs such as silver, gold, copper, titanium, magnesium, and zinc have shown potential inhibition activity against *Staphylococcus aureus*, *Escherichia coli*, *Salmonella typhi*, *Klebsiella pneumoniae*, and *Bacillus subtilis*. The antibacterial activities of these nanoparticles are effective against antibiotic-resistant bacteria. The antimicrobial activity is shown to be dependent on the particle size of the NPs. Smaller particles have better potential as antimicrobials than larger particles [42, 43]. This is due to the higher penetration efficiency of smaller particles and their larger surface-to-volume ratio [44–46]. Nanoparticles synthesized from aluminum (Al), copper (Cu), and titanium dioxide (TiO_2) are used as surface disinfectants and to eliminate microorganisms in foodstuffs [47, 48]. Treatment of viral diseases is challenging throughout the world. MNPs have been explored for antiviral activities and they have emerged as a new opportunity for researchers in the treatment of various viral diseases in plants and animals. A few MNPs, including silver (Ag), gold (Au), and cerium (Ce), have proven activity against bacteriophages and HIV [49, 50]. Antifungal activity of silver nanoparticles (AgNPs) exhibited fungicidal action against pathogenic *Candida spp.* and *Trichophylon mentagrophytes* [51]. The fungicidal activity of AgNPs was the result of interaction between nanoparticles and the cellular structure of *C. albicans*, which leads to the death of the microbial cell [52]. AgNPs have been also elucidated to exert anti-HIV activity through interaction at an early stage of viral replication [53]. Ligand-coated multivalent AuNPs have been considered for interaction with the adsorption and fusion process of virus replication. AuNPs bind specifically with the glycoprotein (gp120) and inhibit HIV infections under in in-vitro [54]. MNPs also show effects against influenza and herpes simplex viruses [55, 56]. MNPs have been used for targeted delivery of chemotherapeutic agents to malignant cells with a very small effect on normal cells. A group of researchers reported that AuNPs inhibits vascular endothelial growth factor (VEGF)-induced angiogenesis in mouse ears injected with VEGF [57].

The quality of selective targeting by MNPs illustrates the influence of nanoparticles at the time of synthesis [58]. Capping agents such as surfactants, polymers, solid support, or ligands with suitable functional groups bind with the surface of the NPs and they act as a stabilizer or modifier of nanoparticles [27]. MNPs with different capping agents can behave as cytotoxic or non-cytotoxic [58, 59]. Thus, the use of an appropriate capping agent is required to establish the maximum desired effect. Biosynthesized MNPs can exhibit anticancer activity or they can be used as drug delivery carriers exclusively depending upon the capping agent used in the biosynthesis [60]. In a study, AuNPs and AgNPs synthesized from guava and clove extract have different effects on cancer cell lines. AuNPs have shown anticancer effects on HeLa, HT-29, and HEK-293 cancer cell lines, while AgNPs did not show any cytotoxic effect on cancer cell lines [60].

The targeted delivery of a drug is effectively achieved by MNPs as they bind with various peptides, antibodies, and RNA/DNA specifically present on different cells [61, 62]. A polyethylene glycol (PEG) coating on MNPs increases the circulation time of nanoparticles and is useful for drug and gene delivery [63, 64]. AuNPs have shown efficient utility in photothermal therapy (PTT) and radiotherapy, which are other aspects of cancer treatment [27].

Another example is bacterial magnetosomes, which are easy to prepare and able to deliver a larger amount of the drug than artificial magnetic NPs [65]. Biosynthesized AuNPs using *Porphyra vietnamensis* as the reducing and capping agent can be used as carriers for DOX (doxorubicin) and results in higher cytotoxicity to LN-229 cells than the free DOX [66].

Biosynthesized MNPs attribute their contribution to imaging and diagnostic areas. They can be easily incorporated into chemical sensors and are quite helpful in the detection of biological compounds or molecules in the human body. AuNPs were successfully used to detect glucose concentration in human blood serum and were found in good agreement with the standard detection spectrophotometric technique [67].

The optical property of biosynthesized NPs includes photoluminescence (PL) properties, high transparency, plasmon resonance, and a low or high refractive index [68]. The photoluminescence properties of a few of the biosynthesized MNPs, AgNPs [69, 70] and the strong near-infrared (NIR) absorption ability of AuNPs using *Maduca longifolia* extract, showed promising medical application [71].

Biosynthesized MNPs also have potential applications in the management of other diseases, such as bleeding disorders [72] and diabetes [73]. AgNPs and AuNPs stabilized by *Brevibacterium casei* have shown anticoagulant properties [74]. It was reported that the enzyme tyrophosphatase type PTP 1B, which is required to dephosphorylate the insulin receptor, was inhibited by biosynthesized AuNPs. In this way, AuNPs can help to efficiently enhance insulin activity [75].

Other than medical and therapeutic applications, metallic nanoparticles are also known for their catalytic properties. They are very useful for heterogeneous catalysis and are preferred due to their high activity and stability. They are also very selective but a bit expensive when it comes to gold (Au), platinum (Pt), and palladium (Pd) NPs [27]. Biosynthesized MNPs have been commercially used for organic synthesis processes such as oxidation or reduction, coupling reactions, and dehalogenation [76, 77]. MNPs have been reportedly used in fuel cell and electrode construction. Biomass-supported MNPs of platinum and palladium were employed as fuel cells for chemical catalysis [78, 79]. The electrical properties of biosynthesized NPs have been well established, and AuNPs when used for electrode construction show efficient electrical transmission [80, 81].

Synthesis of metallic nanoparticles in a biological process involves microorganisms, and the same process can occur in nature and leads to detoxifying the environment. This bioremediation of metals is another way for the formation of metallic nanoparticles [82]. Several microbes like fungi, bacteria, yeast, and algae are able to do the process of binding and concentrating the dissolved metallic ions in the presence of biowaste materials through the process of biosorption. In this way, MNPs offer a cost-effective and natural way to recover dissolved metals from aqueous solutions [83].

The two main approaches for MNP synthesis consist of (1) the top-down approach, where the metal from the bulk state is crushed to a powder form and finally reduced to the nanoscale; whereas in (2) the bottom-up approach, the synthesis arises from the atomic scale with the precursors to the molecular level and finally form nanoparticles as shown in Scheme 2.1 [84].

Different chemical, physical and biological methods have been employed to generate nanoparticles. These methods employ the use of various chemicals, sensitive instrumentation setup, and other physical parameters for the synthesis and

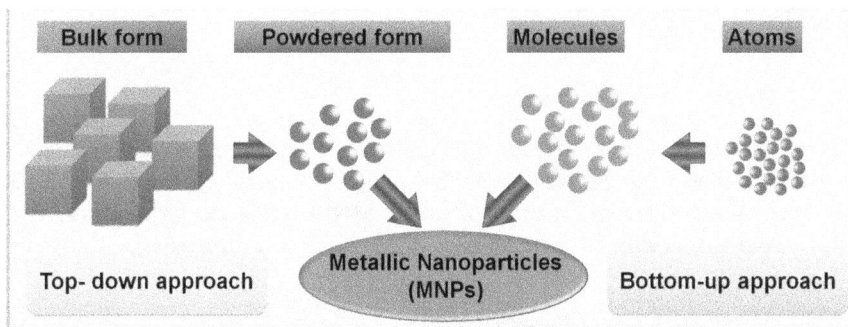

SCHEME 2.1 Top-down and bottom-up approaches for the synthesis of metallic nanoparticles.

characterization purpose. Some of the widely used physical, chemical, and biological methods are explained further.

2.4 CHEMICAL METHODS

2.4.1 CHEMICAL REDUCTION METHOD

Michael Faraday (1857) successfully explored the synthesis and colloidal colors of gold via the chemical reduction method. It has become the widely used method for the synthesis of MNPs by reducing the metals using organic and inorganic reducing solvents. For example, silver can be reduced from the Ag^+ to Ag^0 state that can be obtained by using reducing agents like sodium citrate, sodium borohydride ($NaBH_4$), hydrazine, and phenols [85]. It has been seen that copper is the simplest metal to be reduced by reducing agents, which leads to the production of nanosized copper NPs with a perfect surface morphology [86].

2.4.2 MICROWAVE-ASSISTED SYNTHESIS METHOD

The microwave-assisted synthesis method is more encouraging for MNP synthesis as it consumes microwave heat, which is far better than conventional oil bath heating. As a result, constant synthesis of nanosized MNPs can be achieved with a better dispersity index [87, 88]. The main benefits of this method are a short and precise reaction time, which ultimately leads to a reduction in energy consumption, a better yield of MNPs and eliminates cluster formation [89, 90]. The microwave-assisted method is very popular for the synthesis of noble metals like silver, gold, and platinum NPs. The size can be easily manipulated by the optimization of different physiochemical parameters, namely, reaction time, concentration, temperature, and pH.

2.4.3 IRRADIATION METHODS

In irradiation methods, a laser beam is used to irradiate metal solutions to obtain optimum-sized MNPs. Usually, laser and mercury (Hg) lamps are used as a light source for silver nanoparticle production [91].

2.4.4 PHOTOINDUCED REDUCTION

In photoinduced reduction, the reduction of metals is carried out by adding a reductor to an electron donor species [92]. With the help of this method, it was determined that we can synthesize various morphologies of MNPs, which can have sphere, triangular, cubes, rods, or prisms with an edge of 30–120 nm in length [93].

2.4.5 MICROEMULSION TECHNIQUES

Microemulsion techniques require a two-phase aqueous system for the synthesis of MNPs. It is based on separation of a metal precursor and a reducing agent in a two-phase aqueous system. MNPs were synthesized because of thermostatically stable dispersion obtained from two immiscible solvents using a suitable surfactant [94].

2.5 PHYSICAL METHODS

2.5.1 LASER ABLATION

Laser ablation is one of the most important physical methods for MNP synthesis where a high-power laser beam is used to aim the target material to obtain MNPs in the form of a colloidal solution. The whole process is carried out in a vacuumed chamber [95].

2.5.2 MECHANICAL MILLING METHOD

Mechanical milling is the process of milling solid or bulkier forms of metals to synthesize MNPs. This method requires a heavy machinery setup to process the solid-state material. The type of machine required for the synthesis process is classified as per the capacity and application [96].

2.6 BIOLOGICAL METHODS

2.6.1 BACTERIA

Many researchers have explored whether MNPs can be synthesized by different bacteria. It has been found that reduction of Ag^+ ions from an aqueous solution of silver can be achieved by *Bacillus licheniformis* [97]. Synthesis of gold nanoparticles was also explored using a bacterium, *Shewanella alga*. A study conducted by Nair and team demonstrated that *Lacto bacillus* bacterium from buttermilk can be used to synthesize metallic nanoparticles such as gold, silver, and their alloys [98].

2.6.2 FUNGI

Fungi have been investigated for the synthesis of MNPs because of their unique properties like metal tolerance and bioaccumulation ability. Silver NPs have been successfully synthesized using *Fusarium oxysporum* [99]. Many studies have been conducted on filamentous fungi to produce the extracellular synthesis of MNPs.

2.6.3 ALGAE

A few reports are suggesting that *Sargassum wightii*, marine algae, has been utilized for the synthesis of exceptionally stable gold nanoparticles. In addition, Platinum and Palladium MNPs have also been synthesized by a similar process [100].

2.6.4 PLANTS

Apart from microorganisms, researchers are now moving toward a better way to synthesize high-yield MNPs. In this context, extracts of many plant parts have been explored for their efficiency to reduce metal salts for synthesis purposes [101].

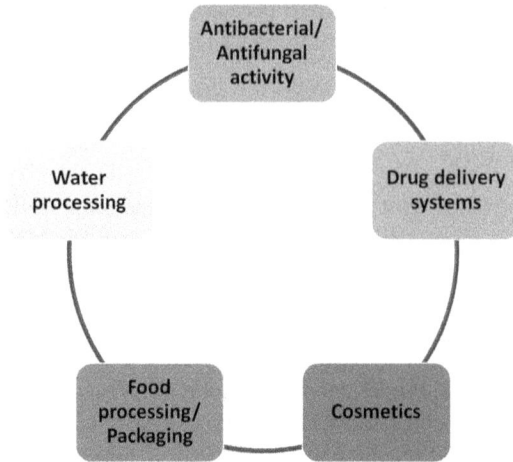

FIGURE 2.1   Biomedical applications of metallic nanoparticles.

Nanotechnology has a strong impact on our day-to-day lives, and it has proven to be beneficial for us in various aspects. It has gotten much attention from the biomedical, industrial, food, electronics, and information sectors, as shown in Figure 2.1. For example, AgNPs have extensively been used in food and packaging industries, cosmetics, and water processing plants [102]. AuNPs are most popular for determining DNA hybridization for the diagnosis of pathogenic and genetic disorders [103]. AuNP-based DNA biosensors with entrapped thiolate have been developed to identify changes in complementary biomolecules [104]. It has been documented that platinum NPs were used as antibacterials and biomedicines [105].

2.7   GREEN SYNTHESIS OF METALLIC NANOPARTICLES

Using natural sources like plants is considered a green approach to synthesize biogenic metallic nanoparticles. Currently, it is the most decisive and ecofriendly approach that eliminates the use of harmful chemicals as needed in physical and chemical approaches. Many studies have been documented suggesting the biogenic synthesis of MNPs using different parts of plants and noble metals [106].

Large-scale productions of nanoparticles are carried out by different means of physical and chemical methods. But there have been certain drawbacks with these conventional methods like the use of heavy machinery or instrumentation, chemicals, huge production cost, and lower yields [107]. To overcome these limitations and for higher yield production, researchers around the world are now moving toward the green approach to synthesize metallic nanoparticles. The idea behind this green approach is to utilize the availability of living organisms provided by nature. Green synthesis utilizes the plant components as reducers, capping, and stabilizing agents. This green approach is most popular due to its low toxicity and environment-friendly properties, as shown in Figure 2.2. Also, this method does not require complex instrumentation, which ultimately results in cost reduction [108].

Less toxicity   No use of harmful chemicals   Cost effective   Natural redcuers

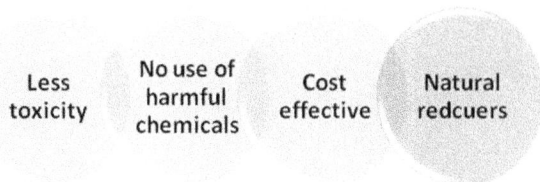

FIGURE 2.2   Advantages of green synthesis of metallic nanoparticles.

The mechanism for the entire synthesis process relies on bioactive compounds that are found in plant extracts primarily acting as metal reducers. Over the past few years, many metals and their derivatives have been used to generate nanoparticles of Ag, Au, Pt, TiO$_2$, copper oxide (CuO), zinc oxide (ZnO), palladium (Pd), nickel (Ni), and iron oxide (FeO) [109–116]. Some plants have been used to produce MNPs especially gold and silver, including *Terminalia arjuna*, *Cordia dichotoma*, *Canarium ovatum*, *Dicoma tomentosa*, *Prosopis julifora*, *Cicer arietinum*, *Trigonella foenum-graecum*, and *Acacia nilotica* [117–124]. Leaves extract from diverse plant species like *Azadirachta indica*, *Cymbopogon* (lemon grass), *Tamarindus indica* (tamarind) and *Aloe barbadensis miller* (aloe vera) have been utilized to synthesize gold nanoparticles [125].

Platinum nanoparticles have been used as catalysts in various fields. The biogenic synthesis of PtNPs was successfully carried under optimal conditions using different aqueous extracts of *X. strumarium* [126]. Shabani et al., for the first time, reported that platinum nanoparticles can be synthesized using sheep milk to become a potential candidate for delivery of drugs and gene therapy [127].

2.8   MECHANISM OF ANTIBACTERIAL ACTION OF METALLIC NANOPARTICLES

Bacteria being the first originated organism on earth have now become highly adaptable and evolved with time. The discovery of antibiotics was acknowledged as one of the greatest accomplishments during the 20th century [128]. The need for antibiotics has significantly increased over the past few decades as new strains of bacteria have drastically emerged [129]. Due to the endless usage of antibiotics to treat infections or diseases caused by bacterial pathogens, a new category is known as antibiotic-resistant bacteria has emerged [130]. Antibiotic resistance is defined as any modifications in the pattern of bacterial resistance that leads to a decrease or inactivation of its therapeutic potential. The occurrence of resistance is a spontaneous process due to genetic mutations and the voluntary use of antibiotics [131]. This uninterrupted development of resistant strains has propelled researchers to establish innovative therapeutic agents. In this regard, metallic nanoparticles are the most prominent innovation to combat the issues related to bacterial resistance. They have shown excellent antibacterial activity against a broad spectrum of gram-positive and gram-negative bacteria [132]. The cellular membrane of bacterium plays an important role in exhibiting toxic effects, as MNPs can directly interact with it [133]. The antibacterial mechanism of MNPs is shown in Figure 2.3. The antibacterial efficacy of AgNPs

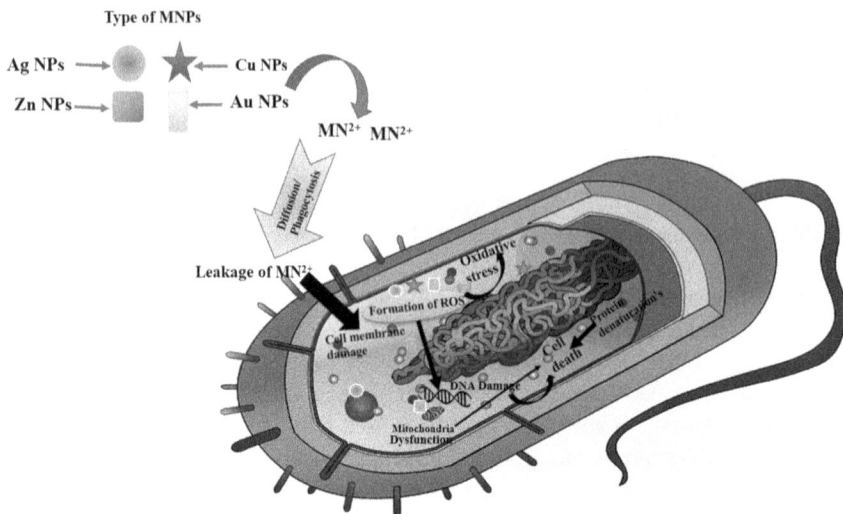

FIGURE 2.3   Antimicrobial mechanism exhibited by metallic nanoparticles.

synthesized using *Cordia dichotoma* leaves was checked against *Escherichia coli* and *Pseudomonas aeruginosa*. The results indicate that these MNPs are very potent as antibacterial agents [134]. The antibacterial effects of palladium nanoparticles (PdNPs) were first investigated by Adams and co-workers. They found that PdNPs with fewer diameters were more lethal than Pd^{2+} ions against *Staphylococcus aureus* [135]. Zinc oxide nanoparticles are widely recognized as antibacterial, as they suppress bacterial growth upon penetration [136]. In another study, zinc oxide and titanium dioxide nanoparticles were analyzed for DNA damage activities. It has been seen that the generation of oxidative stress can lead to damage in bacterial DNA [137]. Arya et al. reported early inhibition of biofilm by silver nanoparticles synthesized using *Acacia nilotica* leaves extract. At the lowest concentration (0.25 μg), AgNPs led to biofilm inhibition to a larger extent as compared to the control (with no treatment) [138–152] (Table 2.1).

2.9   ANTICANCER MECHANISM OF METALLIC NANOPARTICLES

Metallic nanoparticles are of great interest in the field of cancer biology. They are fabricated in such a way that they can be used as either drug delivery systems or therapeutic agents. They overcome the limitations related to the cancer treatments generally used. Less availability of drugs, target site-specificity, and drug resistance by tumors are some of the major issues that are associated with the current treatment system [153]. Among MNPs, silver and gold metals are supremely important because of their properties like flexibility, reliability, and their applications in biomedical science. The green synthesis of silver and gold nanoparticles using plant extracts is known to possess anticancer properties [154]. Goyal et al.

TABLE 2.1

Antimicrobial Activity of Green Synthesis Metallic Nanoparticles

Carrier Name	Characterization Techniques	Particle Size	Mechanism	Reference
Silver Nanoparticles				
Banana peel extract	UV–visible spectra, SEM with EDX, FESEM, TEM, FT-IR spectra, DLS, XRD	20–50 nm, spherical	Banana peels extract silver nanoparticles shown synergistic effect on the antimicrobial activity of the standard antibiotic levofloxacin against gram-positive and gram-negative bacteria under investigation.	[137]
Alternanthera dentata leaf extract	UV–visible spectroscopy, FT-IR, XRD, TEM	50–100 nm	Prepared silver nanoparticles exhibited antibacterial activity against *E. coli, P. aeruginosa, K. pneumonia* and *E. faecalis.*	[138]
Leaf extract of *Pterocarpus santalinus*	UV–visible spectroscopy, FT-IR, XRD, SEM with EDX, AFM	20–50 nm, spherical	MIC of AgNPs exhibited good antibacterial potential against gram-positive and gram-negative bacterial strains and zone of inhibition effect of antibacterial activity depends upon the concentration of AgNPs.	[140]
Phyla dulcis plant extract	UV–visible spectroscopy, DLS, SEM-EDS, ESCA, XPS,	63–76 nm	Green synthesis of antimicrobial effects of AgNPs against four common foodborne pathogens incorporated into food packaging material and in food processing equipment to minimize poisoning related to foodborne pathogens.	[141]
Rhodococcus spp.	UV–visible spectroscopy, TEM, EDX, XRD, FT-IR, DLS	10–50 nm, spherical	AgNPs showed excellent bacteriostatic and bactericidal activity against pathogenic gram-positive and gram-negative microorganisms.	[142]
Gold Nanoparticles				
Aqueous extract of *Caulerpa racemose*	XRD, UV–visible spectroscopy, FT-IR, FESEM, HRTEM, DLS	13.7–85.4 nm, spherical to oval in shape	Cr@AuNPs showed the highest antibacterial activities against gram-negative bacteria than gram-positive bacteria *S. agalactiae.*	[143]

(Continued)

TABLE 2.1 (CONTINUED)
Antimicrobial Activity of Green Synthesis Metallic Nanoparticles

Carrier Name	Characterization Techniques	Particle Size	Mechanism	Reference
Leaf extract of *Nepenthes khasiana*	UV–visible spectroscopy, SEM, XRD, FTIR, TEM	50–80 nm, triangular and spherical	AuNPs showed antibacterial activity against *E. coli*, *Bacillus* species and checked against the standard, i.e. an antibiotic (Ciproflaxin).	[144]
pulp extract of *Abelmoschus esculentus*	DLS, UV–visible spectroscopy, SEM, XRD, HRTEM, EDX, FTIR	13–23 nm, spherical	Pulp extract of AuNP solution exhibits excellent antibacterial activity against the bacteria *Bacillus subtilis*, *Bacillus cereus*, *Escherichia coli*, *Micrococcus luteus* and *Pseudomonas aeruginosa*.	[145]
Fruit of *Ananas comosus*	DLS, UV–visible spectroscopy, TEM, EDX	5–20 nm, tetrahedral	Fruit extract of *A. comosus* of prepared AuNPs reveal good antibacterial activity against gram positive and negative pathogens.	[146]
leaf extracts of *Carica papaya* (CP) and *Catharanthus roseus* (CR)	UV–visible spectroscopy, EDX, FTIR, HRTEM and XRD	2–20 nm, spherical shaped	CP and CR AuNPs was found to be more active against both bacteria and cancer cell lines by confirming the synergistic enhanced effect of both plant extracts.	[147]
Copper Nanoparticles				
Fruits extract of *Momordica charantia*	DLS, UV–vis spectroscopy, FTIR, XRD, TEM, SEM-EDSX	50–60 nm	Green synthesized CuO NPs find potential applications in the field of nanomedicine and could be used to develop targeted therapies against bacteria, fungi and viruses.	[148]
Hageniaabyssinica (Brace) leaf extract	UV–visible, FTIR, TEM, SEM, XRD, HRTEM, EDXA,	34 nm, spherical, hexagonal, triangular, cylindrical, and irregularly shaped NPs	Great potential as a remedy for infectious diseases caused by the tested bacterial pathogens.	[149]

(Continued)

TABLE 2.1 (CONTINUED)

Antimicrobial Activity of Green Synthesis Metallic Nanoparticles

Carrier Name	Characterization Techniques	Particle Size	Mechanism	Reference
Curcuma longa powder extract	XRD, SEM-EDS, TEM, UV-visible, FTIR	5–20 nm, spherical	*C. longa* extract-capped CuNPs have exhibited attractive antibacterial activity with both gram-positive and gram-negative microorganisms.	[150]
Leaf extract of *Adhatodavasica*	XRD, UV–visible, FTIR, FE SEM, EDS, XPS and TGA	7–11 nm, spherical	*A. vasica* leaf extract of CuO/C nanocomposites show significant antimicrobial activities against the gram-negative bacteria *P. aeruginosa*, *E. coli* and *K. pneumoniae* and the gram-positive bacteriae *S. aureus*, and the fungi *A. niger* and *C. albicans*.	[151]
C. vitiginea leaf extract	UV–Vis spectra, XRD, TEM, AFM, FTIR, XPS, SEM-EDX	Spherical shape	*C. vitiginea* leaf extract-based green synthesized copper nanoparticles have strong antioxidant and antibacterial activity against DPPH free radical molecules and urinary tract infection pathogens.	[152]

investigated the anticancer potential of silver nanoparticles biosynthesized using *Trigonella foenum-graecum* (TFG) seed extract in skin cancer cell lines (A431). They have shown that TFG-AgNPs were able to retard the growth of tumor cells up to 50%. Also, anticancer activity was enhanced when the concentrations of TFG-AgNPs increased [155]. Another study depicts the anticancer efficacy of ZnONPs against three types of cancer cell lines, namely, human hepatocellular carcinoma HepG2, human bronchial epithelial BEAS-2B, and human lung adenocarcinoma A549. They conclude that ZnONPs were able to selectively promote apoptosis of tumor cells in the p53 pathway via the generation of reactive oxygen species [156]. Similarly, the anticancer potential of three MNPs – zinc oxide, iron oxide, and gold – were examined using the colon cancer cell line HT 29. ZnONPs along with peptides revealed compelling cytotoxic effects on HT 29 cell lines [157]. A study conducted by Patra et al. demonstrated that AuNPs also possess anticancer activity when studied on the prostate cancer cell line DU145, as they can decline the growth of tumor cells [158–172] (Table 2.2). The anticancer mechanism of metallic nanoparticles is shown in Figure 2.4.

2.10 CATALYTIC ACTIVITY OF METAL NANOPARTICLES (MNPS)

MNPs are used as catalysts for a wide range of reactions, and the enormous literature is hard to ignore. In many applications, the nanoparticles are attached to the colloidal carrier for simple handling throughout the catalysis. Mainly due to their catalytic activity, MNPs have been the subject of deep research in the last two decades. For example, gold behaves as an active catalyst when brought to the nanoscale. We can say that catalysis by NPs is one of the most active fields in modern nanotechnology.

2.11 CATALYTIC CHANGE OF 4-NITROPHENOL (4-NP) INTO 4-AMINOPHENOL (4-AP)

Catalytic as well as photocatalytic activity of MNPs strongly depends upon the morphology and light utilization capability. When MNPs are reduced from bulk to nanoscale, the surface area increases manifold, which facilitates efficient adsorption of 4-nitrophenolate ions that ultimately leads to the enrichment in the catalytic behavior by reducing 4-nitrophenol (4-NP). Due to the increased surface area, MNPs show better catalytic activity and possess more active sites. In our case, MNPs have a large surface area due to the smaller size of NPs and therefore it leads to efficient adsorption of organic contaminants and ultimately it improves photocatalytic activity by reducing 4-NP. Due to the smaller size, MNPs efficiently enhance their catalytic behavior due to the transformation of 4-NP to 4-aminophenol (4-AP).

The observed enhanced photocatalytic performance of MNPs is mainly attributed to their wide surface area facilitating efficient adsorption of organic contaminants and pollutants, which restrain the recombination of photogenerated charge carriers [173].

4-NP and its offshoot are generally used to produce synthetic dyestuffs, insecticides, and herbicides, but they can appreciably harm the ecosystem. 4-NP is

TABLE 2.2
Anticancer Application of Green Synthesis Metallic Nanoparticles

Carrier Name	Characterization Techniques	Morphology	Mechanism	Reference
Silver Nanoparticles				
Leaf of *Azadirachta indica*	UV, SEM, TEM, FTIR	40 nm, spherical	The anticancer potential of the nanoparticles evaluated using MTT assay on HeLa, Hek-293 and MCF-7.	[158] Mittal et al. 2016
Latex of *Calotropis gigantean*	UV, FTIR, SEM, TEM, EDX spectroscopy	5–30 nm, spherical	In vitro anticancer study manifests the cytotoxicity value of synthesized AgNPs against tested HeLa cells.	[159] Rajkuberan et al. 2015
Penicillium decumbens (MTCC 2494)	UV, FTIR, AFM, SEM	30–60 nm, spherical	Ag-NPs showed good anticancer activity at 80 μg·mL^{-1} upon 24 hours of incubation, and toxicity increases upon 48 hours of incubation against A-549 human lung cancer cell lines.	[160] Majeed et al. 2016
Algae, *Sargassum vulgare*	UV, FTIR, AFM, XRD, SEM, TEM	10 nm, spherical	Its preferential ability to kill cancerous human myeloblastic leukemic cells HL60 and cervical cancer cells HeLa as compared with normal peripheral blood mononuclear cells. The biosynthesized AgNPs inhibited the lipid peroxidation-mediated reactive oxygen species generation thus preventing the irradiation-related carcinogenesis. It is endocytosed into the cells and damages the DNA which leads to the process of apoptosis.	[161] Govindaraju et al. 2015
Callus of *Taxus yunnanensis*	FTIR, XRD, SEM, TEM	6.4–27.2 nm, spherical	They exhibited stronger cytotoxic activity against human hepatoma SMMC-7721 cells and induced noticeable apoptosis in SMMC-7721 cells but showed lower cytotoxic activity against normal human liver cells (HL-7702).	[162] Xia et al. 2016
Gold Nanoparticles				
Flower of *Couroupita guianensis*	UV, FTIR, XRD, TEM, SEM	7–48 nm, spherical, triangular, tetragonal, and pentagonal with irregular contours	The cytotoxic effects of biologically synthesized gold nanoparticles, particularly in the context of apoptosis. MTT assay was used to appraise the effect of gold nanoparticles on proliferation of HL-60 cells. DNA fragmentation, apoptosis by DAPI staining, and comet assay for DNA damage.	[163] Geetha et al. 2013

(Continued)

TABLE 2.2 (CONTINUED)
Anticancer Application of Green Synthesis Metallic Nanoparticles

Carrier Name	Characterization Techniques	Morphology	Mechanism	Reference
Plant extracts of *Marsdenia tenacissima*	UV, AFM, FTIR, TEM, energy dispersive x-ray spectroscopy	50 nm, spherical	They performed as *in vitro* anticancer activity against lung cancer cell lines (A549). MTT assay revealed that AuNPs produce toxicity based on the dose dependent A549 cells growth inhibition. AuNPs treatment activates caspase expression and down-regulates the anti-apoptotic protein expression in A549 cells. Our results point out that the AuNPs from *M. tenacissima* extract are apposite stabilizing agents, which serve as an effective anticancer agent against lung cancer cell lines (A549).	[164] Sun et al. 2017
Aqueous extract of *Wedeliatrilobata leaves*	UV, FTIR, TEM, SEM, DLS, XRD	10–50 nm, spherical, crystalline	In vitro cytotoxic efficacy of synthesized gold nanoparticles against HCT 15 human colon cancer cell lines was studied.	[165] Dey et al. 2018
Matured plant of *Scutellaria barbata*	UV, FTIR, TEM, AFM	30–40 nm, spherical	The synthesized gold nanoparticles (AuNPs) possessed effective anticancer activity against pancreatic cancer cell lines (PANC-1).	[166] Wang et al. 2019
Marine bacteria *Enterococcus* sp.	XRD, FTIR, TEM, energy dispersive x-ray spectroscopy	6–13 nm, spherical	Anticancer activity carried out by MTT assay against Hep-G2 and lung cancer cell (A549) lines.	[167] Rajeshkumar et al. 2016
Copper Nanoparticles				
Leaves of *Acalypha indica*	UV, SEM, TEM, FTIR, energy dispersive x-ray spectroscopy	29 nm, spherical	The cytotoxicity activity of *A. indica* mediated copper nanoparticles was evaluated by MTT assay against MCF-7 breast cancer cell lines and confirmed that copper oxide nanoparticles have cytotoxicity activity.	[168] Sivaraj et al. 2014

(Continued)

TABLE 2.2 (CONTINUED)
Anticancer Application of Green Synthesis Metallic Nanoparticles

Carrier Name	Characterization Techniques	Morphology	Mechanism	Reference
Aqueous black bean extract	XRD, FT-IR, XPS, Raman spectroscopy, DLS, TEM, SAED, SEM, and EDX	26–27 nm, spherical	The cytotoxic effect of the CuONPs was determined by sulforhodamine-B assay. In vitro anticancer results indicated that CuONPs induced intracellular ROS generation in a dose-dependent manner and significantly reduced cervical carcinoma colonies. CuO NPs can induce apoptosis and suppress the proliferation of HeLa cells.	[169] Nagajyothi et al. 2017
Using fruit and leaf Andean blackberry (*Rubus glaucus* Benth.)	UV, TEM, XRD, DLS	43–52 nm, spherical	The free radical scavenging activity of CuONPs was measured by using DPPH. Antioxidant was evaluated against DPPH. CuONPs can induce apoptosis and suppress the proliferation of HeLa cells.	[170] Kumar et al. 2017
Aqueous leaf extract of *Ficus religiosa*	UV, SEM, XRD, FTIR, DLS	577 nm, spherical	Biological activities of the synthesized nanoparticles were confirmed based on its stable anticancer effects. The apoptotic effect of copper oxide nanoparticles is mediated by the generation of reactive oxygen species (ROS) involving the disruption of mitochondrial membrane potential (Δψm) in A549 cells.	[171] Sankar et al. 2014
Fungal strains of *Aspergillus niger* strain STA9	UV, FTIR	500 nm, spherical	MTT assay revealed that CuNPs have a significant cytotoxic effect against human hepatocellular carcinoma cell lines (Huh-7).	[172] Noor et al. 2020
Leaves of *Acalypha indica*	UV, SEM, TEM, FTIR, energy dispersive x-ray spectroscopy	29 nm, spherical	The cytotoxicity activity of *A. indica* mediated copper nanoparticles was evaluated by MTT assay against MCF-7 breast cancer cell lines and confirmed that copper oxide nanoparticles have cytotoxicity activity.	[168] Sivaraj et al. 2014

(Continued)

TABLE 2.2 (CONTINUED)
Anticancer Application of Green Synthesis Metallic Nanoparticles

Carrier Name	Characterization Techniques	Morphology	Mechanism	Reference
Aqueous black bean extract	XRD, FT-IR, XPS, Raman spectroscopy, DLS, TEM, SAED, SEM, and EDX	26–27 nm, Spherical	The cytotoxic effect of the CuO NPs was determined by sulforhodamine-B assay. In vitro anticancer results indicated that CuO NPs induced intracellular ROS generation in a dose-dependent manner and significantly reduced cervical carcinoma colonies. CuO NPs can induce apoptosis and suppress the proliferation of HeLa cells.	[174] Nagajyothi et al. 2017
Using fruit and leaf Andean blackberry (*Rubus glaucus* Benth.)	UV, TEM, XRD, DLS	43–52 nm, spherical	The free radical scavenging activity of the CuO NPs was measured by using DPPH. Antioxidant was evaluated against DPPH. CuO NPs can induce apoptosis and suppress the proliferation of HeLa cells.	[175] Kumar et al. 2017
Aqueous leaf extract of *Ficus religiosa*	UV, SEM, XRD, FTIR, DLS	577 nm, spherical	Biological activities of the synthesized nanoparticles were confirmed based on its stable anticancer effects. The apoptotic effect of copper oxide nanoparticles is mediated by the generation of reactive oxygen species (ROS) involving the disruption of mitochondrial membrane potential ($\Delta\psi m$) in A549 cells.	[176] Sankar et al. 2014
Fungal strains of *Aspergillus niger* strain STA9	UV, FTIR	500 nm, spherical	MTT assay revealed that CuNPs have a significant cytotoxic effect against human hepatocellular carcinoma cell lines (Huh-7).	[177] Noor et al. 2020

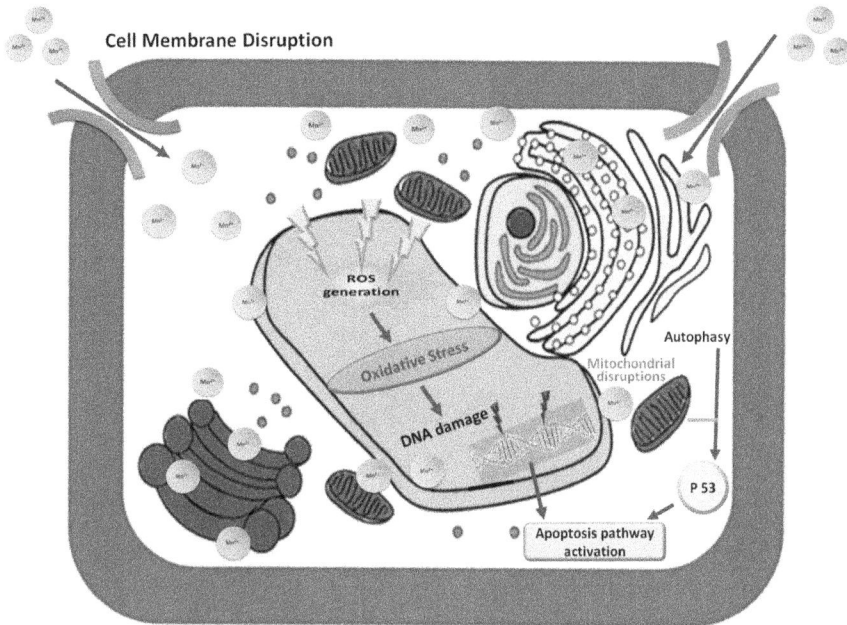

FIGURE 2.4 Anticancer mechanisms of metallic nanoparticles.

an immense ecological concern mainly due to its toxicity. For that reason, the reductions of these pollutants are vital. The reduction product of 4-NP, 4-AP, has been used in various fields as an intermediary, in particular for paracetamol, sulfur dyes, film developers, rubber antioxidants precursors in corrosion inhibitors, and antipyretic and analgesic drugs [174].The easiest way to reduce 4-NP is by simply introducing sodium borohydride ($NaBH_4$) as a reducing agent and a metal catalyst, in particular Ag/Au/Cu/Pd nanoparticles [175–180]. MNPs show splendid catalytic potential mainly due to their wide surface-to-volume ratio and high rate of surface adsorption. On the other hand, the feasibility of the reaction declines because of the extensive potential difference amid the acceptor molecule (nitrophenolate ion) and donor ($NaBH_4/H_3BO_3$), which accounts for the superior activation energy barricade. The rate of reaction is eventually promoted by MNPs by simply escalating the adsorption of reactants on the surface of the MNPs [181]. The absorption spectra of 4-NPs were characterized in the presence of sodium hydroxide (NaOH) by nitrophenolate ions, and a peak was observed at 400 nm. The catalytic properties of the MNPs were studied by the transformation of 4-NP into 4-AP in the aqueous phase using $NaBH_4$. We have observed that 4-NP shows a strong absorption spectrum around 320 nm, which transforms to 400 nm upon the addition of $NaBH_4$.

The C/C_0 plots of prepared MNPs were used as catalysts, and the relative efficiency en route for catalytic conversion of 4-NP can be plotted against degradation time. The rate constants can be evaluated by plotting a graph for $\ln C/C_0$ vs time.

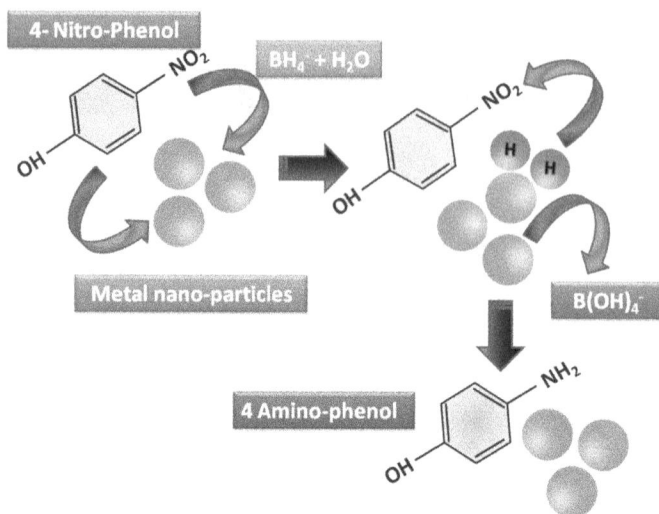

SCHEME 2.2 Catalytic properties of the MNPs show conversion of 4-NP into 4-AP in aqueous medium by NaBH$_4$.

The catalytic reduction mechanism of 4-NP by the prepared MNPs to form 4-AP is as shown in Scheme 2.2. The procedure of reduction of catalyst of 4-NP involves subsequent steps:

1. Adsorption of BH$_4^-$ ions on MNPs
2. Transfer of electrons of MNPs from BH$_4^-$ ions to 4-NP
3. Generation of H$^+$ ions in the solution and forming of 4-AP

In the presence of MNPs, BH$_4^-$ ions are adsorbed on the MNPs' surfaces and electrons are discharged from BH$_4^-$ ions (donor) through MNPs to 4-NP (acceptor). The necessary amount of H$^+$ ions are generated due to the aqueous solution of MNPs, and we observed that an absolute transformation takes place from 4-NP to 4-AP [182].

The surface area of MNPs plays a very significant role in the catalytic conversion of 4-NP. Due to the wide surface area of MNPs, the adsorption rate is improved efficiently by 4-nitrophenolate anions leading to improvement in the catalytic activity. The work reported in this chapter reveals a simple green route preparation of MNPs in aqueous media to be used as an efficient catalyst toward the degradation of toxic 4-NP in wastewater.

2.12 CATALYTIC ACTIVITY OF MNPS BY METHYLENE BLUE (MB) AND METHYL ORANGE (MO)

Photocatalytic behavior of the MNPs was measured through degradation of commonly used dyes, in particular methylene blue (MB), rhodamine blue (R6G), crystal violet (CV), and methyl orange (MO), under solar energy, using MNPs as photocatalysts.

In our study, we have taken MB for degradation in water using MNPs as photocatalysts. We have observed that with increased exposure time, quenching was observed in the absorption spectra. We observed that the characteristic peak of MB ($\lambda = 667$ nm) diminishes with exposure to solar energy. The temporal changes in the peak at 463 nm resemble the characteristic feature of MO when exposed to solar energy was also noted. The absorption spectra of MO decrease significantly with increasing exposure time of solar energy, and the characteristic absorption spectra of MO completely diminish after a certain interval of time, showing the MO degradation kinetics by MNPs and their activities for various photocatalysts for degradation of MO.

The degradation rate constants for MB dye using MNPs as photocatalysts were determined by the standard equation [183]:

$$C = C_0 \exp(-kt) \tag{2.1}$$

where C and C_0 are the concentration of the catalyst in the presence and absence of MB, respectively, and k represents the rate constant.

The reusability of MNPs used as photocatalysts were measured for three runs and it efficiently degrades MB. We show the mechanism for the degradation of MB using MNPs as catalysts in a schematic diagram represented by Scheme 2.3.

Initially, we observe that when solar energy is incident on the system containing MNPs mixed with MB in an appropriate ratio, the electrons are excited from the MNPs. When these photogenerated holes interact with the aqueous solution, it generates hydroxyl radicals ($\bullet$OH), whereas when the photogenerated electrons interact with oxygen (O_2) molecules superoxide radicals ($\bullet O_2$) is formed. When these radicals react with MB, then CO_2 and H_2O are formed, which help in the efficient degradation of the organic contaminants [184]. For a clearer understanding, we describe the whole process step by step using the following equations:

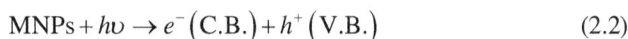

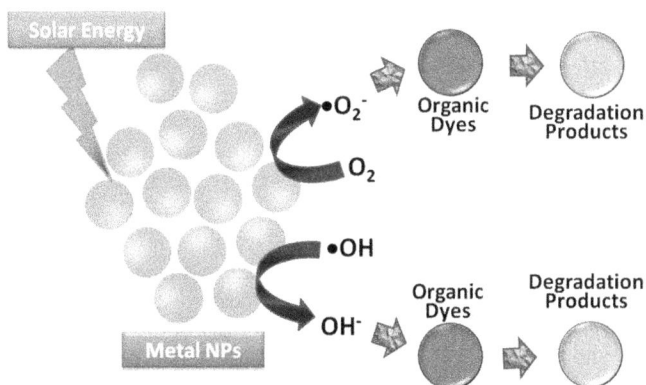

$$\text{MNPs} + h\upsilon \rightarrow e^- \left(\text{C.B.}\right) + h^+ \left(\text{V.B.}\right) \tag{2.2}$$

SCHEME 2.3 Degradation of methylene blue (MB) organic dye using MNPs as catalysts.

$$e^- + O_2 \rightarrow \bullet O_2^-$$ (2.3)

$$h^+ + OH^- \rightarrow \bullet OH$$ (2.4)

$$h^+ + H_2O \rightarrow H^+ + \bullet OH$$ (2.5)

$$\bullet O_2 + MB \, dye \rightarrow degraded \, products$$ (2.6)

$$\bullet OH + MB \, dye \rightarrow degraded \, products$$ (2.7)

2.13 CONCLUSION AND FUTURE PERSPECTIVES

This chapter reports on the green synthesis approach for the preparation of metallic nanoparticles via the green route using biological materials, such as microorganisms (bacteria, fungi, algae, virus), plant extracts, and polymers and their applications in the fields of medical science, biotechnology, pharmaceuticals, material science, and bioelectronics. MNPs are efficiently targeted for drug delivery, cancer treatments, photothermal therapy, and infectious and neurological disorders as well as photocatalysis. MNPs have been efficiently used as catalysts for the degradation of organic contaminants from the textile industry and wastewater. UV-vis spectra depict efficient degradation of MB when exposed to solar energy for various intervals of time. We are planning to use core-shell nanoparticles by varying the core-to-shell ratio with metal-semiconductor nanoparticles for their possible application in the fields of medical sciences.

ACKNOWLEDGMENTS

The authors gratefully acknowledge the contributions of their collaborators and co-workers mentioned in the cited references. NIMS Institute of Pharmacy, NIMS University, Jaipur, Rajasthan, India, for their support. J. R. Ansari gratefully acknowledges the Vice Chancellor, Pro-Vice-Chancellor, and Head, Department of Physics, PDM University, Bahadurgarh, India, for their kind cooperation and support. A. Waziri thankfully acknowledges GGSIPU, New Delhi, for the financial support in the form of STRF.

REFERENCES

1. Ravishankar Rai, V., "Nanoparticles and their potential application as antimicrobials." (2011).
2. Feynman, R. "There's plenty of room at the bottom." *Science* 254 (1991): 1300–1301.
3. Parak, Wolfgang J., Daniele Gerion, Teresa Pellegrino, Daniela Zanchet, Christine Micheel, Shara C. Williams, Rosanne Boudreau, Mark A. Le Gros, Carolyn A. Larabell, and A. Paul Alivisatos. "Biological applications of colloidal nanocrystals." *Nanotechnology* 14(7) (2003): R15.
4. Heiligtag, Florian J., and Markus Niederberger. "The fascinating world of nanoparticle research." *Materials Today* 16(7–8) (2013): 262–271.

5. De, Mrinmoy, Partha S. Ghosh, and Vincent M. Rotello. "Applications of nanoparticles in biology." *Advanced Materials* 20(22) (2008): 4225–4241.

6. Javed, M.N., F.H. Pottoo, A. Shamim, M.S. Hasnain, and M.S. Alam. "Design of experiments for the development of nanoparticles, nanomaterials, and nanocomposites." In *Design of Experiments for Pharmaceutical Product Development*, Beg S. (ed.). Springer, Singapore (2021): 151–169.

7. Chandra, R., A.K. Chawla, and P. Ayyub. "Optical and structural properties of sputter-deposited nanocrystalline Cu2O films: Effect of sputtering gas." *Journal of Nanoscience and Nanotechnology* 6(4) (2006): 1119–1123.

8. Auffan, Mélanie, Jérôme Rose, Mark R. Wiesner, and Jean-Yves Bottero. "Chemical stability of metallic nanoparticles: A parameter controlling their potential cellular toxicity in vitro." *Environmental Pollution* 157(4) (2009): 1127–1133.

9. Samia, Anna C.S., Smita Dayal, and Clemens Burda. "Quantum dot-based energy transfer: Perspectives and potential for applications in photodynamic therapy." *Photochemistry and Photobiology* 82(3) (2006): 617–625.

10. Hasan, Saba. "A review on nanoparticles: Their synthesis and types." *Research Journal of Recent Sciences* 2277 (2015): 2502.

11. Mishra, S., S. Sharma, M.N. Javed, F.H. Pottoo, M.A. Barkat, M.S. Alam, M. Amir, and M. Sarafroz. "Bioinspired nanocomposites: Applications in disease diagnosis and treatment." *Pharmaceutical Nanotechnology* 7(3) (2019): 206–219.

12. Pottoo, F.H., S. Sharma, M.N. Javed, M.A. Barkat, Alam M.S. Harshita, M.J. Naim, O. Alam, M.A. Ansari, G.E. Barreto, and G.M. Ashraf. "Lipid-based nanoformulations in the treatment of neurological disorders." *Drug Metabolism Reviews* 52(1) (2020): 185–204.

13. Chandra, R., A.K. Chawla, and P. Ayyub. "Optical and structural properties of sputter-deposited nanocrystalline Cu2O films: Effect of sputtering gas." *Journal of Nanoscience and Nanotechnology* 6(4) (2006): 1119–1123.

14. Bhattacharya, Resham, and Priyabrata Mukherjee. "Biological properties of "naked" metal nanoparticles." *Advanced Drug Delivery Reviews* 60(11) (2008): 1289–1306.

15. Narayanan, Radha, and Mostafa A. El-Sayed. "Shape-dependent catalytic activity of platinum nanoparticles in colloidal solution." *Nano Letters* 4(7) (2004): 1343–1348.

16. Moura, D., M.T. Souza, L. Liverani, G. Rella, G.M. Luz, J.F. Mano, and A.R. Boccaccini. "Development of a bioactive glass-polymer composite for wound healing applications." *Materials Science and Engineering: C* 76 (2017): 224–232.

17. Banerjee, Kaushik, Satyajit Das, Pritha Choudhury, Sarbari Ghosh, Rathindranath Baral, and Soumitra Kumar Choudhuri. "A novel approach of synthesizing and evaluating the anticancer potential of silver oxide nanoparticles in vitro." *Chemotherapy* 62(5) (2017): 279–289.

18. Gomez-Romero, Pedro. "Hybrid organic–inorganic materials—In search of synergic activity." *Advanced Materials* 13(3) (2001): 163–174.

19. Shaikh, Shoyebmohamad F., Rajaram S. Mane, ByoungKoun Min, Yun Jeong Hwang, and Oh-shim Joo. "D-sorbitol-induced phase control of TiO$_2$ nanoparticles and its application for dye-sensitized solar cells." *Scientific Reports* 6 (2016): 20103.

20. Alam, M.S., A. Garg, F.H. Pottoo, M.K. Saifullah, A.I. Tareq, O. Manzoor, M. Mohsin, and M.N. Javed. Gum ghatti mediated, one pot green synthesis of optimized gold nanoparticles: Investigation of process-variables impact using Box-Behnken based statistical design. *International Journal of Biological Macromolecules* 104(A) (2017): 758–767.

21. Javed, M.N., F.H. Pottoo, and M.S. Alam, *Metallic Nanoparticle Alone and/or in Combination as Novel Agent for the Treatment of Uncontrolled Electric Conductance Related Disorders and/or Seizure, Epilepsy & Convulsions.* Patent acquired on October 10, 2016.

22. Dubchak, S., A. Ogar, J.W. Mietelski, and K. Turnau. "Influence of silver and titanium nanoparticles on arbuscular mycorrhiza colonization and accumulation of radiocaesium in Helianthus annuus." *Spanish Journal of Agricultural Research* 1 (2010): 103–108.

23. Pacioni, Natalia L., Claudio D. Borsarelli, Valentina Rey, and Alicia V. Veglia. "Synthetic routes for the preparation of silver nanoparticles." In *Silver Nanoparticle Applications*. Springer, Cham (2015): 13–46.

24. Rajput, Namita. "Methods of preparation of nanoparticles-a review." *International Journal of Advances in Engineering and Technology* 7(6) (2015): 1806.

25. Chen, Wei, Weiping Cai, Liang Zhang, Guozhong Wang, and Lide Zhang. "Sonochemical processes and formation of gold nanoparticles within pores of mesoporous silica." *Journal of Colloid and Interface Science* 238(2) (2001): 291–295.

26. Geethalakshmi, R., and D.V. Sarada. "Gold and silver nanoparticles from *Trianthema decandra*: Synthesis, characterization, and antimicrobial properties." *International Journal of Nanomedicine* 7 (2012): 5375–5384.

27. Kumar, Harish, Nagasamy Venkatesh, Himangshu Bhowmik, and Anuttam Kuila. "Metallic nanoparticle: A review." *Biomedical Journal of Scientific & Technical Research* 4(2) (2018): 3765–3775.

28. Jamkhande, Prasad Govindrao, Namrata W. Ghule, Abdul Haque Bamer, and Mohan G. Kalaskar. "Metal nanoparticles synthesis: An overview on methods of preparation, advantages and disadvantages, and applications." *Journal of Drug Delivery Science and Technology* 53 (2019): 101174.

29. Willner, Itamar, Ronan Baron, and Bilha Willner. "Growing metal nanoparticles by enzymes." *Advanced Materials* 18(9) (2006): 1109–1120.

30. Klaus, Tanja, Ralph Joerger, Eva Olsson, and Claes-Göran Granqvist. "Silver-based crystalline nanoparticles, microbially fabricated." *Proceedings of the National Academy of Sciences of the United States of America* 96(24) (1999): 13611–13614.

31. Konishi, Yasuhiro, Kaori Ohno, Norizoh Saitoh, Toshiyuki Nomura, Shinsuke Nagamine, Hajime Hishida, Yoshio Takahashi, and Tomoya Uruga. "Bioreductive deposition of platinum nanoparticles on the bacterium Shewanella algae." *Journal of Biotechnology* 128(3) (2007): 648–653.

32. Vigneshwaran, N., N.M. Ashtaputre, P.V. Varadarajan, R.P. Nachane, K.M. Paralikar, and R.H. Balasubramanya. "Biological synthesis of silver nanoparticles using the fungus *Aspergillus flavus*." *Materials Letters* 61(6) (2007): 1413–1418.

33. Shankar, S. Shiv, Akhilesh Rai, Balaprasad Ankamwar, Amit Singh, Absar Ahmad, and Murali Sastry. "Biological synthesis of triangular gold nanoprisms." *Nature Materials* 3(7) (2004): 482–488.

34. Ahmad, Naheed, Seema Sharma, V.N. Singh, S.F. Shamsi, Anjum Fatma, and B.R. Mehta. "Biosynthesis of silver nanoparticles from *Desmodium triflorum*: A novel approach towards weed utilization." *Biotechnology Research International* 2011 (2011).

35. Armendariz, V., and J.L. Gardea-Torresdey. "Jose Yacaman M, Gonzalez J, Herrera I, Parsons JG (2002)." In *Proceedings of Conference on Application of Waste Remediation Technologies to Agricultural Contamination of Water Resources*. Kansas City, MO, USA.

36. Kim, B.Y., Rutka, J.T., and Chan WC. "Nanomedicine." *New England Journal of Medicine* 363(25) (2010): 2434–2443.

37. Kyriacou, Sophia V., William J. Brownlow, and Xiao-Hong Nancy Xu. "Using nanoparticle optics assay for direct observation of the function of antimicrobial agents in single live bacterial cells." *Biochemistry* 43(1) (2004): 140–147.

38. Narayanan, Kannan Badri, and Natarajan Sakthivel. "Biological synthesis of metal nanoparticles by microbes." *Advances in Colloid and Interface Science* 156(1–2) (2010): 1–13.

39. Pottoo, F.H., N. Tabassum, M.N. Javed, S. Nigar, R. Rasheed, A. Khan, M.A. Barkat, et al. The synergistic effect of raloxifene, fluoxetine, and bromocriptine protects against pilocarpine-induced status epilepticus and temporal lobe epilepsy. *Molecular Neurobiology* 56(2) (2019): 1233–1247.

40. Pottoo, F.H., M. Javed, M. Barkat, M. Alam, J.A. Nowshehri, D.M. Alshayban, and M.A. Ansari. Estrogen and serotonin: Complexity of interactions and implications for epileptic seizures and epileptogenesis. *Current Neuropharmacology* 17(3) (2019): 214–231.

41. Pottoo, F.H., S. Sharma, M.N. Javed, M.A. Barkat, Alam M.S. Harshita, M.J. Naim, O. Alam, M.A. Ansari, G.E. Barreto, and G.M. Ashraf. Lipid-based nanoformulations in the treatment of neurological disorders. *Drug Metabolism Reviews* 52(1) (2020): 185–204.

42. Panáček, Aleš, Libor Kvitek, Robert Prucek, Milan Kolář, Renata Večeřová, Virender K. NaděždaPizúrová, Tat Sharma, Jana Nevěčná, and Radek Zbořil. "Silver colloid nanoparticles: Synthesis, characterization, and their antibacterial activity." *The Journal of Physical Chemistry. Part B, Condensed Matter, Materials, Surfaces, Interfaces and Biophysical* 110(33) (2006): 16248–16253.

43. Sau, Tapan K., Arunangshu Biswas, and Parijat Ray. *Metal Nanoparticles: Synthesis and Applications in Pharmaceutical Sciences*, Sreekanth T., Crans D.C. (eds), (2018), 265 pages. ISBN: 978-3-527-82144-0.

44. Cho, Kyung-Hwan, Jong-Eun Park, Tetsuya Osaka, and Soo-Gil Park. "The study of antimicrobial activity and preservative effects of nanosilver ingredient." *Electrochimica Acta* 51(5) (2005): 956–960.

45. Baker, C., A. Pradhan, L. Pakstis, D.J. Pochan, and S. Ismat Shah. "Synthesis and antibacterial properties of silver nanoparticles." *Journal of Nanoscience and Nanotechnology* 5(2) (2005): 244–249.

46. Martínez-Castañon, G.-A., N. Nino-Martinez, F. Martinez-Gutierrez, J.R. Martinez-Mendoza, and F. Ruiz. "Synthesis and antibacterial activity of silver nanoparticles with different sizes." *Journal of Nanoparticle Research* 10(8) (2008): 1343–1348.

47. Sunada, Kayano, Yoshihiko Kikuchi, Kazuhito Hashimoto, and Akira Fujishima. "Bactericidal and detoxification effects of TiO2 thin film photocatalysts." *Environmental Science and Technology* 32(5) (1998): 726–728.

48. Hu, Chun, Yongqing Lan, Jiuhui Qu, Xuexiang Hu, and Aimin Wang. "Ag/AgBr/TiO2 visible light photocatalyst for destruction of azodyes and bacteria." *The Journal of Physical Chemistry. Part B, Condensed Matter, Materials, Surfaces, Interfaces and Biophysical* 110(9) (2006): 4066–4072.

49. Vijayakumar, P.S., and B.L.V. Prasad. "Intracellular biogenic silver nanoparticles for the generation of carbon supported antiviral and sustained bactericidal agents." *Langmuir* 25(19) (2009): 11741–11747.

50. Gusseme, Bart De, Gijs Du Laing, Tom Hennebel, Piet Renard, Dev Chidambaram, Jeffrey P. Fitts, Els Bruneel, et al. "Virus removal by biogenic cerium." *Environmental Science and Technology* 44(16) (2010): 6350–6356.

51. Beauregard, D.A., P. Yong, L.E. Macaskie, and M.L. Johns. "Using non-invasive magnetic resonance imaging (MRI) to assess the reduction of Cr (VI) using a biofilm–palladium catalyst." *Biotechnology and Bioengineering* 107(1) (2010): 11–20.

52. Creamer, N.J., I.P. Mikheenko, P. Yong, K. Deplanche, D. Sanyahumbi, J. Wood, K. Pollmann, M. Merroun, S. Selenska-Pobell, and L.E. Macaskie. "Novel supported Pd hydrogenation bionanocatalyst for hybrid homogeneous/heterogeneous catalysis." *Catalysis Today* 128(1–2) (2007): 80–87.

53. Shahwan, Talal, S. Abu Sirriah, Muath Nairat, Ezel Boyacı, Ahmet E. Eroğlu, Thomas B. Scott, and Keith R. Hallam. "Green synthesis of iron nanoparticles and their application as a Fenton-like catalyst for the degradation of aqueous cationic and anionic dyes." *Chemical Engineering Journal* 172(1) (2011): 258–266.

54. Das, Sujoy K., Md. Motiar, R. Khan, Arun K. Guha, and Nityananda Naskar. "Bio-inspired fabrication of silver nanoparticles on nanostructured silica: Characterization and application as a highly efficient hydrogenation catalyst." *Green Chemistry* 15(9) (2013): 2548–2557.

55. Baram-Pinto, Dana, Sourabh Shukla, Nina Perkas, Aharon Gedanken, and Ronit Sarid. "Inhibition of herpes simplex virus type 1 infection by silver nanoparticles capped with mercaptoethane sulfonate." *Bioconjugate Chemistry* 20(8) (2009): 1497–1502.

56. Papp, Ilona, Christian Sieben, Kai Ludwig, Meike Roskamp, Christoph Böttcher, Sabine Schlecht, Andreas Herrmann, and Rainer Haag. "Inhibition of influenza virus infection by multivalent sialic-acid-functionalized gold nanoparticles." *Small* 6(24) (2010): 2900–2906.

57. Ramezani, Neda, Zeynab Ehsanfar, Fazel Shamsa, Gholamreza Amin, Hamid R. Shahverdi, Hamid R. Monsef Esfahani, Ali Shamsaie, Reza DolatabadiBazaz, and Ahmad Reza Shahverdi. "Screening of medicinal plant methanol extracts for the synthesis of gold nanoparticles by their reducing potential." *ZeitschriftfürNaturforschung B* 63(7) (2008): 903–908.

58. Valodkar, Mayur, Ravirajsinh N. Jadeja, Menaka C. Thounaojam, Ranjitsinh V. Devkar, and Sonal Thakore. "Biocompatible synthesis of peptide capped copper nanoparticles and their biological effect on tumor cells." *Materials Chemistry and Physics* 128(1–2) (2011): 83–89.

59. Moulton, Michael C., Laura K. Braydich-Stolle, Mallikarjuna N. Nadagouda, Samantha Kunzelman, Saber M. Hussain, and Rajender S. Varma. "Synthesis, characterization and biocompatibility of "green" synthesized silver nanoparticles using tea polyphenols." *Nanoscale* 2(5) (2010): 763–770.

60. Raghunandan, Deshpande, Bhat Ravishankar, D. GanachariSharanbasava, Bedre Mahesh, Vasanth Harsoor, Manjunath S. Yalagatti, M. Bhagawanraju, and A. Venkataraman. "Anti-cancer studies of noble metal nanoparticles synthesized using different plant extracts." *Cancer Nanotechnology* 2(1–6) (2011): 57–65.

61. Schröfel, Adam, Gabriela Kratošová, Ivo Šafařík, Mirka Šafaříková, Ivan Raška, and Leslie M. Shor. "Applications of biosynthesized metallic nanoparticles–A review." *Acta Biomaterialia* 10(10) (2014): 4023–4042.

62. Sperling, Reisa A., and Wolfgang J. Parak. "Surface modification, functionalization and bioconjugation of colloidal inorganic nanoparticles." *Philosophical Transactions of the Royal Society, Series A: Mathematical, Physical and Engineering Sciences* 368(1915) (2010): 1333–1383.

63. Ghosh, Partha, Gang Han, Mrinmoy De, Chae Kyu Kim, and Vincent M. Rotello. "Gold nanoparticles in delivery applications." *Advanced Drug Delivery Reviews* 60(11) (2008): 1307–1315.

64. Nishiyama, Nobuhiro. "Nanomedicine: Nanocarriers shape up for long life." *Nature Nanotechnology* 2(4) (2007): 203–204.

65. Sun, Jian-Bo, Jin-Hong Duan, Shun-Ling Dai, Jun Ren, Lin Guo, Wei Jiang, and Ying Li. "Preparation and anti-tumor efficiency evaluation of doxorubicin-loaded bacterial magnetosomes: Magnetic nanoparticles as drug carriers isolated from *Magnetospirillum gryphiswaldense*." *Biotechnology and Bioengineering* 101(6) (2008): 1313–1320.

66. Venkatpurwar, Vinod, Anjali Shiras, and Varsha Pokharkar. "Porphyran capped gold nanoparticles as a novel carrier for delivery of anticancer drug: In vitro cytotoxicity study." *International Journal of Pharmaceutics* 409(1–2) (2011): 314–320.

67. Zheng, Baozhan, Shunping Xie, Lei Qian, Hongyan Yuan, Dan Xiao, and Martin M.F. Choi. "Gold nanoparticles-coated eggshell membrane with immobilized glucose oxidase for fabrication of glucose biosensor." *Sensors and Actuators. Part B: Chemical* 152(1) (2011): 49–55.

68. Iskandar, Ferry. "Nanoparticle processing for optical applications–A review." *Advanced Powder Technology* 20(4) (2009): 283–292.

69. Fayaz, M., C.S. Tiwary, P.T. Kalaichelvan, and R. Venkatesan. "Blue orange light emission from biogenic synthesized silver nanoparticles using *Trichoderma viride.*" *Colloids and Surfaces, Part B: Biointerfaces* 75(1) (2010): 175–178.

70. Sarkar, R., P. Kumbhakar, and A.K. Mitra. "Green synthesis of silver nanoparticles and its optical properties." Digest Journal of Nanomaterials and Biostructures 5(2) (2010): 491–496.

71. Fayaz, A. Mohammed, M. Girilal, R. Venkatesan, and P.T. Kalaichelvan. "Biosynthesis of anisotropic gold nanoparticles using *Maduca longifolia* extract and their potential in infrared absorption." *Colloids and Surfaces, Part B: Biointerfaces* 88(1) (2011): 287–291.

72. Kalishwaralal, K., V. Deepak, S. Ram Kumar Pandian, M. Kottaisamy, S. BarathmaniKanth, B. Kartikeyan, and S. Gurunathan. "Biosynthesis of silver and gold nanoparticles using *Brevibacterium casei.*" *Colloids and Surfaces, Part B: Biointerfaces* 77(2) (2010): 257–262.

73. Basha, Sabjan Khaleel, Kasivelu Govindaraju, Ramar Manikandan, Jong SeogAhn, Eun Young Bae, and Ganesan Singaravelu. "Phytochemical mediated gold nanoparticles and their PTP 1B inhibitory activity." *Colloids and Surfaces, Part B: Biointerfaces* 75(2) (2010): 405–409.

74. De Corte, Simon, Tom Hennebel, Bart De Gusseme, Willy Verstraete, and Nico Boon. "Bio-palladium: From metal recovery to catalytic applications." *Microbial Biotechnology* 5(1) (2012): 5–17.

75. Hennebel, Tom, Simon De Corte, Willy Verstraete, and Nico Boon. "Microbial production and environmental applications of Pd nanoparticles for treatment of halogenated compounds." *Current Opinion in Biotechnology* 23(4) (2012): 555–561.

76. Yong, P., M. Paterson-Beedle, I.P. Mikheenko, and L.E. Macaskie. "From bio-mineralisation to fuel cells: Biomanufacture of Pt and Pd nanocrystals for fuel cell electrode catalyst." *Biotechnology Letters* 29(4) (2007): 539–544.

77. Dimitriadis, S., N. Nomikou, and A.P. McHale. "Pt-based electro-catalytic materials derived from biosorption processes and their exploitation in fuel cell technology." *Biotechnology Letters* 29(4) (2007): 545–551.

78. Wang, Yonghong, Xiaoxiao He, Kemin Wang, Xiaorong Zhang, and Weihong Tan. "BarbatedSkullcup herb extract-mediated biosynthesis of gold nanoparticles and its primary application in electrochemistry." *Colloids and Surfaces, Part B: Biointerfaces* 73(1) (2009): 75–79.

79. Ghoreishi, Sayed M., Mohsen Behpour, and Maryam Khayatkashani. "Green synthesis of silver and gold nanoparticles using *Rosa damascena* and its primary application in electrochemistry." *Physica E: Low-Dimensional Systems and Nanostructures* 44(1) (2011): 97–104.

80. Hennebel, Tom, Bart De Gusseme, Nico Boon, and Willy Verstraete. "Biogenic metals in advanced water treatment." *Trends in Biotechnology* 27(2) (2009): 90–98.

81. Das, Nilanjana. "Recovery of precious metals through biosorption—A review." *Hydrometallurgy* 103(1–4) (2010): 180–189.

82. Patra, Jayanta Kumar, and Kwang-Hyun Baek. "Green nanobiotechnology: Factors affecting synthesis and characterization techniques." *Journal of Nanomaterials* 2014 (2014).

83. Evanoff, David D., and George Chumanov. "Size-controlled synthesis of nanoparticles. 2. Measurement of extinction, scattering, and absorption cross sections." *The Journal of Physical Chemistry. Part B* 108(37) (2004): 13957–13962.

84. Ghorbani, Hamid Reza. "A review of methods for synthesis of Al nanoparticles." *Oriental Journal of Chemistry* 30(4) (2014): 1941–1949.

85. Sastry, M., A. Ahmad, M.I. Khan, and R. Kumar. Biosynthesis of metal nanoparticles using fungi and actinomycete. *Current Science* 85 (2003): 162–170.
86. Korbekandi, H., S. Iravani, and S. Abbasi. Production of nanoparticles using organisms. *Critical Reviews in Biotechnology* 29(4) (2009): 279–306.
87. Nadagouda, Mallikarjuna N., Thomas F. Speth, and Rajender S. Varma. "Microwave-assisted green synthesis of silver nanostructures." *Accounts of Chemical Research* 44(7) (2011): 469–478.
88. Eustis, Susie, Galina Krylova, Anna Eremenko, Natalie Smirnova, Alexander W. Schill, and Mostafa El-Sayed. "Growth and fragmentation of silver nanoparticles in their synthesis with a fs laser and CW light by photo-sensitization with benzophenone." *Photochemical and Photobiological Sciences* 4(1) (2005): 154–159.
89. Polshettiwar, Vivek, Mallikarjuna N. Nadagouda, and Rajender S. Varma. "Microwave-assisted chemistry: A rapid and sustainable route to synthesis of organics and nanomaterials." *Australian Journal of Chemistry* 62(1) (2009): 16–26.
90. Kirstein, Stefan, Hans von Berlepsch, and Christoph Böttcher. "Photo-induced reduction of Noble metal ions to metal nanoparticles on tubular J-aggregates." *International Journal of Photoenergy* 2006 (2006).
91. Jin, Rongchao, Y. Charles Cao, Encai Hao, Gabriella S. Métraux, George C. Schatz, and Chad A. Mirkin. "Controlling anisotropic nanoparticle growth through plasmon excitation." *Nature* 425(6957) (2003): 487–490.
92. Krutyakov, Yu A., A.Yu. Olenin, A.A. Kudrinskii, P.S. Dzhurik, and G.V. Lisichkin. "Aggregative stability and polydispersity of silver nanoparticles prepared using two-phase aqueous organic systems." *Nanotechnologies in Russia* 3(5–6) (2008): 303–310.
93. Kim, Myungjoon, Saho Osone, Taesung Kim, Hidenori Higashi, and Takafumi Seto. "Synthesis of nanoparticles by laser ablation: A review." *KONA Powder and Particle Journal* 34 (2017): 2017009.
94. Yadav, T.P., R. Manohar Yadav, and D. Pratap Singh. "Mechanical milling: A top down approach for the synthesis of nanomaterials and nanocomposites." *Nanoscience and Nanotechnology* 2(3) (2012): 22–48.
95. Kalishwaralal, K., V. Deepak, S. Ramkumarpandian, H. Nellaiah, and G. Sangiliyandi. "Extracellular biosynthesis of silver nanoparticles by the culture supernatant of *Bacillus licheniformis*." *Materials Letters* 62(29) (2008): 4411–4413.
96. Nair, B., and T. Pradeep. "Preparation of gold nanoparticles from *Mirabilis jalapa* flowers." *Crystal Growth and Design* 2(4) (2002): 293–298.
97. Ahmad, Absar, Priyabrata Mukherjee, Satyajyoti Senapati, Deendayal Mandal, M. Islam Khan, Rajiv Kumar, and Murali Sastry. "Extracellular biosynthesis of silver nanoparticles using the fungus Fusarium oxysporum." *Colloids and Surfaces, Part B: Biointerfaces* 28(4) (2003): 313–318.
98. Singaravelu, G., J.S. Arockiamary, V. Ganesh Kumar, and K. Govindaraju. "A novel extracellular synthesis of monodisperse gold nanoparticles using marine alga, *Sargassum wightii* Greville." *Colloids and Surfaces, Part B: Biointerfaces* 57(1) (2007): 97–101.
99. Shankar, S. Shiv, Akhilesh Rai, Absar Ahmad, and Murali Sastry. "Rapid synthesis of Au, Ag, and bimetallic Au core–Ag shell nanoparticles using Neem (Azadirachta indica) leaf broth." *Journal of Colloid and Interface Science* 275(2) (2004): 496–502.
100. Hasnain, M.S., M.N. Javed, M.S. Alam, P. Rishishwar, S. Rishishwar, S. Ali, A.K. Nayak, and S. Beg. Purple Heart plant leaves extract-mediated silver nanoparticle synthesis: Optimization by Box-Behnken design. *Materials Science and Engineering C* 99 (2019): 1105–1114.
101. Burduşel, Alexandra-Cristina, Oana Gherasim, Alexandru Mihai Grumezescu, Laurenţiu Mogoantă, Anton Ficai, and Ecaterina Andronescu. "Biomedical applications of silver nanoparticles: An up-to-date overview." *Nanomaterials* 8(9) (2018): 681.

102. Daraee, Hadis, Ali Eatemadi, Elham Abbasi, Sedigheh Fekri Aval, Mohammad Kouhi, and Abolfazl Akbarzadeh. "Application of gold nanoparticles in biomedical and drug delivery." *Artificial Cells, Nanomedicine, and Biotechnology* 44(1) (2016): 410–422.

103. Ebrahimnezhad, Zohreh, Nosratollah Zarghami, Manoutchehr Keyhani, Soumaye Amirsaadat, Abolfazl Akbarzadeh, Mohammad Rahmati, Zohreh Mohammad Taheri, and Kazem Nejati-Koshki. "Inhibition of hTERT gene expression by silibinin-loaded PLGA-PEG-Fe3O4 in T47D breast cancer cell line." *BioImpacts: BI* 3(2) (2013): 67.

104. Puja, Patel, and Ponnuchamy Kumar. "A perspective on biogenic synthesis of platinum nanoparticles and their biomedical applications." *Spectrochimica Acta. Part A: Molecular and Biomolecular Spectroscopy* 211 (2019): 94–99.

105. Akhtar, Mohd Sayeed, Jitendra Panwar, and Yeoung-Sang Yun. "Biogenic synthesis of metallic nanoparticles by plant extracts." *ACS Sustainable Chemistry and Engineering* 1(6) (2013): 591–602.

106. Raghupathi, K.R., R.T. Koodali, and A.C. Manna. Size-dependent bacterial growth inhibition and mechanism of antibacterial activity of zinc oxide nanoparticles." *Langmuir* 27(7) (2011): 4020–4028.

107. Jha, Anal K., and K. Prasad. "Green synthesis of silver nanoparticles using Cycas leaf." *International Journal of Green Nanotechnology: Physics and Chemistry* 1(2) (2010): 110-P117.

108. Fouda, Amr, Saad El-Din Hassan, Abdullah M. Abdo, and Mamdouh S. El-Gamal. "Antimicrobial, antioxidant and larvicidal activities of spherical silver nanoparticles synthesized by endophytic Streptomyces spp.." *Biological Trace Element Research* 195(2) (2020): 707–724.

109. Hassan, Saad El-Din, Salem S. Salem, Amr Fouda, Mohamed A. Awad, Mamdouh S. El-Gamal, and Abdullah M. Abdo. "New approach for antimicrobial activity and biocontrol of various pathogens by biosynthesized copper nanoparticles using endophytic actinomycetes." *Journal of Radiation Research and Applied Sciences* 11(3) (2018): 262–270.

110. Kumar, P. Vijaya, and S.M.J. Kala. Green synthesis, characterisation and biological activity of platinum nanoparticle using *Croton caudatus* Geisel leaf extract. International Journal of Recent Research Aspects. *(Special Issue: Conscientious Computing Technologies)* (2018): 608–612.

111. Molaei, Rahim, Khalil Farhadi, Mehrdad Forough, and SalaheddinHajizadeh. "Green biological fabrication and characterization of highly monodisperse palladium nanoparticles using *Pistacia atlantica* fruit broth." *Journal of Nanostructures* 8(1) (2018): 47–54.

112. Muthukumar, Harshiny, Samsudeen Naina Mohammed, Nivedhini Iswarya Chandrasekaran, Aiswarya Devi Sekar, Arivalagan Pugazhendhi, and Manickam Matheswaran. "Effect of iron doped zinc oxide nanoparticles coating in the anode on current generation in microbial electrochemical cells." *International Journal of Hydrogen Energy* 44(4) (2019): 2407–2416.

113. Ahmad, Absar, Priyabrata Mukherjee, Satyajyoti Senapati, Deendayal Mandal, M. Islam Khan, Rajiv Kumar, and Murali Sastry. "Extracellular biosynthesis of silver nanoparticles using the fungus *Fusarium oxysporum*." *Colloids and Surfaces, Part B: Biointerfaces* 28(4) (2003): 313–318.

114. Muthukumar, Harshiny, SamsudeenNaina Mohammed, NivedhiniIswarya Chandrasekaran, Aiswarya Devi Sekar, Arivalagan Pugazhendhi, and Manickam Matheswaran. "Effect of iron doped zinc oxide nanoparticles coating in the anode on current generation in microbial electrochemical cells." *International Journal of Hydrogen Energy* 44(4) (2019): 2407–2416.

115. Goutam, Surya Pratap, Gaurav Saxena, Varunika Singh, Anil Kumar Yadav, Ram Naresh Bharagava, and Khem B. Thapa. "Green synthesis of TiO2 nanoparticles using leaf extract of Jatropha curcas L. for photocatalytic degradation of tannery wastewater." *Chemical Engineering Journal* 336 (2018): 386–396.

116. Ahmed, Qadruddin, Nidhi Gupta, Ajeet Kumar, and Surendra Nimesh. "Antibacterial efficacy of silver nanoparticles synthesized employing Terminalia Arjuna bark extract." *Artificial Cells, Nanomedicine, and Biotechnology* 45(6) (2017): 1192–1200.

117. Kumari, R. Mankamna, Nikita Thapa, Nidhi Gupta, Ajeet Kumar, and Surendra Nimesh. "Antibacterial and photocatalytic degradation efficacy of silver nanoparticles biosynthesized using Cordia dichotoma leaf extract." *Advances in Natural Sciences: Nanoscience and Nanotechnology* 7(4) (2016): 045009.

118. Arya, Geeta, Nitin Kumar, Nidhi Gupta, Ajeet Kumar, and Surendra Nimesh. "Antibacterial potential of silver nanoparticles biosynthesised using *Canarium ovatum* leaves extract." *IET Nanobiotechnology* 11(5) (2016): 506–511.

119. Arya, Geeta, R. Mankamna Kumari, Nidhi Gupta, Ajeet Kumar, Ramesh Chandra, and Surendra Nimesh. "Green synthesis of silver nanoparticles using *Prosopis juliflora* bark extract: Reaction optimization, antimicrobial and catalytic activities." *Artificial Cells, Nanomedicine, and Biotechnology* 46(5) (2018): 985–993.

120. Arya, Geeta, Ashish K. Malav, Nidhi Gupta, Ajeet Kumar, and Surendra Nimesh. "Biosynthesis and in vitro antimicrobial potential of silver nanoparticles prepared using Dicoma tomentosa Plant extract." *Nanoscience and Nanotechnology-Asia* 8(2) (2018): 240–247.

121. Arya, Geeta, Nikita Sharma, Jahangir Ahmed, Nidhi Gupta, Ajeet Kumar, Ramesh Chandra, and Surendra Nimesh. "Degradation of anthropogenic pollutant and organic dyes by biosynthesized silver nano-catalyst from Cicer arietinum leaves." *Journal of Photochemistry and Photobiology, Part B: Biology* 174 (2017): 90–96.

122. Goyal, Shivangi, Nidhi Gupta, Ajeet Kumar, Sreemoyee Chatterjee, and Surendra Nimesh. "Antibacterial, anticancer and antioxidant potential of silver nanoparticles engineered using *Trigonella foenum*-Graecum seed extract." *IET Nanobiotechnology* 12(4) (2018): 526–533.

123. Arya, Geeta, R. Mankamna Kumari, Richa Pundir, Sreemoyee Chatterjee, Nidhi Gupta, Ajeet Kumar, Ramesh Chandra, and Surendra Nimesh. "Versatile biomedical potential of biosynthesized silver nanoparticles from *Acacia nilotica* bark." *Journal of Applied Biomedicine* 17(2) (2019): 115–124.

124. Akhtar, Mohd Sayeed, Jitendra Panwar, and Yeoung-Sang Yun. "Biogenic synthesis of metallic nanoparticles by plant extracts." *ACS Sustainable Chemistry and Engineering* 1(6) (2013): 591–602.

125. Naseer, A., A. Ali, S. Ali, A. Mahmood, H.S. Kusuma, A. Nazir, M. Yaseen, M.I. Khan, A. Ghaffar, M. Abbas, and M. Iqbal. "Biogenic and eco-benign synthesis of platinum nanoparticles (Pt NPs) using plants aqueous extracts and biological derivatives: Environmental, biological and catalytic applications." *Journal of Materials Research and Technology* 9(4) (2020): 9093–9107.

126. Gholami-Shabani, Mohammadhassan, Z. Gholami-Shabani, M. Shams-Ghahfarokhi, Azim Akbarzadeh, GhRiazi, and Mehdi Razzaghi-Abyaneh. "Biogenic approach using sheep milk for the synthesis of platinum nanoparticles: The role of milk protein in platinum reduction and stabilization." *International Journal of Nanoscience and Nanotechnology* 12(4) (2016): 199–206.

127. Coates, Anthony R.M., Gerry Halls, and Yanmin Hu. "Novel classes of antibiotics or more of the same?." *British Journal of Pharmacology* 163(1) (2011): 184–194.

128. Tanwar, Jyoti, Shrayanee Das, Zeeshan Fatima, and Saif Hameed. "Multidrug resistance: An emerging crisis." *Interdisciplinary Perspectives on Infectious Diseases* 2014 (2014).

129. Aslam, Bilal, Wei Wang, Muhammad Imran Arshad, Mohsin Khurshid, Saima Muzammil, Muhammad Hidayat Rasool, Muhammad Atif Nisar et al. "Antibiotic resistance: A rundown of a global crisis." *Infection and Drug Resistance* 11 (2018): 1645.
130. Slavin, Yael N., Jason Asnis, Urs O. Häfeli, and Horacio Bach. "Metal nanoparticles: Understanding the mechanisms behind antibacterial activity." *Journal of Nanobiotechnology* 15(1) (2017): 1–20.
131. Wang, Linlin, Chen Hu, and Longquan Shao. "The antimicrobial activity of nanoparticles: Present situation and prospects for the future." *International Journal of Nanomedicine* 12 (2017): 1227.
132. Kumari, R. Mankamna, Nikita Thapa, Nidhi Gupta, Ajeet Kumar, and Surendra Nimesh. "Antibacterial and photocatalytic degradation efficacy of silver nanoparticles biosynthesized using *Cordia dichotoma* leaf extract." *Advances in Natural Sciences: Nanoscience and Nanotechnology* 7(4) (2016): 045009.
133. Adams, Clara P., Katherine A. Walker, Sherine O. Obare, and Kathryn M. Docherty. "Size-dependent antimicrobial effects of novel palladium nanoparticles." *PLOS ONE* 9(1) (2014): e85981.
134. Pati, Rashmirekha, Ranjit Kumar Mehta, Soumitra Mohanty, Avinash Padhi, Mitali Sengupta, Baskarlingam Vaseeharan, Chandan Goswami, and Avinash Sonawane. "Topical application of zinc oxide nanoparticles reduces bacterial skin infection in mice and exhibits antibacterial activity by inducing oxidative stress response and cell membrane disintegration in macrophages." *Nanomedicine: Nanotechnology, Biology and Medicine* 10(6) (2014): 1195–1208.
135. Kumar, Ashutosh, Alok K. Pandey, Shashi S. Singh, Rishi Shanker, and Alok Dhawan. "Engineered ZnO and TiO2 nanoparticles induce oxidative stress and DNA damage leading to reduced viability of Escherichia coli." *Free Radical Biology and Medicine* 51(10) (2011): 1872–1881.
136. Arya, Geeta, R. Mankamna Kumari, Nikita Sharma, Sreemoyee Chatterjee, Nidhi Gupta, Ajeet Kumar, and Surendra Nimesh. "Evaluation of antibiofilm and catalytic activity of biogenic silver nanoparticles synthesized from Acacia nilotica leaf extract." *Advances in Natural Sciences: Nanoscience and Nanotechnology* 9(4) (2018): 045003.
137. Ibrahim, Haytham M.M. "Green synthesis and characterization of silver nanoparticles using banana peel extract and their antimicrobial activity against representative microorganisms." *Journal of Radiation Research and Applied Sciences* 8(3) (2015): 265–275.
138. Kumar, Deenadayalan Ashok, V. Palanichamy, and Selvaraj Mohana Roopan. "Green synthesis of silver nanoparticles using *Alternanthera dentata* leaf extract at room temperature and their antimicrobial activity." *Spectrochimica Acta. Part A: Molecular and Biomolecular Spectroscopy* 127 (2014): 168–171.
139. Gopinath, Kasi, Shanmugam Gowri, and Ayyakannu Arumugam. "Phytosynthesis of silver nanoparticles using *Pterocarpus santalinus* leaf extract and their antibacterial properties." *Journal of Nanostructure in Chemistry* 3(1) (2013): 68.
140. Carson, Laura, Subhani Bandara, Marshall Joseph, Tony Green, Tony Grady, Godson Osuji, Aruna Weerasooriya, Peter Ampim, and Selamawit Woldesenbet. "Green synthesis of silver nanoparticles with antimicrobial properties using *Phyla dulcis* plant extract." *Foodborne Pathogens and Disease* 17(8) (2020). https://doi.org/10.1089/fpd.2019.2714.
141. Otari, S.V., R.M. Patil, S.J. Ghosh, N.D. Thorat, and S.H. Pawar. "Intracellular synthesis of silver nanoparticle by Actinobacteria and its antimicrobial activity." *Spectrochimica Acta. Part A: Molecular and Biomolecular Spectroscopy* 136(B) (2015): 1175–1180.
142. Manikandakrishnan, Muthushanmugam, Subramanian Palanisamy, Manoharan Vinosha, Baskaran Kalanjiaraja, Sonaimuthu Mohandoss, Ramar Manikandan, Mehdi

Tabarsa, SangGuan You, and Narayanasamy Marimuthu Prabhu. "Facile green route synthesis of gold nanoparticles using *Caulerpa racemosa* for biomedical applications." *Journal of Drug Delivery Science and Technology* 54 (2019): 101345.

143. Bhau, B.S., Sneha Ghosh, Sangeeta Puri, B. Borah, D.K. Sarmah, and Raju Khan. "Green synthesis of gold nanoparticles from the leaf extract of *Nepenthes khasiana* and antimicrobial assay." *Advanced Materials Letters* 6(1) (2015): 55–58.

144. Mollick, Md., Masud Rahaman, Biplab Bhowmick, Dibyendu Mondal, Dipanwita Maity, Dipak Rana, Sandeep Kumar Dash, Sourav Chattopadhyay, et al. "Anticancer (in vitro) and antimicrobial effect of gold nanoparticles synthesized using *Abelmoschus esculentus* (L.) pulp extract via a green route." *RSC Advances* 4(71) (2014): 37838–37848.

145. Bindhu, M.R., and M. Umadevi. "Antibacterial activities of green synthesized gold nanoparticles." *Materials Letters* 120 (2014): 122–125.

146. Muthukumar, Thangavelu, Balaji Sambandam, Adithan Aravinthan, Thotapalli Parvathaleswara Sastry, and Jong-Hoon Kim. "Green synthesis of gold nanoparticles and their enhanced synergistic antitumor activity using HepG2 and MCF7 cells and its antibacterial effects." *Process Biochemistry* 51(3) (2016): 384–391.

147. Qamar, Hina, Sumbul Rehman, Dushyant Kumar Chauhan, Ashok Kumar Tiwari, and Vikramaditya Upmanyu. "Green synthesis, characterization and antimicrobial activity of copper oxide nanomaterial derived from *Momordica charantia*." *International Journal of Nanomedicine* 15 (2020): 2541.

148. Murthy, H.C., Tegene Desalegn, Mebratu Kassa, Buzuayehu Abebe, and Temesgen Assefa. "Synthesis of green copper nanoparticles using medicinal plant *Hagenia abyssinica* (Brace) JF. Gmel. Leaf extract: Antimicrobial properties." *Journal of Nanomaterials* 2020 (2020).

149. Jayarambabu, N., A. Akshaykranth, T. Venkatappa Rao, K. Venkateswara Rao, and R. Rakesh Kumar. "Green synthesis of Cu nanoparticles using *Curcuma longa* extract and their application in antimicrobial activity." *Materials Letters* 259 (2020): 126813.

150. Bhavyasree, P.G., and T.S. Xavier. "Green synthesis of copper oxide/carbon nanocomposites using the leaf extract of *AdhatodavasicaNees*, their characterization and antimicrobial activity." *Heliyon* 6(2) (2020): e03323.

151. Wu, Shuang, Shanmugam Rajeshkumar, Malini Madasamy, and Vanaja Mahendran. "Green synthesis of copper nanoparticles using *Cissus vitiginea* and its antioxidant and antibacterial activity against urinary tract infection pathogens." *Artificial Cells, Nanomedicine, and Biotechnology* 48(1) (2020): 1153–1158.

152. Hu, Che-Ming Jack, Santosh Aryal, and Liangfang Zhang. "Nanoparticle-assisted combination therapies for effective cancer treatment." *Therapeutic Delivery* 1(2) (2010): 323–334.

153. Rao, PasupuletiVisweswara, Devi Nallappan, Kondeti Madhavi, Shafiqur Rahman, Lim Jun Wei, and Siew Hua Gan. "Phytochemicals and biogenic metallic nanoparticles as anticancer agents." *Oxidative Medicine and Cellular Longevity* 2016 (2016).

154. Goyal, Shivangi, Nidhi Gupta, Ajeet Kumar, Sreemoyee Chatterjee, and Surendra Nimesh. "Antibacterial, anticancer and antioxidant potential of silver nanoparticles engineered using *Trigonella foenum*-Graecum seed extract." *IET Nanobiotechnology* 12(4) (2018): 526–533.

155. Akhtar, MohdJaved, Maqusood Ahamed, Sudhir Kumar, M.A. Majeed Khan, Javed Ahmad, and Salman A. Alrokayan. "Zinc oxide nanoparticles selectively induce apoptosis in human cancer cells through reactive oxygen species." *International Journal of Nanomedicine* 7 (2012): 845.

156. Bai Aswathanarayan, Jamuna, Ravishankar Rai Vittal, and Umashankar Muddegowda. "Anticancer activity of metal nanoparticles and their peptide conjugates against human colon adenorectal carcinoma cells." *Artificial Cells, Nanomedicine, and Biotechnology* 46(7) (2018): 1444–1451.

157. Patra, Nabanita, Dattatreya Kar, Abhisek Pal, and Anindita Behera. "Antibacterial, anticancer, anti-diabetic and catalytic activity of bio-conjugated metal nanoparticles." *Advances in Natural Sciences: Nanoscience and Nanotechnology* 9(3) (2018): 035001.

158. Mittal, Amit Kumar, Kaushik Thanki, Sanyog Jain, and Uttam Chand Banerjee. "Comparative studies of anticancer and antimicrobial potential of bioinspired silver and silver-selenium nanoparticles." *Journal of Materials Nanoscience* 3(2) (2016): 22–27.

159. Rajkuberan, Chandrasekaran, Kannaiah Sudha, Gnanasekar Sathishkumar, and Sivaperumal Sivaramakrishnan. "Antibacterial and cytotoxic potential of silver nanoparticles synthesized using latex of *Calotropis gigantea* L." *Spectrochimica Acta. Part A: Molecular and Biomolecular Spectroscopy* 136(B) (2015): 924–930.

160. Majeed, Shahnaz, Mohd Syafiq bin Abdullah, Gouri Kumar Dash, Mohammed Tahir Ansari, and Anima Nanda. "Biochemical synthesis of silver nanoprticles using filamentous fungi *Penicillium decumbens* (MTCC-2494) and its efficacy against A-549 lung cancer cell line." *Chinese Journal of Natural Medicines* 14(8) (2016): 615–620.

161. Govindaraju, Kasivelu, Karthikeyan Krishnamoorthy, Suliman A. Alsagaby, Ganesan Singaravelu, and Mariappan Premanathan. "Green synthesis of silver nanoparticles for selective toxicity towards cancer cells." *IET Nanobiotechnology* 9(6) (2015): 325–330.

162. Xia, Qian Hua, Yan Jun Ma, and Jian Wen Wang. "Biosynthesis of silver nanoparticles using *Taxus yunnanensis* callus and their antibacterial activity and cytotoxicity in human cancer cells." *Nanomaterials* 160(9) (2016): 6.

163. Geetha, Ravi, Thirunavukkarasu Ashokkumar, Selvaraj Tamilselvan, Kasivelu Govindaraju, Mohamed Sadiq, and Ganesan Singaravelu. "Green synthesis of gold nanoparticles and their anticancer activity." *Cancer Nanotechnology* 4(4–5) (2013): 91–98.

164. Sun, Butong, Nanjun Hu, Leng Han, Yanan Pi, Yu Gao, and Kang Chen. "Anticancer activity of green synthesised gold nanoparticles from *Marsdenia tenacissima* inhibits A549 cell proliferation through the apoptotic pathway." *Artificial Cells, Nanomedicine, and Biotechnology* 1(1) (2019): 4012–4019.

165. Dey, Arpita, A. Yogamoorthi, and S. Sundarapandian. "Green synthesis of gold nanoparticles and evaluation of its cytotoxic property against colon cancer cell line." *Research Journal of Life Sciences, Bioinformatics, Pharmaceutical and Chemical Sciences* 4 (2018): 1–17.

166. Wang, Lei, Jianwei Xu, Ye Yan, Han Liu, Thiruventhan Karunakaran, and Feng Li. "Green synthesis of gold nanoparticles from *Scutellaria barbata* and its anticancer activity in pancreatic cancer cell (PANC-1)." *Artificial Cells, Nanomedicine, and Biotechnology* 47(1) (2019): 1617–1627.

167. Rajeshkumar, S. "Anticancer activity of eco-friendly gold nanoparticles against lung and liver cancer cells." *Journal of Genetic Engineering and Biotechnology* 14(1) (2016): 195–202.

168. Sivaraj, Rajeshwari, Pattanathu K.S.M. Rahman, P. Rajiv, S. Narendhran, and R. Venckatesh. "Biosynthesis and characterization of *Acalypha indica* mediated copper oxide nanoparticles and evaluation of its antimicrobial and anticancer activity." *Spectrochimica Acta. Part A: Molecular and Biomolecular Spectroscopy* 129 (2014): 255–258.

169. Nagajyothi, P.C., Muthuraman Pandurangan, Doo Hwan Kim, T. V. M. Sreekanth, and Jaesool Shim. "Green synthesis of iron oxide nanoparticles and their catalytic and in vitro anticancer activities." *Journal of Cluster Science* 28(1) (2017): 245–257.

170. Kumar, Brajesh, Kumari Smita, Luis Cumbal, Alexis Debut, and Yolanda Angulo. "Biofabrication of copper oxide nanoparticles using Andean blackberry (*Rubus glaucus* Benth.) fruit and leaf." *Journal of Saudi Chemical Society* 21 (2017): S475–S480.

171. Sankar, Renu, Ramasamy Maheswari, Selvaraju Karthik, Kanchi Subramanian Shivashangari, and Vilwanathan Ravikumar. "Anticancer activity of *Ficus religiosa* engineered copper oxide nanoparticles." *Materials Science and Engineering C: Biomimetic Materials Sensors and Systems* 44 (2014): 234–239.

172. Noor, Sadaf, Ziaullah Shah, Aneela Javed, Amjad Ali, Syed Bilal Hussain, Sidra Zafar, Hazrat Ali, and Syed Aun Muhammad. "A fungal based synthesis method for copper nanoparticles with the determination of anticancer, antidiabetic and antibacterial activities." *Journal of Microbiological Methods* 174 (2020): 105966.

173. Xie, Juan, Yinyan Xu, Xinyan Huang, Yanni Chen, Jing Fu, Mingming Xi, and Li Wang. "Berberine-induced apoptosis in human breast cancer cells is mediated by reactive oxygen species generation and mitochondrial-related apoptotic pathway." *Tumor Biology* 36(2) (2015): 1279–1288.

174. Panigrahi, Sudipa, Soumen Basu, Snigdhamayee Praharaj, Surojit Pande, Subhra Jana, Anjali Pal, Sujit Kumar Ghosh, and Tarasankar Pal. "Synthesis and size-selective catalysis by supported gold nanoparticles: Study on heterogeneous and homogeneous catalytic process." *The Journal of Physical Chemistry C* 111(12) (2007): 4596–4605.

175. Woo, Yin-Tak, and David Y. Lai. "Aromatic amino and nitro–amino compounds and their halogenated derivatives." *Patty's Toxicology* (2001): 1–96.

176. Lim, Soo Hyeon, Eun-Young Ahn, and Youmie Park. "Green synthesis and catalytic activity of gold nanoparticles synthesized by *Artemisia capillaris* water extract." *Nanoscale Research Letters* 11(1) (2016): 474.

177. Rostami-Vartooni, Akbar, Mahmoud Nasrollahzadeh, and Mohammad Alizadeh. "Green synthesis of perlite supported silver nanoparticles using Hamamelis virginiana leaf extract and investigation of its catalytic activity for the reduction of 4-nitrophenol and Congo red." *Journal of Alloys and Compounds* 680 (2016): 309–314.

178. Sharma, Jitendra Kumar, M. Shaheer Akhtar, S. Ameen, Pratibha Srivastava, and Gurdip Singh. "Green synthesis of CuO nanoparticles with leaf extract of *Calotropis gigantea* and its dye-sensitized solar cells applications." *Journal of Alloys and Compounds* 632 (2015): 321–325.

179. Belard, Arnaud, Timothy Buchman, Jonathan Forsberg, Benjamin K. Potter, Christopher J. Dente, Allan Kirk, and Eric Elster. "Precision diagnosis: A view of the clinical decision support systems (CDSS) landscape through the lens of critical care." *Journal of Clinical Monitoring and Computing* 31(2) (2017): 261–271.

180. Gangula, A., R. Podila, M. Ramakrishna, L. Karanam, C. Janardhana, A.M. Rao. "Catalytic reduction of 4-nitrophenol using biogenic gold and silver nanoparticles derived from Breynia rhamnoides." *Langmuir* 27(24) (2011): 15268–15274.

181. Singh, Jagpreet, Preeti Kukkar, Heena Sammi, Mohit Rawat, Gurjinder Singh, and Deepak Kukkar. "Enhanced catalytic reduction of 4-nitrophenol and Congo red dye by silver nanoparticles prepared from *Azadirachta indica* leaf extract under direct sunlight exposure." *Particulate Science and Technology* 37(4) (2019): 434–443.

182. Alam, M.S., M.N. Javed, F.H. Pottoo, A. Waziri, F.A. Almalki, M.S. Hasnain, A. Garg, and M.K. Saifullah. QbD approached comparison of reaction mechanism in microwave synthesized gold nanoparticles and their superior catalytic role against hazardous nirto-dye. *Applied Organometallic Chemistry* 33(9) (2019): e5071.

183. Singh, Dinesh Pratap, Animesh Kumar Ojha, and Onkar Nath Srivastava. "Synthesis of different Cu (OH) 2 and CuO (nanowires, rectangles, seed-, belt-, and sheetlike) nanostructures by simple wet chemical route." *The Journal of Physical Chemistry C* 113(9) (2009): 3409–3418.

184. Zuo, Yong, Ji-Ming Song, He-Lin Niu, Chang-Jie Mao, Sheng-Yi Zhang, and Yu-Hua Shen. "Synthesis of TiO2-loaded CoO. 85Se thin films with heterostructure and their enhanced catalytic activity for p-nitrophenol reduction and hydrazine hydrate decomposition." *Nanotechnology* 27(14) (2016): 145701.

3 Synthesis Approaches for Higher Yields of Nanoparticles

Md. Sabir Alam, Md. Farhan Naseh,
Jamilur R. Ansari, Aafrin Waziri,
Md. Noushad Javed, Amirhossein Ahmadi,
Muhammad Khalid Saifullah, and Arun Garg

CONTENTS

3.1 Introduction 52
3.2 Advantages of Metallic Nanoparticles 52
3.3 Disadvantages of Metallic Nanoparticles 53
3.4 Examples of Metallic Nanoparticles 53
    3.4.1 Gold and Silver NPs 53
    3.4.2 Palladium and Platinum NPs 54
    3.4.3 Copper and Copper-Oxide NPs 54
    3.4.4 Iron and Iron-Oxide NPs 54
    3.4.5 Zinc-based NPs 54
    3.4.6 Titanium Dioxide NPs 54
    3.4.7 Other Metallic NPs 54
3.5 Synthesis Methods of Metallic Nanoparticles 55
3.6 Green Synthesis of Metallic Nanoparticles 57
    3.6.1 Using Plants and Their Parts for Synthesis of Metallic
           Nanoparticles 58
    3.6.2 Using Biopolymers for Synthesis of Metallic Nanoparticles 59
    3.6.3 Using Microorganisms for Synthesis of Metallic Nanoparticles........59
    3.6.4 Bacteria-based Synthesis of Metallic Nanoparticles 59
    3.6.5 Fungi-based Synthesis of Metallic Nanoparticles 60
    3.6.6 Algae-based Synthesis of Metallic Nanoparticles...............................60
    3.6.7 Plants-based Synthesis of Metallic Nanoparticles...............................60
3.7 Factors Influencing Reaction Parameters in the Biological Synthesis
    of Nanoparticles...62
3.8 Reaction Parameters Influencing the Synthesis of Metallic Nanoparticles 62
    3.8.1 Effect of pH 62
    3.8.2 Effect of Reaction Temperature..64
    3.8.3 Effect of Reaction Time...65
    3.8.4 Effect of Reactant Concentration 66

DOI: 10.1201/9781003126256-3

3.9   Characterization of Metallic Nanoparticles 66
3.10 Conclusion and Future Perspectives 71
Acknowledgments...71
References..71

3.1   INTRODUCTION

Nanoparticles (NPs) are nanomaterials that focus on the distinctive physicochemical properties of materials in the nano range, typically between 1 and 100 nm. Owing to properties like large surface area, low melting point, enhanced mechanical strength, good stability, and good penetrability make NPs suitable candidates for pharmaceutical and clinical applications like drug delivery, antimicrobial activity, X-ray imaging, cancer therapy, medicines, and treatment. The ability of NPs to hold both hydrophobic as well as hydrophilic substances makes them important drug carriers. Metallic nanoparticles (MNPs) can be synthesized effectively by any of the two traditional techniques, namely the top-down approach or the bottom-up approach. In the top-down method, the bulk material is greatly reduced in size to very tiny equivalents through physical or mechanical means like ball milling, grinding, thermal ablation, laser ablation, or sputtering. The reduction in size can also be achieved chemically using different reducing reagents and/or by chemical etching. On the other hand, the bottom-up technique includes chemical or biological synthesis of NPs through the process of nucleation, growth, and self-assembly using techniques like atomic/molecular condensation, sol-gel method, spray pyrolysis, or electrochemical precipitation.

Due to the unavoidable toxic effects of many conventionally synthesized MNPs on human health and the environment, a much safer, greener, and sustainable approach holds great importance starting right from the synthesis method up until the final utilization of NPs in various applications. The green strategies include the use of natural substances like plant metabolites, enzymes, vitamins, microbes, parasites, etc. for facilitating the biosynthesis of eco-friendly MNPs. The higher biostability of NPs and ease in scaling-up adds more fiscal value to green approaches. Since NPs can interact with biomolecules like enzymes/DNA and cellular organelles like lysosomes/ribosomes, they can significantly influence the activation of proteins and enzymes, gene expansion, oxidative stress, and cell membrane permeability. And as MNPs can target multiple biological moieties at the same time, the harmful bacterial cells cannot resist such multiple attacks. Singh et al. [1] reported a review on the green synthesis of MNPs demonstrating their bactericidal effects against the antibiotic-resistant bacterial strains.

3.2   ADVANTAGES OF METALLIC NANOPARTICLES

Metallic nanoparticles play a significant role in nanoscience and nanotechnology mainly due to the following facts:

- MNPs have the most stable properties at the nanoscale level as compared to the other nanocarriers.

- MNPs possess unique optical properties. For example, gold NPs show strong surface plasmon resonance (SPR) of electromagnetic waves that lie in the visible region. When electromagnetic waves are irradiated on the metallic surface, a strong SPR effect is observed that occurs due to the oscillation of the conductive electrons. The wavelength of the absorption peak depends upon few factors, in particular the morphology of the MNPs, the dielectric constant of the surrounding media, and the inter-particle distance between the MNPs.
- The synthesis protocol for MNPs is user-friendly so they can be easily synthesized having various morphologies in varying media, according to the desired applications.
- MNPs provide unique properties at the nanoscale range and they can be functionalized with polymers, drugs, and biomaterials for various potential applications.
- MNPs have good photostability and are less toxic.

3.3   DISADVANTAGES OF METALLIC NANOPARTICLES

- High cost.
- Not easily available.
- Stored at low temperature.
- Not feasible in normal temperature.
- Stability, not more than six months.

3.4   EXAMPLES OF METALLIC NANOPARTICLES

Some commonly used metallic nanoparticles are gold (Au), silver (Ag), copper (Cu), zinc (Zn), palladium (Pd), platinum (Pt), and titanium (Ti).

MNPs that have been synthesized via green synthesis includes metals (Ag, Au, Fe, Cu, Pd, Pt), metal oxides (CuO, ZnO, MnO_2, SiO_2, TiO_2, CeO_2, In_2O_3, Sb_2O_3, Fe_2O_3, Fe_3O_4), and metal chalcogenides (ZnS, PbS, CdS, CdSe) nanoparticles. Some examples related to the synthesis and application of these MNPs are specified in the subsequent sections.

3.4.1   GOLD AND SILVER NPS

Among all MNPs, Au and Ag NPs are the two most popularly used nanomaterials for biogenic applications. Biosynthesized Au and Ag NPs have been reported based on plant parts, natural polymers, and microorganisms used for cellular toxicity studies, anticancer therapy, neuro-degenerative disorders, catalysis, biosensing, and anticorrosive and antimicrobial agents [2–8]. Huang et al. have reported biosynthesized Ag and Au NPs using *Cinnamomum camphora* leaf extract and reported triangular- to spherical-shaped NPs [9]. Many other examples of Au and Ag NPs have been cited in the section discussing the influence of reaction parameters on the synthesis of MNPs.

3.4.2 PALLADIUM AND PLATINUM NPs

Baruwati and Varma reported the fast synthesis of gold, silver, platinum, and palladium NPs. They reported the use of red-grape pomace as the reducing and capping agent under microwave irradiation at 50W [10]. In yet another publication Varma along with Nadagouda, reported a simple one-pot, template-/surfactant-/capping-free method for the biosynthesis of Pd and Ag nanocrystals using tea and coffee extracts. The authors suggested that their method could also be applied for synthesizing Pt and Au NPs [11].

3.4.3 COPPER AND COPPER-OXIDE NPs

For studying the antimicrobial effects of copper-based nanomaterials, Cu NPs of 40–100 nm and CuO NPs of 4.8 nm average size were synthesized using magnolia leaf extract and *Sterculia urens* gum extract, respectively [12–13]. Manoj and co-workers reported the green synthesis of Cu NPs/CNTs nanocomposite for sensitive and selective electrochemical sensing of nitrite species over a large linear range of 5–1260 µM [14].

3.4.4 IRON AND IRON-OXIDE NPs

Farshchi et al. reported that *Rosmarinus officinalis* plant extract was used to synthesize Fe NPs for evaluating the effects of cytotoxicity on cancer cell lines [15]. Khalil et al. and Sathish Kumar et al. reported that Fe_2O_3- and Fe_3O_4-based nanohybrid structures were biosynthesized using plant extracts for pharmacogenetic, and cytotoxicity activity for antibacterial studies, respectively [16–17].

3.4.5 ZINC-BASED NPs

Vimala et al. reported on green synthesized ZnO NPs that were used for drug delivery of doxorubicin for the treatment of tumors. The drug was loaded on ZnO NPs, and their performance was tested for treating breast and intestinal colon tumors. ZnS NPs were extra cellularly synthesized using sulfate-reducing bacteria *Desulforibrio caledoiensis*. The photocatalytic property of the prepared NPs was also examined [18].

3.4.6 TITANIUM DIOXIDE NPs

Raja Kumar et al. reported on TiO_2 NPs in the range of 36–68 nm, which were biosynthesized using leaf extracts of *Eclipta prostrata* [19]. Spherical TiO_2 NPs in the range of 20–90 nm were green synthesized using leaf extracts of *Trigonella foenumgraecum* for monitoring their antimicrobial activity against *E. coli* and many other pathogens [20]. Manikandan et al. studied the bactericidal and photocatalytic properties of TiO_2 NPs synthesized using extracts of *Prunus yedoensis* leaf [21].

3.4.7 OTHER METALLIC NPs

Salunke et al. [22] and Xia et al. [23] reported that MnO_2 NPs were obtained extra cellularly using *Saccharophagus degradans* bacteria and *Nanochloropsis oculata* microalgae, respectively. Royston et al. [24] reported that various shaped SiO_2 NPs were synthesized extra cellularly using the TMV virus. Arumugam et al. [25] and Rocca et al. [26] investigated CeO_2 NPs for antibacterial activity and treatment of obesity, respectively. Shah et al. [27] and Jha et al. [28] reported that spherical NPs of In_2O_3 in the range of 5–50 nm and Sb_2O_3 in the range of 3–10 nm were obtained using an extract of *Aloe vera* plant and *Saccharomycetes cerevisiae* yeast, respectively. Rajesh Kumar et al. [29] reported that CdS NPs were extracellularly synthesized using *Enterococcus sp.* bacteria for testing their antimicrobial activity. Seshadri et al. [30] reported that ultra-small (2–5 nm) PbS NPs were prepared intracellularly using *Rhodosporidium diobovatum* yeast. Rose et al. [31] reported that CdSe NPs were used for checking the efficiency of solar cells.

3.5   SYNTHESIS METHODS OF METALLIC NANOPARTICLES

Synthesis methods of metallic nanoparticles are physical, chemical or biological.

1. By and large, the combination of nanoparticles has been completed utilizing three well-defined physical methodologies: physical, concoction, and natural techniques. In the physical method, the nanoparticles are dissipated by the build-up technique, which uses a cylinder heater at the climatic variation. As far as regular physical techniques are involved, flash releasing and pyrolysis are utilized for the amalgamation of MNPs. The advantages of using physical strategies are generally their speed, radiation is utilized as diminishing specialists, and the synthetic procedure includes non-toxic substances. A few drawbacks are low yield, high vitality utilization, dissolvable defilement, and presence of unvarying dispersion [32].

2. Chemical methods are the most common and frequently used methods for the preparation of various types of MNPs as stable, colloidal dispersions in water as well as different organic solvents. These methods use various types of weak and strong reducing agents (borohydrate, citrate, hydrazine hydrate) and stabilizing agents (CTAB, SDS) to reduce and stabilize the metal in order to form MNPs. In this process, metals and ions get ionized, followed by agglomeration, which finally turns them into oligomeric clusters [33–34].

3. In the present scenario of MNP synthesis, it is quite complicated to expand the chemical and physical techniques for multiscale productions mainly due to several drawbacks and disadvantages. In general, some of the drawbacks involve hazardous toxicity in the solvent and by-products. Due to the high level of toxicity and lack of predictability, it is difficult to use the products prepared via chemical synthesis techniques in human beings in the present state without any functionalization.

4. To functionalize and to control the level of toxicity, biological approach has been implemented, as it is eco-friendly and has insignificant toxicity. It does not imply any hazardous effects on the human beings [35]. The various methods/techniques for the preparation of MNPs by conventional/biological methods are shown in Figure 3.1.

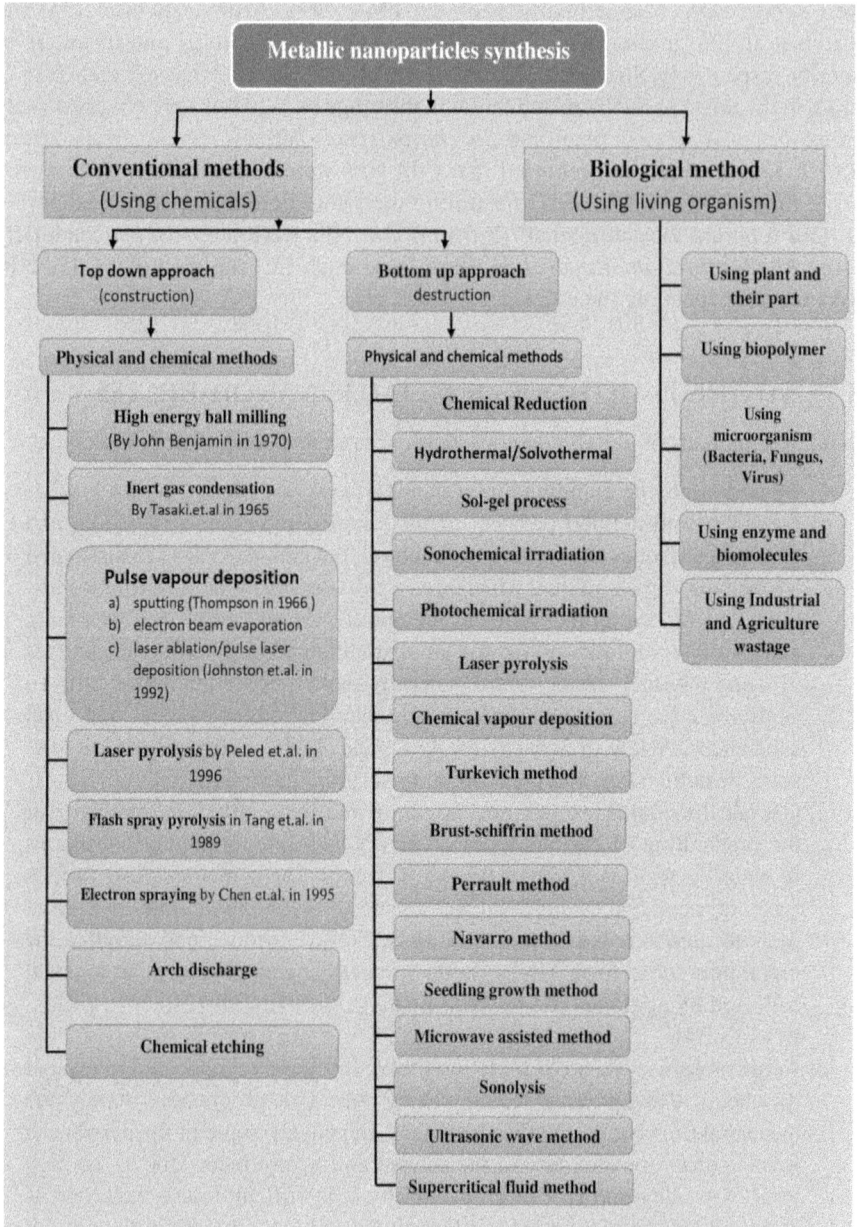

FIGURE 3.1 Synthesis of metallic nanoparticles by conventional/biological methods.

3.6 GREEN SYNTHESIS OF METALLIC NANOPARTICLES

Green synthesis techniques are reliable, sustainable, and eco-friendly synthesis techniques. Also, they are one of the prominent and most frequently applicable techniques for the synthesis of MNPs. By the green synthesis route, the presence of hazardous and toxic compounds can easily be avoided. The green synthesis technique has been adopted to synthesize MNPs using numerous biological materials (e.g., plant extracts, fungi, bacteria, algae). The scheme used for the synthesis of MNPs is shown in Scheme 3.1. Due to the availability of the green synthesis technique, which is simple and user-friendly for the preparation of MNPs at a large scale, it is widely utilized in the manufacturing of metal- and metal-oxide-based nanoparticles. In this process, the produced nanoparticles are alternatively known as biogenic nanoparticles [36].

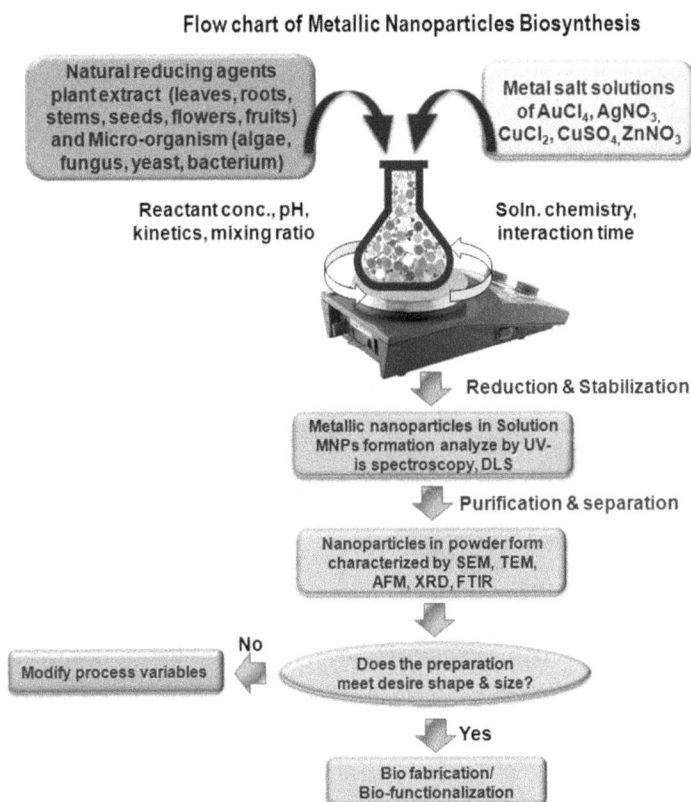

Flow chart of Metallic Nanoparticles Biosynthesis

SCHEME 3.1   Flow chart of metallic nanoparticle biosynthesis.

3.6.1 Using Plants and Their Parts for Synthesis of Metallic Nanoparticles

There are various techniques that have been reported for the synthesis of MNPs, like physical, chemical, and biological methods. Physical and chemical techniques are a bit expensive and are hazardous to the environment. So, there was a need of using the emerging technologies, in particular the green synthesis route, which is non-toxic, eco-friendly, cost-effective, and most reliable for MNP production and thereby extending their applications in the fields of medicine, cosmeceutical preparation, agriculture, biomedical, and pharmaceutical. Currently, there is a need to develop sustainable methods for the production of various types of nanoparticles, which discards the hazardous chemical synthesis technique. For these reasons, biological-mediated techniques are preferred over others, for the synthesis of metallic nanoparticles. In this method, biological materials can be used like parts of plants (leaves, flowers, seeds, stem, bark, fruit, etc.) and plant extracts, as shown in Figure 3.2 [37–38], can be used as bio-sources to synthesize MNPs. Fungi, bacteria, and algae may also be used for the synthesis of MNPs [39].

FIGURE 3.2 Plants extracts as bio-sources to synthesize metallic nanoparticles.

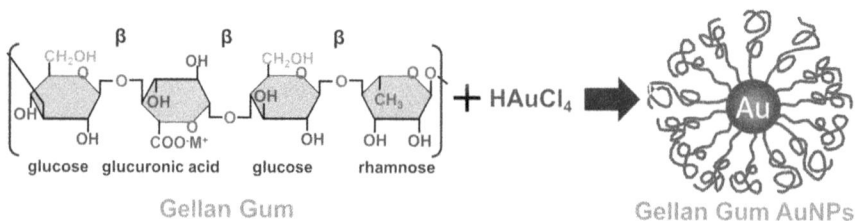

FIGURE 3.3 Synthesis of gold nanoparticles using gellan gum.

3.6.2   Using Biopolymers for Synthesis of Metallic Nanoparticles

Due to numerous advantages, biopolymer-based synthesis has attracted much attention for the green synthesis methods. The historical developments of polymer-based
nanomaterials were started by Paul Ehrlich and co-workers. Extensive works based on
polymers were done by the group of Peter Speiser in the late 1960s and early 1970s
[40]. Polymer-based MNPs are widely used in the fields of nanotechnology as well
as of biotechnology [41–42]. This is due to their unique physical properties such as
wide surface area and enhanced permeability, which makes them prominent candidates for biomedical and pharmaceutical applications such as diagnosis and therapy of
wound healing, cancer, diabetes, malaria, neurodegenerative disorders, and antimicrobial activity. Bipolymers are efficiently used in applications, particularly to enhance
environmental compatibility and biodegradability. Essential phytochemicals like
alkaloids, terpenoids, flavonoids, glycosides, saponins, steroids, tannins, and other
nutritional compounds can be used for the synthesis of MNPs. Likewise, the use of
polymers derived from renewable resources provide an appropriate way to extend the
non-renewable petrochemical supplies. These biopolymers can be used as a reducing
agent and/or stabilizer for metallic nanoparticles. In recent times, much work has been
published concerning the synthesis of Au NPs using gellan gum as a polymer, which
acts both as a reducing and capping agent, as shown in Figure 3.3 [43].

3.6.3   Using Microorganisms for Synthesis of Metallic Nanoparticles

In recent years, the development of MNPs by different techniques, for example,
physical approach, biological and chemical methods, have been reported. Physical
methods have certain limitations and low production rates with high utilization of
energy and are a bit expensive. However, with chemical processes, the costs are low,
but there is possibility of toxicity in the solvent, which creates chemical contamination. Therefore, biological approaches that are based on microorganisms (bacteria,
fungi, yeast, and viruses) are one of the best alternative approaches for the synthesis
of metallic nanoparticles. Microbial synthesis of MNPs using microorganisms is a
part of green chemistry approaches and is considered as one of the exciting areas of
research for various applications in nanoscience and nanotechnology, as shown in
Figure 3.4. In this approach, high yield, low cost, and environmental friendliness are
observed, as shown in Table 3.1 [44–58].

3.6.4   Bacteria-based Synthesis of Metallic Nanoparticles

Bacteria and actinomycetes are categories under the same roof. However, bacteria are broadly used as compared to actinomycetes in the biogenesis of Ag NPs.
Vaidyanathan et al. reported that bacteria are one of the potential prospects for the
preparation of Ag NPs, as the handling procedure and genetic modification are user-
friendly [59]. In 1984, Haefeli et al. for the first time observed that synthesis of Ag
NPs can be done by the bacteria *Pseudomonas stutzeri* AG259 strain, which was
separated from the silver mines [60]. Since then, many groups across the globe have
followed the procedure reported by Haefeli et al., demonstrating quiet well the use of

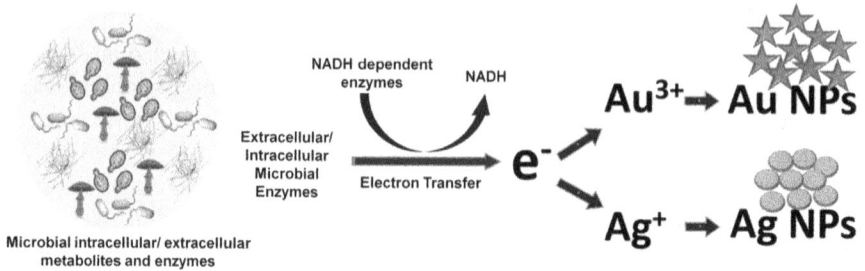

FIGURE 3.4 Synthesis of metallic nanoparticles using microorganisms.

other bacterial strains including various variants like *Escherichia coli* [61], actino-bacteria *Rhodococcus* sp. [62], and *Pseudomonas* sp. [63]. The most widely reported mechanism for the biosynthesis of Ag NPs is the conversion of nitrate into nitrite in the presence of the nitrate reductase enzymes.

3.6.5 FUNGI-BASED SYNTHESIS OF METALLIC NANOPARTICLES

Fungi are widely used for the biosynthesis of Ag NPs. Patil et al. [64] demonstrated *Fusarium semitectum* for the biosynthesis of Ag NPs. Later, Mukherjee et al. [65] found that when *Verticillium* sp. biomass was exposed to silver nitrate solution, it produces Ag NPs. Kathiresan et al. [66] reported on the use of the marine fungus *Penicillium fellutanum* for the production of stable Ag NPs. Mukherjee et al. [67] reported that *Trichoderma asperellum*, when subjected to the solutions of AgNO$_3$, results in the extracellular formation of NPs having diameter of 13–18 nm. As far as fungi is concerned, it can secrete ample amounts of proteins, which randomly increases the productivity of NPs manifold. The biosynthesis mechanism of Ag NPs actually consists of two steps. In the first step, the silver ions are confined at the fungal cells' surface, and in the second step, the reduction of silver ions takes place with the presence of the fungal system in the enzymes.

3.6.6 ALGAE-BASED SYNTHESIS OF METALLIC NANOPARTICLES

Extracellular Ag NP synthesis by several algal species has been reported. It was also reported that with the use of extracts from unicellular green algae *Chlorella vulgaris*, MNPs were synthesized at room temperature. Marine algae such as *Chaetomorpha linum* [68], *Caulerpa resmosa* [69], and *Sargassum polycystum* [70] were also explored to synthesize Ag NPs. As far as the reduction of metal ions is concerned, the presence of hydroxyl groups in tyrosine as well as carboxyl groups in aspartic and/or glutamine residues of the proteins plays a significant role [71].

3.6.7 PLANTS-BASED SYNTHESIS OF METALLIC NANOPARTICLES

For the synthesis of nanoparticles via the green route, plants and their extracts are considered as green nano industries. The green synthesis route for the preparation of Ag NPs using plant extracts and their respective applications in the allied field have caught the

TABLE 3.1
Synthesis of Metallic Nanoparticles Using Microorganisms

Microorganism	Particle Size	Characterization	Morphology	Reference
Silver NPs				
Euplotes focardii	20–70 nm	SEM, TEM, EDX, FTIR, XRD	Spherical	[44]
Bacillus pumilus, B. persicus, and B. licheniformis	77–92 nm	FTIR, UV–Vis spectroscopy, DLS, TEM, EDX	Triangular, hexagonal, and spherical	[45]
Pantoea ananatis	8.06–91.32 nm	FTIR, UV–Vis spectroscopy, TEM, SEM-EDX, DLS,	Spherical	[46]
Bacillus brevis	41–68 nm	UV–Vis spectroscopy, FTIR, TLC, AFM, SEM	Spherical	[47]
Sinomonas mesophila MPKL 26	4–50 nm	UV–Vis spectroscopy, FTIR, TEM	Spherical	[48]
Gold NPs				
Stenotrophomonas sp.	10–50 nm and 40–60 nm	DLS, UV–Vis spectroscopy, TEM, DLS, EDAX	Multi-shaped	[49]
Pseudomonas aeruginosa	15–30 nm	TEM, UV–visible spectrophotometer, fluorescence spectra	Spherical	[50]
Pseudomonas fluorescens 417	5–50 nm	FTIR, UV–visible spectrophotometer, XRD, TEM	Spherical	[51]
Blue green algae	5 nm	FTIR, UV–visible spectrophotometer, HRTEM, EDAX, Raman spectra	Spherical	[52]
Magnetospirillum gryphiswaldense MSR-1	10–40 nm	TEM, XRD, XPS	Spherical	[53]
Copper NPs				
Pseudomonas fluorescens	49 nm	UV–visible spectroscopy, SEM-EDS, TEM, SAED	Spherical and hexagonal	[54]
Stereum hirsutum	5–20 nm	TEM, XRD, UV–visible spectroscopy, FTIR, DLS	Spherical	[55]
Aspergillus niger strain STA9	5–100 nm	FTIR, UV–visible spectroscopy, DLS, SEM, TEM	Spherical	[56]
Endophytic actinomycete *Streptomyces capillispiralis*	50–100 nm	FTIR, XRD, UV–visible spectroscopy, TEM	Spherical	[57]
Rhodotorula mucilaginosa	10.5 nm	SEM-EDS, XPS, TEM	Spherical	[58]

attention of scientists, engineers, biologists, and chemists around the globe. The benefits of using plants and their extracts for the biogenesis of Ag NPs is that they are easily acquirable and non-toxic. Also they are quite safe and have plenty of metabolites that play a significant role in the reduction of silver ions, and are quick as compared to microbes in the synthesis process. The introduction of a photobiological approach has an edge over other synthesis protocols of MNPs because it is eco-friendly, cost-effective, less time-consuming, and has nominal wastage [72]. Green nanotechnology is also known as the photobiological approach because plants and their extracts are utilized as capping as well as reducing agents during the preparation of Ag NPs (Table 3.2) [73–87].

3.7 FACTORS INFLUENCING REACTION PARAMETERS IN THE BIOLOGICAL SYNTHESIS OF NANOPARTICLES

When we talk about green route synthesis, or, in other words, biological synthesis for the formation of nanoparticles, there are numerous controlling factors involved in the nucleation and subsequent growth/formation of stabilized nanoparticles. These factors include the effects of (a) pH, (b) reaction temperature, (c) reaction time, (d) concentrations, and (e) environment.

3.8 REACTION PARAMETERS INFLUENCING THE SYNTHESIS OF METALLIC NANOPARTICLES

During the synthesis of MNPs, we need to emphasize the controlling parameters that are involved in the nucleation and growth/formation of stable MNPs. The effect of various parameters, which include pH, reaction temperature, reaction time, and concentration of the reactants, is shown in Figure 3.5. A brief discussion of each parameter is elaborated in the following sections and in Table 3.3 [88–111].

3.8.1 Effect of pH

The pH of the reaction medium is very significant for the formation and nucleation of MNPs. The number of nucleation centers varies with the variation of pH, facilitating the generation of NPs. It has been observed that pH also affects the size and surface morphology of MNPs. Okitsu et al. [112] studied the pH effects on the size and the aspect ratio of Au nanorods and Au NPs. They concluded that the aspect ratio and size decreased with the increase in pH. Armendariz and co-workers checked the pH-dependent response of Au NPs synthesized using *Avena sativa* biomass. They noticed that at a pH of 2, few NPs having a large size of 25–85 nm were formed, and upon increasing the pH to 3 and 4, smaller sized (5–20 nm) Au NPs were formed. They reasoned that due to lesser amounts of nucleation centers at a lower pH, the NPs aggregated to form larger sized Au NPs [113]. Similar results were obtained by Konishi et al. for intracellular synthesis of Au NPs using *Shewanella algae*. Large nanoparticles in the range of 15–200 nm were observed at pH 2–6, and small particles in the range of 10–20 nm at pH 7 [114]. He et al. biosynthesized Au NPs extracellularly using *Rhodopseudomonas capsulata* bacteria by varying the pH. The shape of the particles changed from nanoplates to spherical upon increasing the pH from 4 to 7 [115].

TABLE 3.2
Green Synthesis of Metallic Nanoparticles Using Plants and Their Extracts

Metallic NP	Characterizations	Morphology	Particle Size	Reference
Gold NPs				
Aloe vera leaf extract Au and Ag NPs	TEM, UV–Vis–NIR spectroscopy, AFM, FTIR, EDAX	Spherical	50–350 nm	[73]
Azadirachta indica leaf extract Au/Ag NPs	UV–vis spectra, XRD, TEM, FTIR	Spherical, triangular, hexagonal	5–35 nm and 50–100 nm	[74]
Camellia sinensis leaf extracts Au/Ag NPs	TEM, UV–Vis spectroscopy, cyclic voltammetry	Spherical, prism	20–100 nm	[75]
Coriandrum sativum Au NPs	UV–Vis spectroscopy, FTIR, EDAX, TEM, XRD	Spherical, triangular, decahedral	6.75–57.91 nm	[76]
Syzygium aromaticumbuds Au NPs	UV–Vis spectroscopy, FESEM, TEM, AFM, XRD, EDAX, FTIR	Irregular shaped	5–100 nm	[77]
Silver NPs				
Acalypha indica leaf extracts Ag NPs	UV–Vis spectrum, SEM, XRD, EDS, HRTEM	Spherical	20–30 nm	[78]
Carica papaya callus extract	UV–Vis spectrum, FTIR, SEM	Spherical	60–80 nm	[79]
Citrus limon	UV–Vis, DLS, XRD, FTIR, AFM, TEM	Spherical, spheroidal	50 nm	[80]
Eucalyptus citriodora and *Ficus bengalensis*	UV–Vis spectra, DSC, TGA, SEM, TEM	Spherical	20–50 nm	[81]
Garcinia mangostana leaf extract	UV–Vis spectra, FTIR, TEM	Spherical	35 nm	[82]
Zinc NPs				
Calliandra haematocephala leaves	UV–visible spectra, FESEM, EDS, XRD, FTIR	Hexagonally formed	19.45 nm	[83]
Parthenium hysterophorus leaves	UV–visible spectra, XRD, FTIR, SEM, TEM, EDX	Spherical and hexagonal	28–84 nm	[84]
Lycopersicon esculentum fruit	UV–visible spectra, SEM, TEM, FTIR, XRD, EDX	Crystalline	49.8–191 nm	[85]
Agathosma betulina dry leaves	HRTEM, SAED, XRD, FTIR, Raman, EDX	Quasi-spherical agglomerates	12–26 nm	[86]
Coptidis Rhizoma	UV–Vis spectroscopy, FTIR, TGA, SEM, XRD, EDX, SAED	Spherical, rod shaped	2.9–25.2 nm	[87]

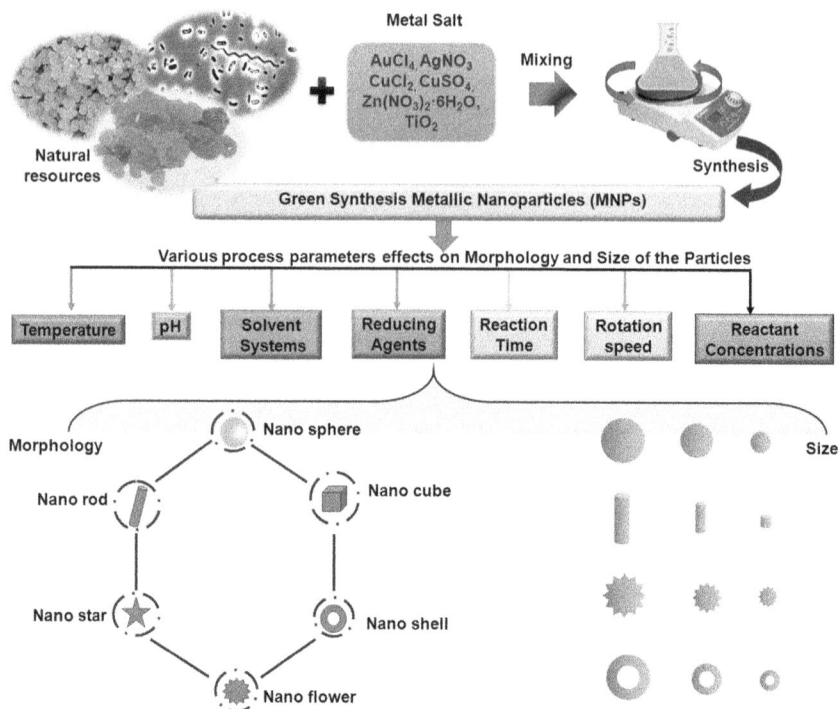

FIGURE 3.5 Effect of various parameters on the synthesis of metallic nanoparticles.

3.8.2 EFFECT OF REACTION TEMPERATURE

Temperature plays a vital role and influences the rate of synthesis and affects the shape and size of MNPs. With a rise in temperature, the rate of reaction increases resulting in the formation of more nucleation centers. Sneha and co-workers observed the variations in surface morphology of Au NPs with the change in the temperature. Moving from 20°C to 30°C–40°C and increasing to 50°C–60°C, the shape of the NPs changed from triangular to octahedral to spherical, correspondingly [116]. Iravani and Zolfaghari biosynthesized Ag NPs using bark extract of *Pinus eldarica* at 25°C, 40°C, 50°C, 100°C, and 200°C. They reported that the size of the NPs reduced with the increase in the synthesis temperature [117]. Fayaz et al. reported that a decrease in particle size is accompanied by a change in the shape of fungal-based Ag NPs with rising reaction temperature. At 10°C, the particles were in the form of nanoplates and in the range of 80–100 nm. Increasing the temperature to 27°C reduced the size to 10–40 nm with spherical and rod-shaped NPs, and finally at 40°C, ultra-small spherical nanoparticles in the range of 2–4 nm were obtained [118]. A reverse trend was observed by Kumar et al. with Mn-doped cobalt ferrite NPs where the size of the NPs grew with the consequent rise in the reaction temperature [119]. Islam and Mukherjee synthesized Ag NPs from a silver–ammonia

TABLE 3.3

Factors Influencing Synthesis of Metallic Nanoparticles

Controlling Factors	Influence on Biological Synthesis of MNPs	Morphology	Particle Size	Reference
pH	Variability in size and shape	Fcc, tetrahedral, decahedral, icosahedral, irregular shaped	20–100 nm	[88–92]
Reactant concentration	Variability in shape	Quasi-spherical, triangular, spherical shapes, hexagonal, decahedral	55–80 nm	[93–95]
Reaction time	Increase in reaction time increases the size of MNPs	Oval, elliptical	10–30 nm	[90, 96–97]
Reaction temperature	Size, shape, yield, and stability	Triangles, pentagons, hexagons, and spherical	5–300 nm, 35–10 nm	[90, 98–99]
Solvent systems	Different solvent systems effect on morphology as well as particle size	Spherical	4.2 nm ± 1.0 nm; and 14.1 nm ± 1.0 nm	[100–104]
Reducing agents	Effect on morphology and particle size	Spherical, hexagonal, triangle, quasi-spherical, rod	20–400 nm	[105–108]
Rotation speed	Increase in rotation speed decreases size of MNPs	Spherical, hexagonal	3–12 nm	[109–111]

complex in triblock copolymer micellar solution at different temperatures. They concluded that the particle size at lower temperatures was controlled by variations in the morphology of the polymer, while at higher temperatures it was due to the change in the chemical composition of the polymer [120].

3.8.3 EFFECT OF REACTION TIME

Alongside pH and temperature, the reaction time also controls the morphology of MNPs to a great extent. Karade et al. demonstrated the effect of reaction time on the structural properties of magnetic NPs. Its effect on magnetic behavior was also noted. They took Fe NPs prepared using an iron salt solution and green tea extract, and found that with increasing the reaction time, the size increased from 7.5 to 12 nm enhancing the saturation magnetization of Fe NPs [121]. Flor and co-workers also reported an increase in the size of ZnO NPs and Ce-doped ZnO NPs, with the increase in reaction time. They also mentioned that at a constant reaction time, the morphology of the Ce-doped ZnO NPs was larger than the undoped ZnO NPs [122]. Contrary to the aforementioned results, a group of researchers also reported a gradual decrease in the particle size of CdSe NPs, with the increase in the time of reaction. From 15.8 nm at 4 hours to 10.5 nm at 8 hours of reaction time, the particle size again reduced to 6.7 nm at the end of 12 hours. They also noticed that at further increase of reaction time to 16 hours, the size of particles increased suddenly to 111.7 nm, possibly due to aggregation [123]. Ahmad et al. investigated the effect of reaction time on the reduction and stability of Au NPs prepared using oil palm leaves and reported that the rate of reduction leading to particle formation increased with the increase in reaction time [124].

3.8.4   EFFECT OF REACTANT CONCENTRATION

The concentration of reactants that includes metal salt solutions, biomolecules, and biomass solutions has a predominant effect on the synthesis of metallic NPs. During the reduction of NPs, solvents and reducing agents have a correlative effect on the physicochemical and structural properties relating to the surface morphology, size, and shape of NPs. These influences impact in the manner the NPs are utilized for application purposes. Mishra et al. [125] studied the influence of concentration on the size and shape of Au NPs biosynthesized using the fungus *Penicillium brevicompactum*. At 1 mM of gold salt solution, the particles were spherical having a size of 10–50 nm, while at 2 mM concentration, the size of particles reached 70 nm. Spherical, triangular, and diamond-shaped particles having the size of 50–120 nm were observed at 3 mM of salt solution. Similar results of spherical NPs at low metal salt concentration, and triangular and hexagonal NPs at high concentration were reported with Au NPs synthesized using the marine yeast named *Yerrowia lipolytica* [126]. With respect to biomass concentration, Owais et al. [127] observed that smaller Au NPs were formed when the concentration of yeast extract was increased during the process of synthesis. The influence of the concentration of plant extract-based biomolecules was also observed during the synthesis of MNPs. Huang et al. [128] synthesized Ag and Au NPs and reported the transformation in particle shape from triangular to spherical when the amount of *Cinnamomum camphora* leaf extract was increased in the precursor solution. Au NPs of various morphology, namely spherical, triangular, hexagonal, and decahedral, were obtained upon varying the concentration of *Coleus amboinicus* leaf extract during synthesis [129].

3.9   CHARACTERIZATION OF METALLIC NANOPARTICLES

There are various methods that are used to characterize the metallic nanoparticles, as described in this section and shown in Figure 3.6.

1. UV–visible spectrophotometry
2. Fluorescence spectroscopy

Characterization Techniques

Microscopic Techniques

Instrument Name	Principle
SEM	Surface morphology, Particle morphology and dispersion
TEM	Shape, Size, Size distribution and 2D Image
HRTEM	Shape, Size, Crystalline nature, Crystalline Structure down to atomic level
AFM	Surface topography, Particle shape, 3D image.
SAED	Crystal information & Planes.

Spectroscopic Techniques

Instrument Name	Principle
UV-Visible spectra	Absorption spectra & Size of the metal Nanoparticles
FTIR	Surface chemistry & Functional group analysis
DLS	Methods for sizing & Potential study
XRD	Purity and Crystalline nature with their structure present & grain nature
EDX	Chemical composition
XPS	Surface composition chemical state
ICP-MS	Elemental analysis & state quantification.

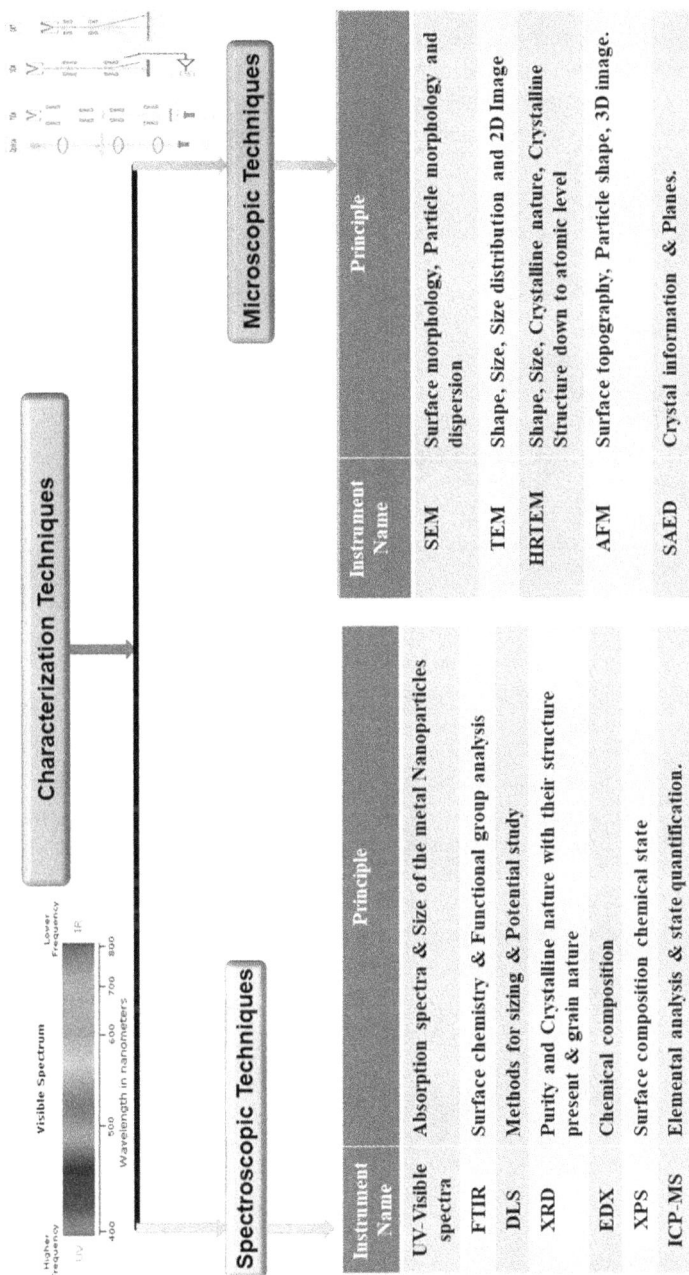

FIGURE 3.6 Characterization techniques used for metallic nanoparticles.

3. DLS (Dynamic light scattering)
  3.1. Zeta sizing
  3.2. Zeta potential
4. FTIR (Fourier-transform infrared spectroscopy)
5. NMR (Nuclear magnetic resonance)
  5.1. CNMR (Carbon-13 nuclear magnetic resonance)
  5.2. HNMR (Hydrogen-1 nuclear magnetic resonance)
6. XRD (X-ray diffraction)
7. Contact angle
8. SEM (Scanning electron microscopy) with EDX (Energy-dispersive X-ray spectroscopy)
9. TEM (Transmission electron microscopy) with SAED (selected area electron diffraction)
10. AFM (Atomic force microscopy)
11. TGA (Thermo-gravimetric analysis)
12. XPS (X-ray photoelectron spectroscopy)
13. MPAES (Microwave plasma atomic emission spectroscopy)

Instrument	Principles	
UV–visible spectrophotometry	The UV–visible spectroscopy technique is based on the light absorption and reflection ions mechanism in the UV–visible electromagnetic spectral regions, ranging between 200 nm and 800 nm, in which the molecules and other chemical compounds undergo electronic transitions [130].	
Fluorescence spectroscopy	Fluorescence spectroscopy (also known as spectrofluorometry) is one of the most reliable, simple, fast, and inexpensive technique to calculate the concentration of the NPs in the solution based on their fluorescent characteristic features. They can be used to study the type of the compound which has to be analyzed (analyte) as well as to calculate the unknown concentration of the compounds. This technique is mainly used for measuring the fluorescent compounds present in the solution. In spectrofluorometry, a beam of light with a wavelength range between 180 nm and 800 nm is passed through a solution kept in a cuvette inside the holder [131].	
DLS (Dynamic light scattering)	Zeta sizing	DLS (also known as photon correlation spectroscopy or quasi-elastic light scattering) is user-friendly, easy to use, and it is one of the most reliable techniques used for the determination of nanoparticle size distribution in an aqueous medium. It provides an overall measurement of the hydrodynamic diameter of the NPs as well as the PDI (polydispersity index) [132].

	Zeta potential	The zeta potential analyzer is a sophisticated technique used to determine the electric charge in the aqueous/colloidal media due to the presence of functional groups on the surface of the particles. The net charge on the surface of the NPs develops because of the ionization of the electrons. This may be due to the surface groups or adsorbed species. The presence of an electrical double layer in the vicinity of the NPs is due to the presence of a surface charge that affects the overall distribution of the ions. The aqueous layer in and around the NPs can be bifurcated into two parts, namely "stern layer in the core" and "diffusive layer in the shell" [133].
FTIR (Fourier-transform infrared spectroscopy)	FTIR is a technique based on vibrational spectroscopy and is widely used for characterizing the organic and inorganic materials in the presence of infrared light. When the sample is moved through the electromagnetic radiation (at wavelength range 2.5 µm to 25 µm, or wavenumbers 400 cm^{-1} to 4000 cm^{-1}), the energy is transmitted or absorbed. At specific wavelengths, rise and variation in the rotational or vibrational movement is observed, which occurs mainly due to absorbance. The wavelength of the molecules at which this phenomenon occurs is quite important, as it indicates the nature of the molecule which is under consideration. Recently, FTIR has been used to study the interactions between different types of NPs and organic molecules: both the binding sites and conformational changes [134].	
NMR (Nuclear magnetic resonance)	CNMR	Also known as Carbon-13 (C-13) nuclear magnetic resonance. It helps in the characterization of carbon atoms in an organic molecule or prepared NPs, and also helps in chemical structure elucidation in the prepared organic compounds (NPs) [135–136].
	HNMR	Also known as proton NMR. It is one of the most common NMRs used. The proton (Hydrogen-1 nucleus) is a hypersensitive nucleus (apart from tritium) and it determines the prominent signals, which makes proton NMR very beneficial having resonance energy 400 MHz [135, 137].
XRD (X-ray diffraction)	This analytical technique determines the crystalline structure of the compound (particles) or atomic arrangement of the prescribed material or chemicals. Their electromagnetic wavelength ranges between 0.01 and 10 nm and energies range between 120 eV and 120 KeV [138].	
Contact angle	Also known as hydrophobicity test of the compounds. This technique, for the measurement of nanoparticle behavior, is a traditional method, but it is used for the advancement of technology mainly to describe the behavior of materials either the hydrophobic (water evading) or hydrophilic (water loving) nature of a material. The principle of the contact angle provides precise information regarding the wettability of an ideal surface. In most cases, the intrinsic property of the contact angle describes the surface porosity, roughness, and heterogeneity [139].	

SEM (Scanning electron microscopy)	A scanning electron microscope is an electron microscope technique that scans the surface of the desired material, and the focused electron beam scans over the surface of the materials to create an image. The electrons interact with the secondary atoms in the sample producing various signals depending upon the morphologies of the NPs to provide the desired information of the surface and their composition [140].
EDX (Energy-dispersive X-ray spectroscopy)	EDX is commonly known as energy-dispersive X-ray analysis (EDXA) or energy-dispersive X-ray microanalysis (EDXMA). It is a micro-analytical technique used for identifying and quantifying elemental compositions present in the samples or chemical characterization of the samples [141].
TEM (Transmission electron microscopy)	TEM is a very powerful technique in the field of material science and biological science to analyze the detailed information about the samples. It mainly includes seeing the morphology of the prepared NPs and the various components involved in the formation of the NPs. Basically, in TEM, the whole setup is enclosed in a vacuum chamber, and a very high energy beam of electrons is allowed to pass through the sample kept on the TEM grid. The electron beam interacts with the atoms and various characteristic features of a particular crystal (structures, planes, dislocations, and grain boundaries) are observed. TEM can also be used to study the growth of material layers, their composition, and defects in semiconductors. Their high-resolution power helps to analyze the various properties of materials like quality, shape, size, and density of quantum wells, wires, dots, and other characteristic features [142].
XPS (X-ray photoelectron spectroscopy)	XPS falls under photo-emission spectroscopy (PES) or electron spectroscopy for chemical analysis (ESCA). It is a surface analyzer technique that provides information of all the elements, composition, binding energies, and chemical states related to the outermost layers of the surface as well as both quantitative and qualitative information of the surface. The XPS spectrum is obtained by irradiating the sample surface with an X-ray beam, and the sample surface depth of about 2–5 nm penetrates the sample surface by X-ray [143].
AFM (Atomic force microscopy)	The AFM technique analyzes the surface morphology of a sample in three dimensions (3D). By examining the interaction between the cantilever tip and the sample surface, AFM generates the exact morphological structure of the samples. This characterization method is complementary to SEM and HRTEM techniques [144]. These techniques allow us to analyze and characterize the samples from a few micrometers to nanometer-scale levels. It also allows us to make analysis of insulators, semiconductors, and electrical conductors.
TGA (Thermogravimetric analysis)	TGA is a technique used for the thermal characterization of the NPs, and it determines the sample's weight variations at a certain range of temperature and atmosphere. This analytical technique allows determining the bonding strength of the ligand conjugated to the nanoparticle surface and its thermal stability. For example, this technique provides direct information about the quantity of organic components on gold NPs, by measuring the total mass loss under gradual heating of dry NPs [145].

3.10 CONCLUSION AND FUTURE PERSPECTIVES

Metallic nanoparticles can be synthesized by various physical, chemical, and biological methods. Scientists are working on most of these methods and the problems they are facing are related to stability, morphology, size distribution, and aggregation. It has been observed that MNPs synthesized via parts of plants, plant extracts, and microorganisms are more stable, and their rate of reaction is faster compared to other routes. Researchers are focusing to understand the mechanism behind such a transformation. So, scientists advocate utilizing various natural resources like plant parts (e.g., roots, shoots, leaves, flowers, stems, barks, fruits, and seeds), essential phytochemicals (e.g., alkaloids, terpenoids, flavonoids, glycosides, saponins, steroids, tannins, and other nutritional compounds), microorganisms (bacteria, fungi, viruses, seaweeds, microalgae, etc.) and natural biopolymers to synthesize MNPs via the green route. A clear-cut idea about the biochemical process involves plant and organism detoxification, resistance, and accumulation, which will help in the production of MNPs. Hence, green synthesis-based approaches have been sought after in the recent years as another efficient, reasonable, and environmentally safe method for producing MNPs with desired properties and applications. We plan and suggest using core-shell MNPs by varying either the core or the shell materials with metal-semiconductor nanoparticles for their possible applications in the field of medical science.

ACKNOWLEDGMENTS

The authors gratefully acknowledge the contributions of their collaborators and co-workers mentioned in the cited references. NIMS Institute of Pharmacy, NIMS University, Jaipur, Rajasthan, India, for their support. J. R. Ansari gratefully acknowledges the Vice-Chancellor, Pro-Vice-Chancellor, and Head, Department of Physics, PDM University, Bahadurgarh, India, for their kind cooperation and support. A. Waziri thankfully acknowledges GGSIPU, New Delhi, for the financial support in the form of STRF.

REFERENCES

1. Singh, Anirudh, Pavan Kumar Gautam, Arushi Verma, Vishal Singh, Pingali M. Shivapriya, Saurabh Shivalkar, Amaresh Kumar Sahoo, and Sintu Kumar Samanta. "Green synthesis of metallic nanoparticles as effective alternatives to treat antibiotics resistant bacterial infections: A review." *Biotechnology Reports* 25 (2020): e00427.
2. Shah, Monaliben, Derek Fawcett, Shashi Sharma, Suraj Kumar Tripathy, and Gérrard Eddy Jai Poinern. "Green synthesis of metallic nanoparticles via biological entities." *Materials* 8, no. 11 (2015): 7278–7308.
3. Patil, Maheshkumar Prakash, and Gun-Do Kim. "Marine microorganisms for synthesis of metallic nanoparticles and their biomedical applications." *Colloids and Surfaces B: Biointerfaces* 172 (2018): 487–495.
4. Gour, Aman, and Narendra Kumar Jain. "Advances in green synthesis of nanoparticles." *Artificial Cells, Nanomedicine, and Biotechnology* 47, no. 1 (2019): 844–851.

5. Pottoo, Faheem Hyder, Nahida Tabassum, Md Noushad Javed, Shah Nigar, Rouqia Rasheed, Ayash Khan, Md Abul Barkat, et al. "The synergistic effect of raloxifene, fluoxetine, and bromocriptine protects against pilocarpine-induced status epilepticus and temporal lobe epilepsy." *Molecular Neurobiology* 56, no. 2 (2019): 1233–1247.

6. Pottoo, Faheem H., Md Javed, Md Barkat, Md Alam, Javaid Ashraf Nowshehri, Dhafer Mahdi Alshayban, and Mohammad Azam Ansari. "Estrogen and serotonin: Complexity of interactions and implications for epileptic seizures and epileptogenesis." *Current Neuropharmacology* 17, no. 3 (2019): 214–231.

7. Pottoo, Faheem Hyder Shrestha Sharma, Md Noushad Javed, Md Abul Barkat, Harshita, Mohd Sabir Alam, Mohd Javed Naim, et al. "Lipid-based nanoformulations in the treatment of neurological disorders." *Drug Metabolism Reviews* 52, no. 1 (2020): 185–204.

8. Pottoo, Faheem Hyder, Nahida Tabassum, Md Noushad Javed, Shah Nigar, Shrestha Sharma, Md Abul Barkat, Md Sabir Alam, Mohammad Azam Ansari, George E. Barreto, and Ghulam Md Ashraf. "Raloxifene potentiates the effect of fluoxetine against maximal electroshock induced seizures in mice." *European Journal of Pharmaceutical Sciences* 146 (2020):105261.

9. Huang, Jiale, Qingbiao Li, Daohua Sun, Yinghua Lu, YuanboSu, Xin Yang, Huixuan Wang, et al. "Biosynthesis of silver and gold nanoparticles by novel sundried *Cinnamomum camphora* leaf." *Nanotechnology* 18, no. 10 (2007): 105104.

10. Baruwati, Babita, and Rajender S Varma. "High value products from waste: Grape pomace extract-a three-in-one package for the synthesis of metal nanoparticles." *ChemSusChem* 2, no. 11 (2009): 1041–1044.

11. Nadagouda, M. N. and R. S. Varma. "Green synthesis of silver and palladium nanoparticles at room temperature using coffee and tea extract." *Green Chemistry* 10, no. 8 (2008): 859.

12. Tardif, F., M. Malarde, L. Golanski, A. Guiot, and M. Pras. "Nanotechnology 2011: Advanced materials, CNTs, particles, films and composites." In *Proceedings of the Nanotech 2011 Conference*, Boston, vol. 495 (2011).

13. Padil, Vinod VelloraThekkae, and Miroslav Černík. "Green synthesis of copper oxide nanoparticles using gum karaya as a biotemplate and their antibacterial application." *International Journal of Nanomedicine* 8 (2013): 889.

14. Manoj, Devaraj, R. Saravanan, Jayadevan Santhanalakshmi, Shilpi Agarwal, Vinod Kumar Gupta, and Rabah Boukherroub. "Towards green synthesis of monodisperse Cu nanoparticles: An efficient and high sensitive electrochemical nitrite sensor." *Sensors and Actuators B: Chemical* 266 (2018): 873–882.

15. Farshchi, HelaleKaboli, Majid Azizi, Mahmoud Reza Jaafari, SeyydHossien Nemati, and Amir Fotovat. "Green synthesis of iron nanoparticles by Rosemary extract and cytotoxicity effect evaluation on cancer cell lines." *Biocatalysis and Agricultural Biotechnology* 16 (2018): 54–62.

16. Khalil, Ali Talha, Muhammad Ovais, Ikram Ullah, Muhammad Ali, Zabta Khan Shinwari, and Malik Maaza. "Biosynthesis of iron oxide (Fe2O3) nanoparticles via aqueous extracts of *Sageretiathea* (Osbeck.) and their pharmacognostic properties." *Green Chemistry Letters and Reviews* 10, no. 4 (2017): 186–201.

17. Sathishkumar, G., V. Logeshwaran, S. Sarathbabu, Pradeep K. Jha, M. Jeyaraj, C. Rajkuberan, N. Senthilkumar, and S. Sivaramakrishnan. "Green synthesis of magnetic Fe3O4 nanoparticles using *Couroupita guianensis* Aubl. fruit extract for their antibacterial and cytotoxicity activities." *Artificial Cells, Nanomedicine, and Biotechnology* 46, no. 3 (2018): 589–598.

18. Qi, Peng, Dun Zhang, and Yi Wan. "Sulfate-reducing bacteria detection based on the photocatalytic property of microbial synthesized ZnS nanoparticles." *Analytica Chimica Acta* 800 (2013): 65–70.

19. Rajakumar, G., A. Abdul Rahuman, B. Priyamvada, V. Gopiesh Khanna, D. Kishore Kumar, and P. J. Sujin. "*Eclipta prostrata* leaf aqueous extract mediated synthesis of titanium dioxide nanoparticles." *Materials Letters* 68 (2012): 115–117.
20. Subhapriya, S., and P. Gomathipriya. "Green synthesis of titanium dioxide (TiO2) nanoparticles by *Trigonella foenum*-graecum extract and its antimicrobial properties." *Microbial Pathogenesis* 116 (2018): 215–220.
21. Manikandan, Velu, Palanivel Velmurugan, Palaniyappan Jayanthi, Jung-Hee Park, Woo-Suk Chang, Yool-Jin Park, Min Cho, and Byung-Taek Oh. "Biogenic synthesis from *Prunus × yedoensis* leaf extract, characterization, and photocatalytic and anti-bacterial activity of TiO_2 nanoparticles." *Research on Chemical Intermediates* 44, no. 4 (2018): 2489–2502.
22. Salunke, Bipinchandra K., Shailesh S. Sawant, Sang-Ill Lee, and Beom Soo Kim. "Comparative study of MnO_2 nanoparticle synthesis by marine bacterium *Saccharophagus degradans* and yeast *Saccharomyces cerevisiae*." *Applied Microbiology and Biotechnology* 99, no. 13 (2015): 5419–5427.
23. Xia, Yang, Zhen Xiao, Xiao Dou, Hui Huang, Xianghong Lu, Rongjun Yan, Yongping Gan, et al. "Green and facile fabrication of hollow porous MnO/C microspheres from microalgaes for lithium-ion batteries." *ACS Nano* 7, no. 8 (2013): 7083–7092.
24. Royston, Elizabeth S., Adam D. Brown, Michael T. Harris, and James N. Culver. "Preparation of silica stabilized Tobacco mosaic virus templates for the production of metal and layered nanoparticles." *Journal of Colloid and Interface Science* 332, no. 2 (2009): 402–407.
25. Arumugam, Ayyakannu, Chandrasekaran Karthikeyan, Abdulrahman SyedahamedHaja Hameed, Kasi Gopinath, Shanmugam Gowri, and Viswanathan Karthika. "Synthesis of cerium oxide nanoparticles using *Gloriosa superba* L. leaf extract and their structural, optical and antibacterial properties." *Materials Science and Engineering: C* 49 (2015): 408–415.
26. Rocca, Antonella, Stefania Moscato, Francesca Ronca, Simone Nitti, Virgilio Mattoli, Mario Giorgi, and Gianni Ciofani. "Pilot in vivo investigation of cerium oxide nanoparticles as a novel anti-obesity pharmaceutical formulation." *Nanomedicine: Nanotechnology, Biology and Medicine* 11, no. 7 (2015): 1725–1734.
27. Shah, Monaliben, Derek Fawcett, Shashi Sharma, Suraj Kumar Tripathy, and Gérrard Eddy Jai Poinern. "Green synthesis of metallic nanoparticles via biological entities." *Materials* 8, no. 11 (2015): 7278–7308.
28. Jha, Anal K., K. Prasad, Kamlesh Prasad, and A. R. Kulkarni. "Plant system: Nature's nanofactory." *Colloids and Surfaces B: Biointerfaces* 73, no. 2 (2009): 219–223.
29. Rajeshkumar, S., M. Ponnanikajamideen, C. Malarkodi, M. Malini, and G. Annadurai. "Microbe-mediated synthesis of antimicrobial semiconductor nanoparticles by marine bacteria." *Journal of Nanostructure in Chemistry* 4, no. 2 (2014): 96.
30. Seshadri, Sachin, K. Saranya, and Meenal Kowshik. "Green synthesis of lead sulfide nanoparticles by the lead resistant marine yeast, Rhodosporidium diobovatum." *Biotechnology Progress* 27, no. 5 (2011): 1464–1469.
31. Singh, Anirudh, Pavan Kumar Gautam, Arushi Verma, Vishal Singh, Pingali M. Shivapriya, Saurabh Shivalkar, Amaresh Kumar Sahoo, and Sintu Kumar Samanta. "Green synthesis of metallic nanoparticles as effective alternatives to treat antibiotics resistant bacterial infections: A review." *Biotechnology Reports* 25 (2020): e00427.
32. Kruis, Frank Einar, Heinz Fissan, and Bernd Rellinghaus. "Sintering and evaporation characteristics of gas-phase synthesis of size-selected PbS nanoparticles." *Materials Science and Engineering: B* 69 (2000): 329–334.
33. Tao, Andrea, Prasert Sinsermsuksakul, and Peidong Yang. "Polyhedral silver nanocrystals with distinct scattering signatures." *Angewandte Chemie International Edition* 45, no. 28 (2006): 4597–4601.

34. Wiley, Benjamin, Yugang Sun, Brian Mayers, and Younan Xia. "Shape-controlled synthesis of metal nanostructures: The case of silver." *Chemistry–A European Journal* 11, no. 2 (2005): 454–463.

35. Thakkar, Kaushik N., Snehit S. Mhatre, and Rasesh Y. Parikh. "Biological synthesis of metallic nanoparticles." *Nanomedicine: Nanotechnology, Biology and Medicine* 6, no. 2 (2010): 257–262.

36. Singh, Anirudh, Pavan Kumar Gautam, Arushi Verma, Vishal Singh, Pingali M. Shivapriya, Saurabh Shivalkar, Amaresh Kumar Sahoo, and Sintu Kumar Samanta. "Green synthesis of metallic nanoparticles as effective alternatives to treat antibiotics resistant bacterial infections: A review." *Biotechnology Reports* 25 (2020): e00427.

37. Krishnaraj, C., E. G. Jagan, S. Rajasekar, P. Selvakumar, P. T. Kalaichelvan, and N. J. C. S. B. B. Mohan. "Synthesis of silver nanoparticles using Acalypha indica leaf extracts and its antibacterial activity against water borne pathogens." *Colloids and Surfaces B: Biointerfaces* 76, no. 1 (2010): 50–56.

38. Hasnain, M. Saquib, Md Noushad Javed, Md Sabir Alam, Poonam Rishishwar, Sanjay Rishishwar, Sadath Ali, Amit Kumar Nayak, and Sarwar Beg. "Purple heart plant leaves extract-mediated silver nanoparticle synthesis: Optimization by Box-Behnken design." *Materials Science and Engineering: C* 99 (2019): 1105–1114.

39. Mukunthan, K. S., and S. Balaji. "Cashew apple juice (*Anacardium occidentale* L.) speeds up the synthesis of silver nanoparticles." *International Journal of Green Nanotechnology* 4, no. 2 (2012): 71–79.

40. Khanna, S. C., M. Soliva, and P. Speiser. "Epoxy resin beads as a pharmaceutical dosage form II: Dissolution studies of epoxy-amine beads and release of drug." *Journal of Pharmaceutical Sciences* 58, no. 11 (1969): 1385–1388.

41. Alam, Md Sabir, Arun Garg, Faheem Hyder Pottoo, Mohammad Khalid Saifullah, Abu Izneid Tareq, Ovais Manzoor, Mohd Mohsin, and Md Noushad Javed. "Gum ghatti mediated, one pot green synthesis of optimized gold nanoparticles: Investigation of process-variables impact using Box-Behnken based statistical design." *International Journal of Biological Macromolecules* 104 (2017): 758–767.

42. Alam, Md Sabir, Md Noushad Javed, Faheem Hyder Pottoo, Aafrin Waziri, Faisal A. Almalki, Md Saquib Hasnain, Arun Garg, and Md Khalid Saifullah. "QbD approached comparison of reaction mechanism in microwave synthesized gold nanoparticles and their superior catalytic role against hazardous nirto-dye." *Applied Organometallic Chemistry* 33, no. 9 (2019): e5071.

43. Javed, M. N., F. H. Pottoo, and M. S. Alam. "Metallic nanoparticle alone and/or in combination as novel agent for the treatment of uncontrolled electric conductance related disorders and/or seizure, epilepsy & convulsions." (2016). Patent registered on October 10, 2016.

44. John, Maria Sindhura, Joseph Amruthraj Nagoth, Kesava Priyan Ramasamy, Alessio Mancini, Gabriele Giuli, Antonino Natalello, Patrizia Ballarini, Cristina Miceli, and Sandra Pucciarelli. "Synthesis of bioactive silver nanoparticles by a Pseudomonas strain associated with the Antarctic psychrophilic protozoon *Euplotes focardii.*" *Marine Drugs* 18, no. 1 (2020): 38.

45. Elbeshehy, Essam KF, Ahmed M. Elazzazy, and George Aggelis. "Silver nanoparticles synthesis mediated by new isolates of *Bacillus* spp., nanoparticle characterization and their activity against Bean Yellow Mosaic Virus and human pathogens." *Frontiers in Microbiology* 6 (2015): 453.

46. Monowar, Tahmina, Md Rahman, Subhash J. Bhore, Gunasunderi Raju, and Kathiresan V. Sathasivam. "Silver nanoparticles synthesized by using the endophytic bacterium *Pantoea ananatis* are promising antimicrobial agents against multidrug resistant bacteria." *Molecules* 23, no. 12 (2018): 3220.

47. Saravanan, Muthupandian, Sisir Kumar Barik, Davoodbasha Mubarak Ali, Periyakaruppan Prakash, and Arivalagan Pugazhendhi. "Synthesis of silver nanoparticles from *Bacillus brevis* (NCIM 2533) and their antibacterial activity against pathogenic bacteria." *Microbial Pathogenesis* 116 (2018): 221–226.

48. Manikprabhu, Deene, Juan Cheng, Wei Chen, Anil Kumar Sunkara, Sunilkumar B. Mane, Ram Kumar, Wael N. Hozzein, Yan-Qing Duan, and Wen-Jun Li. "Sunlight mediated synthesis of silver nanoparticles by a novel actinobacterium (*Sinomonas mesophila* MPKL 26) and its antimicrobial activity against multi drug resistant *Staphylococcus aureus*." *Journal of Photochemistry and Photobiology B: Biology* 158 (2016): 202–205.

49. Malhotra, Ankit, Kunzes Dolma, Navjot Kaur, Yogendra S. Rathore, S. Mayilraj, and Anirban Roy Choudhury. "Biosynthesis of gold and silver nanoparticles using a novel marine strain of Stenotrophomonas." *Bioresource Technology* 142 (2013): 727–731.

50. Husseiny, M. I., M. Abd El-Aziz, Y. Badr, and M. A. Mahmoud. "Biosynthesis of gold nanoparticles using *Pseudomonas aeruginosa*." *Spectrochimica Acta Part A: Molecular and Biomolecular Spectroscopy* 67, no. 3–4 (2007): 1003–1006.

51. Syed, Baker, Nagendra M. N. Prasad, and Sreedharamurthy Satish. "Endogenic mediated synthesis of gold nanoparticles bearing bactericidal activity." *Journal of Microscopy and Ultrastructure* 4, no. 3 (2016): 162–166.

52. Suganya, K. S., K. Uma, V. Govindaraju Kumar Ganesh, T. Stalin Dhas, V. Karthick, G. Singaravelu, and M. Elanchezhiyan. "Blue green alga mediated synthesis of gold nanoparticles and its antibacterial efficacy against Gram positive organisms." *Materials Science and Engineering: C* 47 (2015): 351–356.

53. Cai, Fang, Jing Li, Jinsheng Sun, and Yulan Ji. "Biosynthesis of gold nanoparticles by biosorption using *Magnetospirillum gryphiswaldense* MSR-1." *Chemical Engineering Journal* 175 (2011): 70–75.

54. Shantkriti, S., and P. Rani. "Biological synthesis of copper nanoparticles using *Pseudomonas fluorescens*." *International Journal of Current Microbiology and Applied Sciences* 3, no. 9 (2014): 374–383.

55. Cuevas, R., Nelson Durán, M. C. Diez, G. R. Tortella, and Olga Rubilar. "Extracellular biosynthesis of copper and copper oxide nanoparticles by *Stereum hirsutum*, a native white-rot fungus from Chilean forests." *Journal of Nanomaterials* 2015 (2015).

56. Noor, Sadaf, Ziaullah Shah, Aneela Javed, Amjad Ali, Syed Bilal Hussain, Sidra Zafar, Hazrat Ali, and Syed Aun Muhammad. "A fungal based synthesis method for copper nanoparticles with the determination of anticancer, antidiabetic and antibacterial activities." *Journal of Microbiological Methods* (2020): 105966.

57. Hassan, Saad El-Din, Salem S. Salem, Amr Fouda, Mohamed A. Awad, Mamdouh S. El-Gamal, and Abdullah M. Abdo. "New approach for antimicrobial activity and bio-control of various pathogens by biosynthesized copper nanoparticles using endophytic actinomycetes." *Journal of Radiation Research and Applied Sciences* 11, no. 3 (2018): 262–270.

58. Salvadori, Marcia R., Rômulo A. Ando, Cláudio A. Oller do Nascimento, and Benedito Corrêa. "Intracellular biosynthesis and removal of copper nanoparticles by dead biomass of yeast isolated from the wastewater of a mine in the Brazilian Amazonia." *PLoS One* 9, no. 1 (2014): e87968.

59. Vaidyanathan, Ramanathan, Shubaash Gopalram, Kalimuthu Kalishwaralal, Venkataraman Deepak, Sureshbabu Ram Kumar Pandian, and Sangiliyandi Gurunathan. "Enhanced silver nanoparticle synthesis by optimization of nitrate reductase activity." *Colloids and Surfaces B: Biointerfaces* 75, no. 1 (2010): 335–341.

60. Haefeli, C., C. Franklin, and K. Hardy. "Plasmid-determined silver resistance in *Pseudomonas stutzeri* isolated from a silver mine." *Journal of Bacteriology* 158, no. 1 (1984): 389–392.

61. El-Shanshoury, Abd El-Raheem R., Sobhy E. ElSilk, and Mohamed E. Ebeid. "Extracellular biosynthesis of silver nanoparticles using *Escherichia coli* ATCC 8739, *Bacillus subtilis* ATCC 6633, and *Streptococcus thermophilus* ESh1 and their antimicrobial activities." *ISRN Nanotechnology* 2011 (2011).

62. Otari, S. V., R. M. Patil, N. H. Nadaf, S. J. Ghosh, and S. H. Pawar. "Green biosynthesis of silver nanoparticles from an actinobacteria *Rhodococcus* sp." *Materials Letters* 72 (2012): 92–94.

63. Thomas, R., A. Viswan, J. Mathew, and E. K. Radhakrishnan. "Evaluation of antibacterial activity of silver nanoparticles synthesized by a novel strain of marine *Pseudomonas* sp." *Nano Biomedicine and Engineering* 4, no. 3 (2012): 139–143.

64. Patil, Sarvamangala R. "Antibacterial activity of silver nanoparticles synthesized from *Fusarium semitectum* and green extracts." *International Journal of Scientific and Engineering Research* 2 (2014): 140–145.

65. Mukherjee, Priyabrata, Absar Ahmad, Deendayal Mandal, Satyajyoti Senapati, Sudhakar R. Sainkar, Mohammad I. Khan, Renu Parishcha, et al. "Fungus-mediated synthesis of silver nanoparticles and their immobilization in the mycelial matrix: A novel biological approach to nanoparticle synthesis." *Nano Letters* 1, no. 10 (2001): 515–519.

66. Kathiresan, K., S. Manivannan, M. A. Nabeel, and B. Dhivya. "Studies on silver nanoparticles synthesized by a marine fungus, *Penicillium fellutanum* isolated from coastal mangrove sediment." *Colloids and Surfaces B: Biointerfaces* 71, no. 1 (2009): 133–137.

67. Mukherjee, 4P, M. Roy, B. P. Mandal, G. K. Dey, P. K. Mukherjee, J. Ghatak, A. K. Tyagi, and S. P. Kale. "Green synthesis of highly stabilized nanocrystalline silver particles by a non-pathogenic and agriculturally important fungus *T. asperellum*." *Nanotechnology* 19, no. 7 (2008): 075103.

68. Kannan, R. Ragupathi Raja, R. Arumugam, D. Ramya, K. Manivannan, and P. Anantharaman. "Green synthesis of silver nanoparticles using marine macroalga *Chaetomorpha linum*." *Applied Nanoscience* 3, no. 3 (2013): 229–233.

69. Kathiraven, T., A. Sundaramanickam, N. Shanmugam, and T. Balasubramanian. "Green synthesis of silver nanoparticles using marine algae *Caulerpa racemosa* and their antibacterial activity against some human pathogens." *Applied Nanoscience* 5, no. 4 (2015): 499–504.

70. Asha, K. S., M. Johnson, P. K. Chandra, T. Shibila, and I. Revathy. "Extracellular synthesis of silver nanoparticles from a marine alga, *Sargassum polycystum* C. agardh and their biopotentials." *WJPPS* 4 (2015): 1388–1400.

71. Xie, Jianping, Jim Yang Lee, Daniel I. C. Wang, and Yen Peng Ting. "Silver nanoplates: From biological to biomimetic synthesis." *ACS Nano* 1, no. 5 (2007): 429–439.

72. Reddy, N. Jayachandra, D. Nagoor Vali, M. Rani, and S. Sudha Rani. "Evaluation of antioxidant, antibacterial and cytotoxic effects of green synthesized silver nanoparticles by *Piper longum* fruit." *Materials Science and Engineering: C* 34 (2014): 115–122.

73. Chandran, S. Prathap, Minakshi Chaudhary, Renu Pasricha, Absar Ahmad, and Murali Sastry. "Synthesis of gold nanotriangles and silver nanoparticles using *Aloe vera* plant extract." *Biotechnology Progress* 22, no. 2 (2006): 577–583.

74. Shankar, S. Shiv, Akhilesh Rai, Absar Ahmad, and Murali Sastry. "Rapid synthesis of Au, Ag, and bimetallic Au core–Ag shell nanoparticles using neem (*Azadirachta indica*) leaf broth." *Journal of Colloid and Interface Science* 275, no. 2 (2004): 496–502.

75. Mondal, Samiran, Nayan Roy, Rajibul A. Laskar, Ismail Sk, Saswati Basu, Debabrata Mandal, and Naznin Ara Begum. "Biogenic synthesis of Ag, Au and bimetallic Au/Ag alloy nanoparticles using aqueous extract of mahogany (*Swietenia mahogani* JACQ.) leaves." *Colloids and Surfaces B: Biointerfaces* 82, no. 2 (2011): 497–504.

76. Narayanan, K. Badri, and N. Sakthivel. "Coriander leaf mediated biosynthesis of gold nanoparticles." *Materials Letters* 62, no. 30 (2008): 4588–4590.
77. Raghunandan, Deshpande, Mahesh D. Bedre, S. Basavaraja, Balaji Sawle, S. Y. Manjunath, and A. Venkataraman. "Rapid biosynthesis of irregular shaped gold nanoparticles from macerated aqueous extracellular dried clove buds (*Syzygium aromaticum*) solution." *Colloids and Surfaces B: Biointerfaces* 79, no. 1 (2010): 235–240.
78. Krishnaraj, C., E. G. Jagan, S. Rajasekar, P. Selvakumar, P. T. Kalaichelvan, and N. J. C. S. B. B. Mohan. "Synthesis of silver nanoparticles using *Acalypha indica* leaf extracts and its antibacterial activity against water borne pathogens." *Colloids and Surfaces B: Biointerfaces* 76, no. 1 (2010): 50–56.
79. Mude, Namrata, Avinash Ingle, Aniket Gade, and Mahendra Rai. "Synthesis of silver nanoparticles using callus extract of Carica papaya—a first report." *Journal of Plant Biochemistry and Biotechnology* 18, no. 1 (2009): 83–86.
80. Prathna, T. C., N. Chandrasekaran, Ashok M. Raichur, and Amitava Mukherjee. "Biomimetic synthesis of silver nanoparticles by *Citrus limon* (lemon) aqueous extract and theoretical prediction of particle size." *Colloids and Surfaces B: Biointerfaces* 82, no. 1 (2011): 152–159.
81. Ravindra, S., Y. Murali Mohan, N. Narayana Reddy, and K. Mohana Raju. "Fabrication of antibacterial cotton fibres loaded with silver nanoparticles via "Green Approach"." *Colloids and Surfaces A: Physicochemical and Engineering Aspects* 367, no. 1–3 (2010): 31–40.
82. Veerasamy, Ravichandran, Tiah Zi Xin, Subashini Gunasagaran, Terence Foo Wei Xiang, Eddy Fang Chou Yang, Nelson Jeyakumar, and Sokkalingam Arumugam Dhanaraj. "Biosynthesis of silver nanoparticles using mangosteen leaf extract and evaluation of their antimicrobial activities." *Journal of Saudi Chemical Society* 15, no. 2 (2011): 113–120.
83. Vinayagam, Ramesh, Raja Selvaraj, Pugazhendhi Arivalagan, and Thivaharan Varadavenkatesan. "Synthesis, characterization and photocatalytic dye degradation capability of *Calliandra haematocephala*-mediated zinc oxide nanoflowers." *Journal of Photochemistry and Photobiology B: Biology* 203 (2020): 111760.
84. Rajiv, P., Sivaraj Rajeshwari, and Rajendran Venckatesh. "Bio-fabrication of zinc oxide nanoparticles using leaf extract of *Parthenium hysterophorus* L. and its size-dependent antifungal activity against plant fungal pathogens." *Spectrochimica Acta Part A: Molecular and Biomolecular Spectroscopy* 112 (2013): 384–387.
85. Ogunyemi, Solabomi Olaitan, Yasmine Abdallah, Muchen Zhang, Hatem Fouad, Xianxian Hong, Ezzeldin Ibrahim, Md Mahidul Islam Masum, Afsana Hossain, Jianchu Mo, and Bin Li. "Green synthesis of zinc oxide nanoparticles using different plant extracts and their antibacterial activity against *Xanthomonas oryzae* pv. *oryzae*." *Artificial Cells, Nanomedicine, and Biotechnology* 47, no. 1 (2019): 341–352.
86. Thema, F. T., E. Manikandan, M. S. Dhlamini, and M. Maaza. "Green synthesis of ZnO nanoparticles via *Agathosma betulina* natural extract." *Materials Letters* 161 (2015): 124–127.
87. Nagajyothi, P. C., T. V. M. Sreekanth, Clement O. Tettey, Yang In Jun, and Shin Heung Mook. "Characterization, antibacterial, antioxidant, and cytotoxic activities of ZnO nanoparticles using Coptidis Rhizoma." *Bioorganic & Medicinal Chemistry Letters* 24, no. 17 (2014): 4298–4303.
88. Gardea-Torresdey, J. L., K. J. Tiemann, G. Gamez, K. Dokken, S. Tehuacanero, and M. Jose-Yacaman. "Gold nanoparticles obtained by bio-precipitation from gold (III) solutions." *Journal of Nanoparticle Research* 1, no. 3 (1999): 397–404.
89. Sathishkumar, Muthuswamy, Krishnamurthy Sneha, and Yeoung-Sang Yun. "Immobilization of silver nanoparticles synthesized using *Curcuma longa* tuber powder and extract on cotton cloth for bactericidal activity." *Bioresource Technology* 101, no. 20 (2010): 7958–7965.

90. Dubey, Shashi Prabha, Manu Lahtinen, and Mika Sillanpää. "Tansy fruit mediated greener synthesis of silver and gold nanoparticles." *Process Biochemistry* 45, no. 7 (2010): 1065–1071.

91. Armendariz, Veronica, Isaac Herrera, Miguel Jose-yacaman, Horacio Troiani, Patricia Santiago, and Jorge L. Gardea-Torresdey. "Size controlled gold nanoparticle formation by *Avena sativa* biomass: Use of plants in nanobiotechnology." *Journal of Nanoparticle Research* 6, no. 4 (2004): 377–382.

92. Armendariz, Veronica, Isaac Herrera, Miguel Jose-yacaman, Horacio Troiani, Patricia Santiago, and Jorge L. Gardea-Torresdey. "Size controlled gold nanoparticle formation by *Avena sativa* biomass: Use of plants in nanobiotechnology." *Journal of Nanoparticle Research* 6, no. 4 (2004): 377–382.

93. Huang, Jiale, Qingbiao Li, Daohua Sun, Yinghua Lu, YuanboSu, Xin Yang, Huixuan Wang, et al. "Biosynthesis of silver and gold nanoparticles by novel sundried *Cinnamomum camphora* leaf." *Nanotechnology* 18, no. 10 (2007): 105104.

94. Chandran, S. Prathap, Minakshi Chaudhary, Renu Pasricha, Absar Ahmad, and Murali Sastry. "Synthesis of gold nanotriangles and silver nanoparticles using Aloevera plant extract." *Biotechnology Progress* 22, no. 2 (2006): 577–583.

95. Narayanan, Kannan Badri, and Natarajan Sakthivel. "Phytosynthesis of gold nanoparticles using leaf extract of *Coleus amboinicus* Lour." *Materials Characterization* 61, no. 11 (2010): 1232–1238.

96. Ahmad, Naheed, and Seema Sharma. "Green synthesis of silver nanoparticles using extracts of *Ananas comosus*." *Green and Sustainable Chemistry* 2 (2012): 141–147.

97. Dwivedi, Amarendra Dhar, and Krishna Gopal. "Biosynthesis of silver and gold nanoparticles using *Chenopodium album* leaf extract." *Colloids and Surfaces A: Physicochemical and Engineering Aspects* 369, no. 1–3 (2010): 27–33.

98. Song, Jae Yong, Hyeon-Kyeong Jang, and Beom Soo Kim. "Biological synthesis of gold nanoparticles using *Magnolia kobus* and *Diopyros kaki* leaf extracts." *Process Biochemistry* 44, no. 10 (2009): 1133–1138.

99. Kaviya, S., J. Santhanalakshmi, B. Viswanathan, J. Muthumary, and K. Srinivasan. "Biosynthesis of silver nanoparticles using *Citrus sinensis* peel extract and its antibacterial activity." *Spectrochimica Acta Part A: Molecular and Biomolecular Spectroscopy* 79, no. 3 (2011): 594–598.

100. Okoli, Celest U., Kurian A. Kuttiyiel, Jesse Cole, J. McCutchen, Hazem Tawfik, Radoslav R. Adzic, and Devinder Mahajan. "Solvent effect in sonochemical synthesis of metal-alloy nanoparticles for use as electrocatalysts." *Ultrasonics Sonochemistry* 41 (2018): 427–434.

101. Jha, Anal K., and K. Prasad. "Green synthesis of silver nanoparticles using Cycas leaf." *International Journal of Green Nanotechnology: Physics and Chemistry* 1, no. 2 (2010): P110–P117.

102. Song, Jae Yong, and Beom Soo Kim. "Rapid biological synthesis of silver nanoparticles using plant leaf extracts." *Bioprocess and Biosystems Engineering* 32, no. 1 (2009): 79.

103. Shanker, Uma, Vidhisha Jassal, Manviri Rani, and Balbir Singh Kaith. "Towards green synthesis of nanoparticles: From bio-assisted sources to benign solvents. A review." *International Journal of Environmental Analytical Chemistry* 96, no. 9 (2016): 801–835.

104. Yoosaf, Karuvath, Binil Itty Ipe, Cherumuttathu H. Suresh, and K. George Thomas. "In situ synthesis of metal nanoparticles and selective naked-eye detection of lead ions from aqueous media." *The Journal of Physical Chemistry C* 111, no. 34 (2007): 12839–12847.

105. Hu, Bo, Shang-Bing Wang, Kan Wang, Meng Zhang, and Shu-Hong Yu. "Microwave-assisted rapid facile "green" synthesis of uniform silver nanoparticles: Self-assembly into multilayered films and their optical properties." *The Journal of Physical Chemistry C* 112, no. 30 (2008): 11169–11174.

106. Hidouri, Slah, Manoubia Ben Yohmes, Ahmed Landoulsi, and Salah Ammar. "Commune propriety between reducing agents implicated in synthesis of metallic nanoparticles." *Review Journal of Chemistry* 9, no. 3 (2019): 153–160.
107. Rivero, Pedro Jose, Javier Goicoechea, Aitor Urrutia, and Francisco Javier Arregui. "Effect of both protective and reducing agents in the synthesis of multicolor silver nanoparticles." *Nanoscale Research Letters* 8, no. 1 (2013): 1–9.
108. Rodríguez-León, Ericka, Ramón Iñiguez-Palomares, Rosa Elena Navarro, Ronaldo Herrera-Urbina, Judith Tánori, Claudia Iñiguez-Palomares, and Amir Maldonado. "Synthesis of silver nanoparticles using reducing agents obtained from natural sources (*Rumex hymenosepalus* extracts)." *Nanoscale Research Letters* 8, no. 1 (2013): 318.
109. Smith, Nigel, Colin L. Raston, Martin Saunders, and Robert Woodward. "Synthesis of magnetic nanoparticles using spinning disc processing." In *Technical Proceedings of the 2006 NSTI Nanotechnology Conference and Trade Show*. Nano Science and Technology Institute (2006): 343–346.
110. Ma, Houyi, Bingsheng Yin, Shuyun Wang, Yongli Jiao, Wei Pan, Shaoxin Huang, Shenhao Chen, and Fanjun Meng. "Synthesis of silver and gold nanoparticles by a novel electrochemical method." *ChemPhysChem* 5, no. 1 (2004): 68–75.
111. Ijaz, Irfan, Ezaz Gilani, Ammara Nazir, and Aysha Bukhari. "Detail review on chemical, physical and green synthesis, classification, characterizations and applications of nanoparticles." *Green Chemistry Letters and Reviews* 13, no. 3 (2020): 223–245.
112. Okitsu, Kenji, Kohei Sharyo, and Rokuro Nishimura. "One-pot synthesis of gold nanorods by ultrasonic irradiation: The effect of pH on the shape of the gold nanorods and nanoparticles." *Langmuir* 25, no. 14 (2009): 7786–7790.
113. Armendariz, Veronica, Isaac Herrera, Miguel Jose-yacaman, Horacio Troiani, Patricia Santiago, and Jorge L. Gardea-Torresdey. "Size controlled gold nanoparticle formation by *Avena sativa* biomass: Use of plants in nanobiotechnology." *Journal of Nanoparticle Research* 6, no. 4 (2004): 377–382.
114. Konishi, Y., T. Tsukiyama, T. Tachimi, N. Saitoh, T. Nomura, and S. Nagamine. "Microbial deposition of gold nanoparticles by the metal-reducing bacterium *Shewanella algae*." *Electrochimica Acta* 53, no. 1 (2007): 186–192.
115. He, Shiying, Zhirui Guo, Yu Zhang, Song Zhang, Jing Wang, and Ning Gu. "Biosynthesis of gold nanoparticles using the bacteria *Rhodopseudomonas capsulata*." *Materials Letters* 61, no. 18 (2007): 3984–3987.
116. Sneha, Krishnamurthy, Muthuswamy Sathishkumar, Sok Kim, and Yeoung-Sang Yun. "Counter ions and temperature incorporated tailoring of biogenic gold nanoparticles." *Process Biochemistry* 45, no. 9 (2010): 1450–1458.
117. Iravani, Siavash, and Behzad Zolfaghari. "Green synthesis of silver nanoparticles using *Pinus eldarica* bark extract." *BioMed Research International* 2013 (2013).
118. Fayaz, A. Mohammed, K. Balaji, P. T. Kalaichelvan, and R. Venkatesan. "Fungal based synthesis of silver nanoparticles—An effect of temperature on the size of particles." *Colloids and Surfaces B: Biointerfaces* 74, no. 1 (2009): 123–126.
119. Kumar, E. Ranjith, R. Jayaprakash, T. ArunKumar, and Sanjay Kumar. "Effect of reaction time on particle size and dielectric properties of manganese substituted CoFe2O4 nanoparticles." *Journal of Physics and Chemistry of Solids* 74, no. 1 (2013): 110–114.
120. Islam, A. K. M. Maidul, and M. Mukherjee. "Effect of temperature in synthesis of silver nanoparticles in triblock copolymer micellar solution." *Journal of Experimental Nanoscience* 6, no. 6 (2011): 596–611.
121. Karade, V. C., T. D. Dongale, Subasa C. Sahoo, P. Kollu, A. D. Chougale, P. S. Patil, and P. B. Patil. "Effect of reaction time on structural and magnetic properties of green-synthesized magnetic nanoparticles." *Journal of Physics and Chemistry of Solids* 120 (2018): 161–166.

122. Flor, Juliana, SA Marques de Lima, and Marian Rosaly Davolos. "Effect of reaction time on the particle size of ZnO and ZnO:Ce obtained by a sol–gel method." In *Surface and Colloid Science. Progress in Colloid and Polymer Science.* Springer, Berlin, Heidelberg Vol. 128 (2004): 239–243.

123. Singh, Anirudh, Pavan Kumar Gautam, Arushi Verma, Vishal Singh, Pingali M. Shivapriya, Saurabh Shivalkar, Amaresh Kumar Sahoo, and Sintu Kumar Samanta. "Green synthesis of metallic nanoparticles as effective alternatives to treat antibiotics resistant bacterial infections: A review." *Biotechnology Reports* 25 (2020): e00427.

124. Ahmad, Tausif, Muhammad Irfan, Mohamad Azmi Bustam, and Sekhar Bhattacharjee. "Effect of reaction time on green synthesis of gold nanoparticles by using aqueous extract of *Elaise guineensis* (oil palm leaves)." *Procedia Engineering* 148 (2016): 467–472.

125. Mishra, Amrita, Suraj Kumar Tripathy, Rizwan Wahab, Song-HoonJeong, Inho Hwang, You-Bing Yang, Young-Soon Kim, Hyung-Shik Shin, and Soon-Il Yun. "Microbial synthesis of gold nanoparticles using the fungus *Penicillium brevicompactum* and their cytotoxic effects against mouse mayo blast cancer C 2 C 12 cells." *Applied Microbiology and Biotechnology* 92, no. 3 (2011): 617–630.

126. Pimprikar, P. S., S. S. Joshi, A. R. Kumar, S. S. Zinjarde, and S. K. Kulkarni. "Influence of biomass and gold salt concentration on nanoparticle synthesis by the tropical marine yeast *Yarrowia lipolytica* NCIM 3589." *Colloids and Surfaces B: Biointerfaces* 74, no. 1 (2009): 309–316.

127. Chauhan, Arun, Swaleha Zubair, Saba Tufail, Asif Sherwani, Mohammad Sajid, Suri C. Raman, Amir Azam, and Mohammad Owais. "Fungus-mediated biological synthesis of gold nanoparticles: Potential in detection of liver cancer." *International Journal of Nanomedicine* 6 (2011): 2305.

128. Huang, Jiale, Qingbiao Li, Daohua Sun, Yinghua Lu, YuanboSu, Xin Yang, Huixuan Wang, et al. "Biosynthesis of silver and gold nanoparticles by novel sundried *Cinnamomum camphora* leaf." *Nanotechnology* 18, no. 10 (2007): 105104.

129. Narayanan, Kannan Badri, and Natarajan Sakthivel. "Phytosynthesis of gold nanoparticles using leaf extract of *Coleus amboinicus* Lour." *Materials Characterization* 61, no. 11 (2010): 1232–1238.

130. Amendola, Vincenzo, and Moreno Meneghetti. "Size evaluation of gold nanoparticles by UV-vis spectroscopy." *The Journal of Physical Chemistry C* 113, no. 11 (2009): 4277–4285.

131. Parang, Z., A. Keshavarz, S. Farahi, S. M. Elahi, M. Ghoranneviss, and S. Parhoodeh. "Fluorescence emission spectra of silver and silver/cobalt nanoparticles." *Scientia Iranica* 19, no. 3 (2012): 943–947.

132. Lim, J., S. Yeap, H. Che, and S. Low. "Characterization of magnetic nanoparticle by dynamic light scattering." *Nanoscale Research Letters* 8, no. 1 (2013): 381.

133. Dougherty, George M., Klint A. Rose, Jeffrey B.-H. Tok, Satinderpall S. Pannu, Frank Y.S. Chuang, Michael Y. Sha, Gabriela Chakarova, and Sharron G. Penn. "The zeta potential of surface-functionalized metallic nanorod particles in aqueous solution." *Electrophoresis* 29, no. 5 (2008): 1131–1139.

134. Faghihzadeh, Fatemeh, Nelson M. Anaya, Laura A. Schifman, and Vinka Oyanedel-Craver. "Fourier transform infrared spectroscopy to assess molecular-level changes in microorganisms exposed to nanoparticles." *Nanotechnology for Environmental Engineering* 1, no. 1 (2016): 1.

135. Marbella, Lauren E., and Jill E. Millstone. "NMR techniques for noble metal nanoparticles." *Chemistry of Materials* 27, no. 8 (2015): 2721–2739.

136. Guo, Chengchen, and Jeffery L. Yarger. "Characterizing gold nanoparticles by NMR spectroscopy." *Magnetic Resonance in Chemistry* 56, no. 11 (2018): 1074–1082.

137. Marbella, Lauren E., and Jill E. Millstone. "NMR techniques for noble metal nanoparticles." *Chemistry of Materials* 27, no. 8 (2015): 2721–2739.
138. Ingham, Bridget. "X-ray scattering characterisation of nanoparticles." *Crystallography Reviews* 21, no. 4 (2015): 229–303.
139. Gao, Nan, and Yuying Yan. "Characterisation of surface wettability based on nanoparticles." *Nanoscale* 4, no. 7 (2012): 2202–2218.
140. Buhr, E., N. Senftleben, T. Klein, D. Bergmann, D. Gnieser, C. G. Frase, and H. Bosse. "Characterization of nanoparticles by scanning electron microscopy in transmission mode." *Measurement Science and Technology* 20, no. 8 (2009): 084025.
141. Herzing, Andrew A., Masashi Watanabe, Jennifer K. Edwards, Marco Conte, Zi-Rong Tang, Graham J. Hutchings, and Christopher J. Kiely. "Energy dispersive X-ray spectroscopy of bimetallic nanoparticles in an aberration corrected scanning transmission electron microscope." *Faraday Discussions* 138 (2008): 337–351.
142. Asadi Asadabad, Mohsen, and Mohammad Jafari Eskandari. "Transmission electron microscopy as best technique for characterization in nanotechnology." *Synthesis and Reactivity in Inorganic, Metal-Organic, and Nano-Metal Chemistry* 45, no. 3 (2015): 323–326.
143. Sublemontier, Olivier, Christophe Nicolas, Damien Aureau, Minna Patanen, Harold Kintz, Xiaojing Liu, Marc-André Gaveau, et al. "X-ray photoelectron spectroscopy of isolated nanoparticles." *The Journal of Physical Chemistry Letters* 5, no. 19 (2014): 3399–3403.
144. Darwich, Samer, Karine Mougin, Akshata Rao, Enrico Gnecco, Shrisudersan Jayaraman, and Hamidou Haidara. "Manipulation of gold colloidal nanoparticles with atomic force microscopy in dynamic mode: Influence of particle–substrate chemistry and morphology, and of operating conditions." *Beilstein Journal of Nanotechnology* 2, no. 1 (2011): 85–98.
145. Mansfield, Elisabeth, Katherine M. Tyner, Christopher M. Poling, and Jenifer L. Blacklock. "Determination of nanoparticle surface coatings and nanoparticle purity using microscale thermogravimetric analysis." *Analytical Chemistry* 86, no. 3 (2014): 1478–1484.

4 Methods for Characterization and Quantitation of Nanomaterials

Kuna Lakshun Naidu, Thirupathi Gadipelly,
Rajakumar Anbazhagan, Shweta Yadav,
and Nisha Gautam

CONTENTS

4.1 Introduction 84
4.2 X-Ray Based Techniques 84
    4.2.1 X-Ray Diffraction (XRD)..84
    4.2.2 X-Ray Absorption Spectroscopy (XAS)..85
4.3 Scanning Electron Microscopy and Energy-Dispersive X-Ray
    Spectroscopy...90
    4.3.1 Electron–Matter Interaction 90
    4.3.2 Scanning Electron Microscopy 91
    4.3.3 Transmission Electron Microscopy (TEM)...92
    4.3.4 Scanning Tunneling Electron Microscopy (STEM)............................92
    4.3.5 Energy-Dispersive X-Ray Spectroscopy (EDS) 92
    4.3.6 Electron Backscatter Diffraction (EBSD) 93
4.4 Raman Spectroscopy and Surface-Enhanced Raman Spectroscopy 93
    4.4.1 Raman Spectroscopy 93
    4.4.2 Surface-Enhanced Raman Spectroscopy (SERS) 96
4.5 UV–Vis–NIR Absorption Spectroscopy...99
4.6 Fourier-Transform Infrared Spectroscopy (FTIR) 102
4.7 Dynamic Light Scattering (DLS) Technique.. 103
4.8 Physical Property Measuring System–Vibrating Sample Magnetometer
    (PPMS-VSM).. 103
4.9 Electron Spin Resonance (ESR) spectroscopy and AC-Susceptibility
    Studies.. 105
4.10 Impedance Spectroscopy 105
4.11 Methods for Physicochemical Characterization.. 106
Acknowledgment 107
References.. 108

DOI: 10.1201/9781003126256-4

4.1 INTRODUCTION

The importance of nanomaterials is not limited to materials science, but the essentiality is there in all engineering and science branches. Different from bulk material characterization, the nanomaterial characterization needs highly sensitive equipment and design. The nanocomposite's stability is known to be nanostructure property range aggregates, composite, crystalline, shape, size, and corrosion (Phan and Haes 2019). By tuning these nanoparticle parameters, the factors can affect the clearance and biodistribution of polymeric nanoparticles (Alexis et al. 2008). In this chapter, various techniques for characterizing nanomaterials are covered. The techniques primarily refer to the structural, magnetic, electrical and optical studies.

4.2 X-RAY BASED TECHNIQUES

4.2.1 X-Ray Diffraction (XRD)

X-ray diffraction (Kim, Kim, and Koo 2004) is a robust technique for the study of a crystal structure. X-rays are electromagnetic waves that lie between ultraviolet and gamma rays in an electromagnetic spectrum. The wavelength of X-rays is on the order of 10^{-10} m. In laboratory diffractometers, applying a high voltage of tens of kilovolts between electrodes causes ejection and acceleration of electrons from the cathode toward the anode target. As the electrons impinge on the anode target, they lose kinetic energy and decelerate, producing X-rays of varying wavelengths, known as continuous X-rays. However, at specific high potentials, the electrons gain enough kinetic energy to knock out the inner shell electrons, which are filled by the valance or outer shell electrons, producing characteristic X-rays, with energy equal to the difference between the shells of the target material. These characteristic X-rays are specific to the target material. Moseley's law gives the relation between the generated characteristic X-ray and the target anode material. It says the square root of the frequency of the characteristic X-ray is directly proportional to the atomic number of the target material.

The crystal structures are classified based on symmetry operations into a 32-point group and 7 crystal systems consisting of 14 Bravais lattices. Further inclusion of space groups via consideration of screw axes and glide planes, then 230 groups are categorized. Every crystal will be in one of these crystal groups. The seven crystal systems are cubic, tetragonal, orthorhombic, rhombohedral, hexagonal, monoclinic, and triclinic.

In the study of the crystal structure of nanomaterials, the X-ray diffraction technique is extensively used to obtain the phase identification, crystallite size, and lattice parameters, apart from the distribution ratio of different sizes of nanomaterial in the sample. The XRD patterns were indexed based on the standards provided by the International Centre for Diffraction Data (ICDD), established by the Joint Committee on Powder Diffraction Standards (JCPDS). The PC-PDF software program uses a search/display system to obtain a comprehensive list of Powder Diffraction File (PDF) (Jenkins and Holomany 1987).

As compared to the bulk, the X-ray diffraction of the nanomaterial requires careful analysis. One of the profound effects of nanosize on the diffraction peaks is the broadening of the peaks, including the particle size effect and instrumental broadening. The Scherrer formula (Patterson 1939) is used to obtain the crystallite size and is given by

$$D = \frac{0.9\ \lambda}{\beta\ Cos\ \theta}$$

where D is the crystallite size, λ is the wavelength of the X-ray, β is the full width at half maximum (FWHM) of the individual peak measured in radians, and θ is the Bragg angle. Depending on composition and instrumental resolution, the typical crystallite size range is 2 nm to 5 μm for powder diffraction. There are complementary techniques to find the size, such as dynamic light scattering and electron microscopy techniques. The X-ray diffraction of the same nanomaterial with different shapes may have different orientation of planes by exhibiting the variation in the intensity at the peaks in the diffraction pattern (Holder and Schaak 2019). The orientation may result in changing the morphology of the nanomaterial.

The average crystallite size (D) and maximum lattice strain ($\mathcal{E}$) are calculated by the following formula (Williamson–Hall plot):

$$\beta \cos\theta = \frac{0.93\lambda}{D} + \sin\theta$$

Apart from the crystallite size, the XRD mainly focuses on the complex structural analysis where multiple phases will be observed in the nanomaterials. The quantitative phase analysis refers to the phase fractions of various crystal structures presented in a material. Also, the order of crystallinity will be calculated by the XRD spectra analysis in a polycrystalline material. A pyrochlore structure evolution with calcination has been reported in a recent study where multiple phases are analyzed by Rietveld refinement (Gadipelly et al. 2020). Table 4.1 lists the different structural parameters.

4.2.2   X-Ray Absorption Spectroscopy (XAS)

Conventional X-ray tubes generate discrete characteristic X-rays: they are widely used for X-ray diffraction in many laboratories. However, by generating synchrotron radiation in large-scale electron accelerators and storage rings, continuous intense X-ray beams can be used for different spectroscopy and advanced investigations.

Synchrotron radiation emits centripetally accelerated charges, like electrons, moving at relativistic speed. The useful properties of the synchrotron radiation are its high intensity, continuous spectrum that can be tuned using a monochromator, high collimation with divergence on the order of mrads leading to good resolution, horizontal polarization, and the pulsed time structure (Shenoy 2003; Mobilio and Vlaic 2003; Hofmann n.d.).

TABLE 4.1
Primary Information Needed for Rietveld Analysis of XRD Pattern

Structural Parameter	Importance
Phase fraction	Multiple structural phases will be quantified
Lattice parameters	Unit cell information
Space group	Symmetry notation
Wyckoff positions or atomic coordinates	Position of atoms substituted or displaced in a particular composition
Thermal parameters	Debye–Waller factor refers to thermal motion of atoms
Average crystallite size	The average extent of the periodic arrangement of atoms in the structure, broader for nanosize
Maximum lattice strain	The strain induced in the lattice structure of the material

The Synchrotron Radiation Facility in the European Synchrotron Radiation Facility (ESRF) (Haensel 1989; Revol et al. 2013) located in Grenoble (France) has an accelerator complex which consists of a pre-injector or linear accelerator (LINAC), a booster or synchrotron, and the storage ring. The electrons are emitted by a 100 KeV electron gun and are accelerated in a LINAC to reach 200 MeV. The accelerated electrons are then transmitted to a circular accelerator (booster synchrotron) where they are further accelerated to reach 6 GeV. The circumference of the booster is about 300 meters. The storage ring stores and maintains the 6 GeV electron beam injected from the booster. The circumference of the storage ring is 844.4 m and consists of 64 bending magnets, 320 quadrupoles, and 224 sextupoles.

The storage ring includes 32 straight and 32 curved sections in alternating order. Each curved section consists of two large bending magnets, which force the path of electron beam into racetrack orbit of storage ring. In the straight sections (each ~6 m), several focusing magnets ensure that electrons remain in their ideal paths. The straight sections also host insertion devices (5 m) that force the electron beam unto sinusoidal trajectory and hence produce a much more intense beam than produced by the bending magnets. The electron energy in the storage ring is given by

$$E = 6\,\mathrm{GeV} = \gamma E_0,$$

where E_0 is 0.511 eV, implying γ is 12000.

Two main insertion devices used are undulators and wigglers (Elleaume 1988, 1989; Clarke 2004). These undulators are magnetic structures, made up of a complex array of small magnets. The radiation emitted at each consecutive bend of trajectory overlaps and interferes with that from the other bends, generating a more focused and intense beam of radiation than that generated by a single magnet. Also, the photons emitted are concentrated at certain energies (called the fundamental and harmonics). The gap between the rows of magnets can be changed to fine-tune the wavelength of the X-rays in the beam. The configuration of the magnets in wigglers and undulators is basically the same, but the devices are operated under different conditions.

The electron beam emits photons in a narrow cone of natural emission angle (γ^{-1}). However, electron beams from the wiggler or undulator emit photons in a narrow cone of natural emission angle ($K\gamma^{-1}$, where K is a deflection parameter). If the deflection parameter $K > 1$, it is a wiggler and emits a continuous spectrum, although more intense than that from a bending magnet. Whereas if $K \leq 1$, it is an undulator and emits a sharp intense beam (Hwang 2014).

Discontinuities (edge jumps) in the X-ray absorption spectra were first observed in 1912 by Fricke (Fricke 1920) and later by Hertz (Hertz 1920). The first theoretical explanation was given by Kronig (1931), by considering the oscillations in the absorption coefficient as effects from the band structure or electronic structure, i.e., as due to the long-range order (LRO) effect. However, LRO failed to explain the experimental results. Kronig (1932) later proposed the short-range order theory (SRO) to explain the absorption structure in the gaseous molecules. The SRO effect was based on the scattering of photoelectrons by the neighboring atoms. After almost 40 years of discussion about the origin of EXAFS oscillation, Sayers, Stern, and Lytle (1971) developed the EXAFS theory on the basis of single-scattering SRO. The single-scattering SRO approach is not only able to reproduce the EXAFS oscillations (that appear in the reciprocal space), but is also able to invert the experimental data to present real-space structural information, such as bond lengths, coordination numbers, and kind of shells around a selected absorbing atom. Sayers showed that the Fourier transform (FT) of the EXAFS function is similar to a radial distribution function, where each peak may be associated with one or more shells of atoms around the absorbing one. This led to the improvement of the SRO approach to explain the XAFS oscillation. Further improvement came from the ability to evaluate the functions describing the amplitude of backscattering and backscattering phase shifts (McKale et al. 1988), together with EXAFS Debye–Waller factors, due to thermal effects. Later, Müller and Schaich (1983) improved the single-scattering SRO theory by introducing the curved photoelectron wave. In 1980, Rehr et al. introduced the effective scattering amplitude to explain the SRO effect with multiple scattering by spherical photoelectron waves (Rehr et al. 1986; Rehr and Albers 1990). Theoretical standards previously tabulated by McKale et al. (1988) were replaced by those calculated by automated ab initio single-scattering EXAFS codes, known as the FEFF code. The modern FEFF code includes both inelastic losses and curved wave effects, as summarized by Rehr et al. (1991). However, the inclusion of whole multiple scattering (MS) terms into the FEFF code is limited by the problem of exponential increase of MS paths, many of which have negligible scattering amplitude. These can be eliminated by path filters, fast path generation and path sorting algorithms developed by Zabinsky et al. (1995). The EXAFS analysis program FEFFIT (Newville et al. 1995), which uses the FEFF code, allows us to obtain reliable information from XAFS data by fitting the experimental data using the outputs of FEFF calculations (Newville 2001b). The advances in theoretical approaches of FEFF code are discussed in detail in some reports (Rehr and Albers 2000; Rehr and Ankudinov 2001). A single package of interactive algorithms for general data manipulation including AUTOBK

(Newville et al. 1993) and FEFFIT has been developed by Newville (2001a) and is named the IFEFFIT library. Two main programs based on IFEFFIT algorithms have been developed: (1) ATHENA (Ravel 2008) for data processing, and (2) ARTEMIS for quantitative analysis of EXAFS using theoretical standards calculated by FEFF (Ravel and Newville 2005). The application ATOMS (Ravel 2001) uses the crystallographic details to generate the input data for the ab initio FEFF code. In most of XAFS work, FEFF6 and FEFF8 have been used to analyze the EXAFS data.

The raw EXAFS spectrum is characterized by (1) the absorption region, (2) the region of energies < absorption edge called the pre-edge region, (3) the region of energies > absorption edge called the post-edge region, and (4) oscillations in the post-edge region. For all the films, at least two scans were carried out at the selected absorption edge. Further scans were also made at different energies close to the selected absorption edge to confirm the stability of experimental conditions and beam parameters. The different scans at the selected absorption edge were then merged to increase the signal-to-noise ratio. The reference samples were measured in transmission mode and the films in fluorescence mode. In order to obtain the structural information from the EXAFS data, the oscillatory part of the X-ray absorption spectra were isolated according to the equation

$$\chi(E) = \left(\frac{\mu(E) - \mu_0(E)}{\mu_0(E_0)} \right)$$

where $\mu_0(E)$ is the absorption coefficient of the isolated central atom, i.e., without the influence of backscattering from neighboring atoms; $\mu(E)$ is absorption coefficient of the probed sample and $\chi(E)$ is called the extracted EXAFS function. E_0 is defined as the threshold energy, representing the $k = 0$ position. The choice of E_0 on experimental spectra influences the accuracy of the structural parameters that will be obtained from the quantitative analysis of EXAFS data. Thus it is quite important to select for reference compounds the correct values. We did the choice of E_0 by a sequence of steps. At first, for each measured reference compound, an initial guess is done, based on one of the following options: (1) at a point where absorption is half of the energy step, or (2) at the first peak of the first derivative of the spectrum, or (3) allowing the ATHENA program to decide about E_0. Then, the experimentally calculated $\chi(E)$ is extracted using the equation

$$\chi(E) = \frac{\mu_{\exp}(E) - \mu_{bkg}(E)}{\Delta\mu_{\exp}(E_0)}$$

The $\mu_0(E)$ term is now approximated by a background function [$\mu_{bkg}(E)$], evaluated through the AUTOBK algorithm of ATHENA, which requires from the user the choice of pre-edge and post-edge range, spline range, and R_{bkg} value (which plays a crucial role in obtaining a smooth background function at low-k and high-k values). The modulated absorption coefficient $\chi(E)$ is then evaluated as (k) in k-space, by using $k = \sqrt{\frac{2m}{\hbar^2}(E - E_0)}$. The $\chi(k)$ is successively Fourier transformed into $\chi(R)$, where we can identify different coordination shells. At this point, the $\chi(R)$ of the first shell and the related EXAFS function can be best-fitted using ARTEMIS on the basis of known lattice parameters of the reference compound. If the best fit is obtained through a forced shift in E_0, then the new E_0 position is used in a successive extraction of EXAFS to be again analyzed. This process is repeated until the shift in E_0 is minimized to zero.

The EXAFS oscillation $\chi(k)$ is related to the structural parameters by the equation

$$\chi(k) = \sum_j N_j S_i(k) F_j(k) e^{-2\sigma_j^2 k^2} e^{-2r_j/\lambda_j(k)} \frac{sin(2kr_j + \Phi_{ij}(k))}{kr_j^2}$$

where $F_j(k)$ is the backscattering amplitude from each jth neighboring atom, $S_i(k)$ is the amplitude reduction factor to account for the shake-up/-off processes of the central atom i during the ejection and backscattering of the photoelectron wave from and toward the central atom, σ_j accounts for the thermal and static disorder at a distance r_j, N_j corresponds to the coordination number, and $\phi_{ij}(k)$ corresponds to the total phase shift observed by the photoelectron near the central atom. The term $exp(-2r_j/\lambda_j(k))$ is for the inelastic losses during the scattering process.

The parameters $F_j(k)$, $\phi_{ij}(k)$, and $exp(-2r_j/\lambda_j(k))$ are determined from the FEFF model. The parameters $S_i(k)$ and r_j are kept as free parameters and N_j as a fixed parameter for the theoretical FEFF fit. The final fit will have the fit values of $S_i(k)$, r_j (or dr; change in distance), and Debye–Waller factor σ_j.

X-ray absorption spectroscopy (XAS) includes both X-ray absorption near edge structure (XANES) and extended X-ray absorption fine structure (EXAFS). The XANES analysis will give information about the density of states (DOS) of empty or partially filled states, to which a core electron is excited, whereas the EXAFS refers to the oscillatory portion above the absorption edge and gives information about the local environment around the selected element. The EXAFS analysis provides information about the structural changes like changes in bond length, bond angle, variance of bond length distribution, appearance of new radial shells, and coordination number around the probed atom.

Lakshun Naidu (2013) reported on the role of the metal layer (Cr or Ni) of thickness (50 to 200 nm) for lowering the crystallization temperature of a-Si, known as metal-induced crystallization of a-Si. XAFS analysis of the films around the metal species indicated the formation of different metal silicides, which act as seeds for the c-Si formation. The driving force for the crystallization of a-Si is the difference in Gibbs free energy between the amorphous and crystalline phases of Si. In the case of silicide forming metals, the metal silicide itself acts as the nucleating center for the crystallization of a-Si, with a 0.04% of mismatch between the silicide and the c-Si for the Ni–Si system. The growth of c-Si is preceded by the diffusion of a-Si atoms across the thickness of $NiSi_2$ to the $NiSi_2$/c-Si interface, which is accompanied by diffusion of Ni atoms from the c-Si/NiSi2 interface to the NiSi2/a-Si interface (Hayzelden and Batstone 1993). Paesler et al. (1983) showed ordering in a-Ge before crystallization by confirming the appearance of a second neighboring shell and a decrease of bond-angle distortion by 7°. Similarly, Wakagi and Maeda (1994) showed an increase in bond length and variations in bond length distributions with an increase in temperature below the crystallization temperature. Note that usually these structural parameters decrease at crystallization, indicating a final structural relaxation. Similar results have been reported in amorphous films with metal dopants to explain the underlying mechanisms for the crystallization process (Tan et al. 1992; Greaves et al. 1993; Ayers et al. 1994; Zanatta and Chambouleyron 2005). In some cases, it was possible to evaluate the crystalline fraction of amorphous-crystalline mixed phases of silicon by a linear combination of individual XAFS spectra:

$$X_c(E) = \frac{C_c}{C_c + C_a}$$

where C_c and C_a are the coefficients of XAFS spectra of the crystalline and amorphous phase of silicon, respectively, as shown in the equation

$$\chi_{mixed} = C_c\chi_c + C_a\chi_a$$

where χ corresponds to the XAFS spectrum (Wakagi et al. 1992). Similar crystal fraction determination is done to obtain the fraction of Ge nanocrystal embedded in an $a\text{-}SiO_2$ matrix (Araujo et al. 2008) or for understanding nanocrystalline structure of an a-Fe (Si) system (Kim, Kim, and Koo 2004). Davidson et al. (2015) studied the evolution of silver nanoparticle in the rat lung by XAFS carried out at 85 K. The size of the Ag nanoparticles used for the 6 h inhalation exposure were 20 and 110 nm diameter, and it was found that the Ag nanoparticles were transformed to other forms of Ag nanoparticles of smaller size over a period of seven days. The rate of this transformation is size dependent on the size of the nanoparticle and the larger 110 nm particles undergo significant changes as compared to the 20 nm particles. However, when similar experiments were performed with silica-coated Au nanoparticles, no modification to the process was observed.

The advantage of XAFS over other mentioned techniques is that it has a large area probe with easy sample preparation and enables non-destructive depth profiling of different compounds and statistical averaging.

4.3 SCANNING ELECTRON MICROSCOPY AND ENERGY-DISPERSIVE X-RAY SPECTROSCOPY

Different from an optical microscope, a beam of accelerated electrons is used as the illuminated source in an electron microscope, where high-resolution images will be captured based on electron–matter interaction. Structural characterization is one of the major areas in finding the properties of the nanomaterials. Mainly, the study involves analysis related to the distribution of size, shape, orientation, and chemical composition.

4.3.1 ELECTRON–MATTER INTERACTION

When the electron beam falls on the specimen, the electron–matter interaction generates various signals that contain the microstructural information of the specimen. The

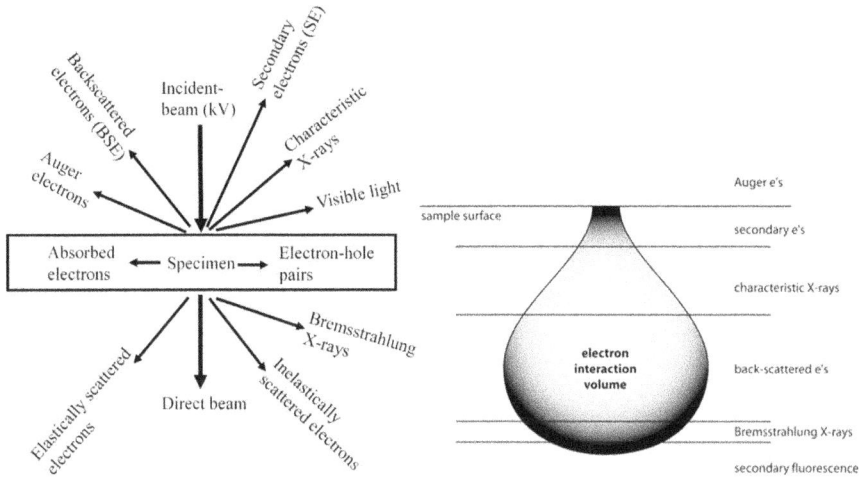

FIGURE 4.1 Schematic representation of electron–matter interaction.

generated events are represented in a schematic diagram as shown in Figure 4.1. The signals will be generated as per the electron–interaction volume within the specimen.

4.3.2 SCANNING ELECTRON MICROSCOPY

The primary signals are detected by the imaging detectors using the backscattered electrons (BSE) and secondary electrons (SE) where the grayscale images of the sample are captured at high magnifications. The selected area will be scanned at selected speed via scanning electron microscopy (SEM). The electrons are emitted and accelerated by a combination of a thermionic and electric field in present techniques known as field emission SEM (FESEM). The source, the field emission gun (FEG), needs a high vacuum greater than 10^{-10} torr to preserve emission tip.

FESEM can provide information on the surface topography, crystalline structure, and chemical composition of the specimen. FESEM supports a large depth of field, so most of the specimen surface is simultaneously in focus independent of the surface roughness. Figure 4.1 shows the schematic representation of an electron microscope (Williams and Carter 2009).

When the specimen is sufficiently thin enough, a significant number of electrons are transmitted. Then we see the elastically scattered (without loss of energy) and inelastically scattered electrons. The X-ray photons will be emitted through inelastic scattering where deceleration of electrons (loss of energy) takes place. These photons will be seen as continuous spectra, called Bremsstrahlung X-rays. Also, we see peaks in the spectra, known as characteristic X-rays, which are produced by the bombarding of electrons with the target. The characteristic X-ray will give the chemical composition in the form of energy dispersion. Figure 4.2 shows a schematic diagram of general FESEM.

FE gun

Condensed lens

Electron beam

BSE detector

Inlens SE detector

objective lens

Magnetic lens

EDS detector

Scan coils

EBSD camera

Sample

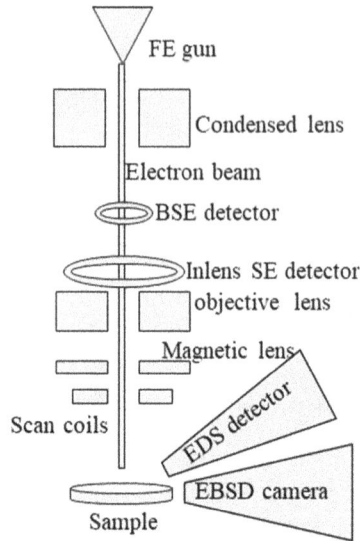

FIGURE 4.2 Schematic diagram of FE-SEM.

4.3.3 TRANSMISSION ELECTRON MICROSCOPY (TEM)

Transmission electron microscopy (TEM) is a technique where the imaging can be done in the transmission mode and the specimen is prepared sufficiently thin to pass the electron beam about 20 nm. The inelastically scattered electron beam contains the microstructural information of the thin specimen. The elastically scattered beam will give the electron diffraction pattern where the local crystal structure will be determined. The combined technique will be helpful to capture the bright field image where the local distribution of particle size can be accurately found.

4.3.4 SCANNING TUNNELING ELECTRON MICROSCOPY (STEM)

Scanning tunneling electron microscopy (STEM) is an advanced technique of TEM in which the scanning is included to improve the imaging technique and to achieve atomic resolution. The STEM lab should be highly isolated from mechanical, magnetic, and acoustic disturbances because atomic-level detection is the focus.

4.3.5 ENERGY-DISPERSIVE X-RAY SPECTROSCOPY (EDS)

Energy-dispersive X-ray spectroscopy (EDS) is a technique where the EDS detector is enabled for use in SEM or TEM for quantifying the microchemical distribution of the specimen. The characteristic X-rays of the material is counted by the EDS detector. The spectra will be helpful in finding the elemental analysis. In the case of nanomaterials, the local chemistry can be found. Moreover, the homogeneity will be quantified. The EDS is useful in the nanostructures, specifically the complex studies such as doping analysis and diffusion studies. Microchemistry is the basis and a tool for all the materials in the field of chemical engineering and biotechnology.

4.3.6 ELECTRON BACKSCATTER DIFFRACTION (EBSD)

Like EDS, the electron backscatter diffraction (EBSD) detector helps in finding the orientational distribution of grains and particles. In the case of nanoengineering, the EBSD technique will help in modeling the required nanomaterials.

4.4 RAMAN SPECTROSCOPY AND SURFACE-ENHANCED RAMAN SPECTROSCOPY

4.4.1 RAMAN SPECTROSCOPY

The scattering of light incident on a solid medium can be elastic or inelastic depending upon the interaction with the molecules. In the process of elastic scattering, also called as Rayleigh scattering (Strutt 1871; Hahn 2006), the light is scattered with no change in energy. However, in the case of inelastic scattering there is a change in energy upon interaction with the medium. The Raman scattering phenomenon (Smith and Dent 2005; Lewis and Edwards 2001) of inelastic light scattering was first postulated by Smekal and observed experimentally in 1928 by Raman and Krishnan.

When the laser light interacts with matter, photons may be excited from the ground state to a virtual state and then return back to a different energy state $h\nu_p$. We have the so-called Stokes lines if photons lose energy, or anti-Stokes lines if they gain energy. Figure 4.3 shows the Stokes, anti-Stokes, and Rayleigh transitions. The intensity of Raman lines are very low compared to the Rayleigh scattered, so different optical arrangements are made to detect the Raman lines.

The experimental setup for the micro-Raman spectrometer includes a monochromatic source, optics to focus the light (lenses and density filter wheel) on the sample, a notch filter to block the excitation wavelength (about 75–90 cm^{-1} Stokes edge), three objective lenses (10×, 50×, 100×), a color camera for the microscope, an asymmetric Czerny–Turner spectrograph with two gratings, and an air-cooled Si CCD detector that gives the measured Raman spectrum. The extraction of the Raman

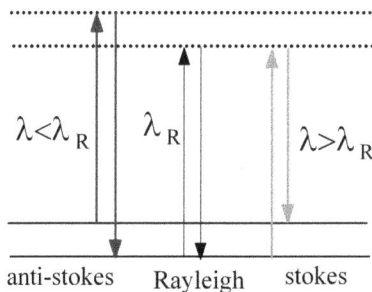

FIGURE 4.3   Stokes, anti-Stokes, and Rayleigh transitions.

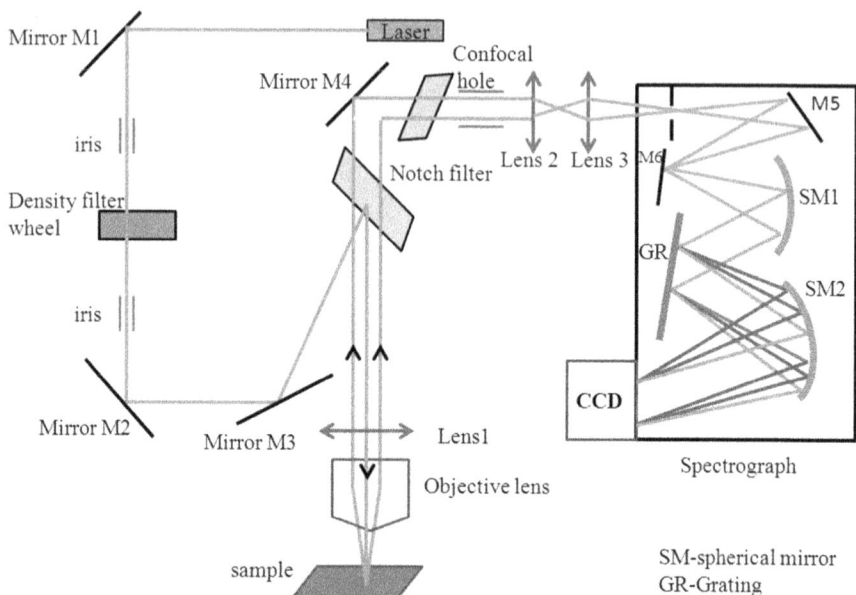

FIGURE 4.4 Typical arrangement for micro-Raman setup.

signals is mainly performed by the edge/notch filters or by the tunable double/triple monochromatic gratings. These filters block the Rayleigh scattered light and allow the Raman signals through the spectrometer. The edge/notch filters block about 2 nm (~75 cm^{-1} above and below) about the central Rayleigh line. The typical schematic of Raman instrumentation is shown in Figure 4.4.

By classical theory, an electric dipole P is induced in the molecule upon interaction with the oscillating electric field of the incident light and is given by

$$P = \alpha E = \alpha E_0 Cos\, 2\pi v_0 t$$

where α is known as the polarizability of the molecule. Further, for small vibration, the polarizability is linearly proportional to the displacement q:

$$\alpha = \alpha_0 + \left(\frac{\partial \alpha}{\partial q}\right)_0 q_0 + \ldots$$

where α_0 and $\left(\dfrac{\partial \alpha}{\partial q}\right)_0$ are the polarizability and rate of change of polarizability with

vibration at equilibrium, respectively. Hence, the dipole moment is

$$P = \left(\alpha_0 + \left(\frac{\partial \alpha}{\partial q}\right)_0 q_0 + \ldots\right) E_0 Cos\, 2\pi v_0 t = \alpha_0 E_0 Cos\, 2\pi v_0 t + \left(\frac{\partial \alpha}{\partial q}\right)_0 q_0 E_0 Cos\, 2\pi v_0 t$$

$$= \alpha_0 E_0 Cos\, 2\pi v_0 t + \frac{1}{2}\left(\frac{\partial \alpha}{\partial q}\right)_0 q_0 E_0 \left[Cos 2\pi \left(v_0 + v_m\right)t + Cos 2\pi \left(v_0 - v_m\right)t \right]$$

where v_m is the vibrational frequency of the molecule. The first term represents the Rayleigh scattering, the second term represents the anti-Stokes line, and the third term is the Stokes line. If the rate of change of polarizability with respect to vibration is zero, then the vibration will be not be Raman active (Ferraro, Nakamoto, and Brown 2003). The molecule, acting as an oscillating dipole, radiates the electromagnetic wave with shifted energy, known as Raman peaks, and these Raman peaks have shown to be of Lorentzian shape. The dependence of the Raman signal on the change of polarizability explains high-intense Raman scattering in aromatic molecules rather than in aliphatic molecules.

In nanomaterials, the modifications in the structure due to changes of the symmetry, changes in short-/long-range order, or due to shape and size of the nanomaterial are imprinted in the Raman spectra (Colomban and Slodczyk 2009). This sensitivity of Raman to the slight changes in structure is useful to study a wide range of carbon nanomaterials, such as carbon nanotubes (CNTs), fullerenes, and graphene-related material like graphene oxide (GO) and reduced graphene oxide (RGO). Such nanomaterials differ in their structure and nature of bonding. Wall reported the Raman characterization for the single, double, triple and multilayers of graphene in terms of the D, G, and 2D bands appearing at 1350, 1587 and 2700 cm^{-1}, respectively (Wall 2011). It is shown that the Raman shift toward lower wavenumbers and an increase in the intensity of the G band of graphene occurs with an increase in the number of layers. Also with the increase in the number of graphene layers, the 2D band shapes change, owing to additional overlapping modes arising due to an decrease of symmetry in the multilayer graphene. A single layer of graphene can accurately be characterized by the peak intensity ratio of 2D and G bands, and is found to be equal to two, i.e., $I_{2D}/I_G = 2$. The two well-known graphene-related materials are GO and rGO. GO is obtained by the chemical oxidation of graphene and can be described as heavily oxygenated with the presence of oxygen-containing functional groups on its basal planes and edges. Also, the GO is known to be hydrophilic in nature, allowing better interactions with solution and is relatively easy to be transferred on a substrate for device fabrication. rGO has similar properties to that of graphene. The Raman spectrum of graphene oxide has two broad peaks represented by D and G bands. The D band is located around 1325 cm^{-1}, assigned to the structural imperfections caused by the hankering of OH and epoxy groups on the basal plane of carbon. The G band is located around 1582 cm^{-1} and is assigned to the sp$_3$ and sp$_2$ hybridization due to distortion in the crystal structure of the carbon atoms, resulting from oxidation (Siyar, Maqsood, and Khan 2014). The additional bands (including D′, 2D, and D + G) arise from the defects present in the graphitic structure of the carbon material. The I_D/I_G ratio (calculated from the intensity of D and G bands) can be used to characterize the disorder of the graphitic structure in carbon nanomaterials (Ajdari et al. 2018). The defect deconvoluted bands D$_1$, D$_2$, D$_3$, and D$_4$ peak positions are at 1229 cm^{-1}, 1484 cm^{-1}, 1565 cm^{-1}, and 1323 cm^{-1}, respectively. The D$_1$ band arises from the disordered graphene layer edges A_{1g} symmetry; the D_2 band from the disordered graphene surface layers E_{2g} symmetry; D_3 from amorphous carbon; and D_4 from the different polyenes, ionic impurities, etc. (Rusi and Majid 2015; Hareesh et al. 2017).

The Raman spectra in nanomaterials is given by the phonon confinement model

$$I(\omega) = I_0 \int \frac{|C(q)|^2 d^3 q}{\left[\omega - \omega(q)\right]^2 + \left(\dfrac{\Gamma_0}{2}\right)^2}$$

where I_0 is the intensity pre-factor, $\omega(q)$ is the phonon dispersion curve, Γ_0 is the full width at half maximum (FWHM) of the peak, $C(q)$ is the confinement function given by

$$|C(q)|^2 = \exp\left(-q^2 d^2 / 16\pi^2\right)$$

where q is the wave vector and d is the diameter of the nanoparticle. Further, the phonon dispersion curve $\omega(q)$ can be expressed as

$$\omega(q) = \omega(0) - \Delta\omega.Sin^2(qa/4)$$

The phonon confinement model is being employed to obtain the size of the nanomaterials. Rajalakshmi et al. reported the size of ZnO nanoparticles to be 8.4 and 4.5 nm from the line shape analysis using the confinement model. Sanjeev et al. reported on the modified confinement model considering the size distribution, improved confinement and phonon dispersion functions explaining the size-dependent Raman shift for Si nanocrystals. It is demonstrated that as the size of nanoparticles increases from 2 nm to 100 nm, the Raman signal for the nc-Si shows a redshift in the spectra. The experimental Raman spectra is deconvoluted to obtain the different components or phases present. The best procedure to obtain the deconvoluted spectra is to remove the baseline and fit the spectra by Voigt function with the Lorentzian to Gaussian ratio selected based on repeated trials (Schuster et al. 2014; Liu et al. 2013; Krylov et al. 2012; Ammar and Rouzaud 2012; Kumar and Krishna 2008; Lei et al. 1998; Zhang et al. 2010; Mangione et al. 2005). The Lorentzian shape in the spectra is due to the natural or collision broadening, whereas the Gaussian shape is due to the instrumental and Doppler broadening.

4.4.2 Surface-Enhanced Raman Spectroscopy (SERS)

Surface-enhanced Raman spectroscopy (SERS) is a powerful technique to characterize low-concentration analytes or ultrathin films where the Raman signal is enhanced with the order of 10^7 times. The enhancement in Raman signal is achieved by depositing the adsorbate layer over metal nanoparticles like gold and silver. The early research on SERS was triggered by the trace detection of chemical and biological samples (Jarvis and Goodacre 2004; Albrecht and Creighton 1977).

The dielectric constant of a metal is a function of plasmon frequency, obtained primarily from the collective oscillation of electrons in response to the electric field of the incident light. These induced collective oscillations are known as surface plasmons and interact with the incident electromagnetic wave to form the surface plasmon polariton (SPP). The corresponding phenomenon is known as surface plasmon resonance (SPR). The frequency of the incident electromagnetic wave is typically low for the propagation of SPP along the interface of the dielectric–metal layer, with decaying electric fields on either medium. The condition for the SPP to exist is

$$k_z = \frac{\omega}{c}\sqrt{\frac{\varepsilon_1\varepsilon_2}{\varepsilon_1 + \varepsilon_2}} \text{ and } \varepsilon_1\varepsilon_2 < 0$$

where ε_1 and ε_2 are the dielectric constants of the dielectric and metal medium, respectively; and k_z is the propagation vector. Any change in the dielectric environment will change the reflectivity position of the wavelength or angle of surface plasmon resonance, and is useful in chemical and biological sensing.

However, when the electromagnetic wave interacts with the nanoparticles smaller than the wavelength of incident light, it causes the plasmon to oscillate around the nanoparticle. When the frequency of incident light is equal to the plasmon frequency of the nanoparticle, it produces the localized surface plasmon (LSP) resonance. The LSP resonance is the main phenomenon responsible for SERS. The SERS mechanism is due to two main mechanisms: the (1) electromagnetic mechanism (EM) and (2) the chemical enhancement mechanism (EM). The EM is due to the resonance of the incident light with the LSP of the nanostructure, whereby the electric field is intense at the hotspots over the surface. These hotspots are obtained by the roughened surfaces or placement of metal nanoparticles over a surface, which are capable of retaining oscillations perpendicular to the surface (Smith and Dent 2019). The CM includes the charge transfer between the metal surface and the biological species.

The wavelength of the LSPR depends on the size, shape, and dielectric properties of the nanomaterial. Willets and Van Duyne (2007) demonstrated that the electric field outside the spherical nanoparticle of size a, irradiated by a z-polarized EM wave of wavelength λ, is given by

$$E_{out} = E_0\hat{z} - \left(\frac{\varepsilon_{in} - \varepsilon_{out}}{\varepsilon_{in} + 2\varepsilon_{out}}\right)a^3 E_0\left[\frac{\hat{z}}{r^3} - \frac{3z}{r^5}\left(x\hat{x} + y\hat{y} + z\hat{z}\right)\right]$$

where ε_{in} and ε_{out} are dielectric constants of the metal nanoparticle and outside dielectric media. The term $\left(\dfrac{\varepsilon_{in} - \varepsilon_{out}}{\varepsilon_{in} + 2\varepsilon_{out}}\right)$ determines the LSPR in the nanoparticle.

When the ε_{in} is equal to $-2\varepsilon_{out}$, the enhancement in the EM field occurs with respect to the incident wave. Further, the presence of a chemical or biological species over the metal nanoparticles changes the LSPR wavelength as per the following relationship:

$$\Delta\lambda_{max} = m\,\Delta n\left[1 - exp\left(\frac{-2d}{l_d}\right)\right]$$

where m is the refractive index of the metal nanoparticles, Δn is the change in refractive index of nanoparticles due to the presence of the chemical or biological species of effective thickness d adsorbed on the surface of nanoparticle, and l_d is the decay length of the LSPR wave. Using the preceding equation, LSPR-based SERS sensing is performed. The SERS enhancement factor of the Raman signal is given by

$$EF = \frac{I_{SERS}/N_{SERS}}{I_{RS}/N_{RS}}$$

where I_{SERS} and I_{RS} are the enhanced Raman intensity from the SERS spectra and normal Raman intensity, respectively. N_{SERS} is the number of molecules given the rise of the SERS Raman signal and N_{RS} is the number of molecules given the Raman signal in absence of the SERS effect.

Further, N_{SERS} and N_{RS} in a case of solution of volume V added on SERS substrate can be determined from the relation

$$N_{SERS} = \eta N_A V C_{SERS} \frac{A_{laser}}{A_{SERS}} \text{ and } N_{RS} = N_A V C_{RS} \frac{A_{laser}}{A_{RS}}$$

where η is the adsorption factor; N_A is Avogadro's number; and A_{SERS}, A_{RS}, and A_{laser} are the areas of the SERS, non-SERS, and laser spots, respectively. C_{SERS} and C_{RS} are concentrations of the chemical or biological species applied over the SERS source materials.

Xia et al. showed the different peaks of LSPR in the extinction spectrum of the Ag nanostructure and related them to the shape of the nanostructures (Wiley et al. n.d.). Most researchers synthesized plasmonic nanomaterials in wet chemical methods and a few people worked on a well-known top-down approach like laser ablation. Laser ablation is a surfactant-free green synthesis that does not require any sophisticated chemical mixtures to synthesize nanomaterials in a single experiment within a short period of time.

Podagatlapalli et al. (2013) utilized silver nanostructures fabricated at different fluences in double distilled water, and thus fabricated Ag nanostructures to detect the hexahydro-1,3,5-trinitroperhydro–1,3,5-triazine (RDX) through SERS using the excitation lines 532 nm and 785 nm. In their studies, a 460-fold enhancement was observed for all the vibrational signatures of RDX fabricated at fluences of 12 J/cm^2 and 16 J/cm^2. In another report, Podagatlapalli et al. (2014) utilized the silver nanostructures fabricated through laser ablation of the silver target immersed in double distilled water at different angles of incidence (5°, 15°, 30° and 45°) of the irradiated laser beam. These Ag nanostructures were utilized to investigate the SERS activity of Rhodamine 6G and an explosive molecule 5-amino, 3-nitro, 1,3,5-nitrozole (ANTA) at micromolar concentrations using both 532 nm and 785 nm. In this study they found that a particular Ag nanostructure fabricated at an incident angle 30° demonstrated an enhancement factor above 10^8. In this particular study the authors attempted to investigate the multiple time utility of SERS active substrates after a proper cleaning protocol. The same group utilized Ag nanostructures fabricated by the ablation silver target immersed in double distilled water/acetone using non-diffracting Bessel beams (Podagatlapalli et al. 2015), in the detection of the explosive molecule CL-20 (2,4,6,8,10,12-hexanitro-2,4,6,8,10,12-hexaazaisowurtzitane) at 5 μM concentration. Bessel beams were produced by allowing the Gaussian profiled laser beam through a conical lens or an axicon. Ag nanostructures fabricated using axicon demonstrated an enhancement of 10^6. In contrast to the SERS activity of the plasmonic metal nanostructures fabricated by ultrafast laser ablation, Podagatlapalli, Hamad, and Rao (2015) investigated the Raman enhancements offered by Ag–Au bimetallic nanoparticles, and nanostructures were utilized to detect secondary explosive molecules such as 1,1-diamino-2,2-dinitroethene (FOX-7, 5 μM concentration) and 1-nitro pyrazole (1NPZ, 20 nM concentration). In this particular study, it was observed that the increment of gold percentage reduced the surface activity of Ag–Au bimetallic nanoparticles or nanostructures. The estimated enhancement factors (EFs) in the case of FOX-7 and 1 NPZ were typically $>10^8$. Furthermore, their study revealed that the bimetallic nanoparticles and nanostructures with the fraction $Ag_{0.65}Au_{0.35}$ exhibited significant EFs compared to other ratios and pure metals of Ag and Au. Recently, Krishna Podagatlapalli et al. (2013) demonstrated a good reproducible SERS activity from the Ag nanoribbons fabricated by picosecond laser ablation for the explosive molecules TNT at 20 nanomolar concentration. The estimated enhancement factor is beyond 10^7. In this study, detection of Au was also reported. Along with silver SERS active platforms, Hamad et al. (2014) demonstrated the SERS activity of copper nanostructures prepared by laser ablation for the detection of many explosive molecules. Shaik et al. (2016) reported the Ag-covered ZnO nanostructures fabricated using thermal evaporation, showing SERS enhancement of an order about 10^9 for Rh6g molecules, which is three orders higher than Ag nanostructures over BSG substrate. Further, the Ag/ZnO nanostructures are shown to detect a very low concentration (~10 μM) of secondary explosives molecules like ANTA, FOX-7, and CL-20.

Ling et al. (2018) introduced graphene as one of the promising material for enhancing the Raman signal via the "charge transfer" effect for some dye molecules, and named it graphene-enhanced Raman spectroscopy (GERS). The molecules can be deposited on a Si/SiO_2 substrate followed by deposition of the graphene layer. It is shown that the graphene suppresses the background photoluminescence and enhances the Raman signal for R6G molecules. Further, the enhancement improves the annealing, allowing better adsorption of the molecule, and is independent of number of graphene layers. However, the enhancement is generally less than 100, but coupling gold nanostructures with graphene causes an additional enhancement and low detection limits (10^{-11}) (Khalil et al. 2016). Banaszak (2017) developed gold-decorated graphene-related materials for GERS platforms for various biomolecule detection.

4.5 UV–VIS–NIR ABSORPTION SPECTROSCOPY

UV–Vis spectroscopy is widely used to characterize materials where electron transitions occur from low to high energy atomic or molecular orbital levels by absorbing energy from the ultraviolet or visible electromagnetic radiation (Skoog, Holler, and Crouch 2007; Yadav 2005). The spectra are usually transmittance, reflectance or absorbance plotted against the wavelength and are useful in quantifying the optical properties of metallic, organic and inorganic bioconjugated nanomaterials.

In general, the absorbance A is given by the Beer–Lambert law,

$$A = -log\left(\frac{I}{I_0}\right) = -\log(T) = \varepsilon cd$$

where I and I_0 are transmitted and initial intensities, T is the transmittance, ε is the extinction coefficient of the sample, c is concentration of the sample, and d is the path length of the light through the sample. In the case of thin films, the Beer–Lambert law defines the absorption coefficient $\alpha(v) = \dfrac{2.303\, A(\lambda)}{d}$, where $A(\lambda)$ is the absorbance and d is the thickness of the thin film.

Naidu and Krishna (2014) reported the optical properties, like the refractive index and extinction coefficient, of the Cr/Si bilayer thin films from the transmission spectra using the Swanepoel method. Further, it was shown that the bandgap is calculated by fitting using Tauc's equation (Tauc, Grigorovici, and Vancu 1966), given by

$$\alpha h v = B\left(h v - E_g\right)^b$$

where α is the absorption coefficient, E_g is the optical bandgap, B is a constant, and b gives the type of the optical transition. For $b = 0.5$ represents the allowed direct band gap, while $b = 2$ indicates the indicate the allowed indirect band gap. The absorption coefficient is obtained using the Beer–Lambert law. The absorption coefficient plot is divided into two regions at the point of inflexion, obtained by the derivative of the absorption coefficient with respect to energy. The low energy is known as the Urbach's region, while the high energy region is the Tauc's region. Figure 4.5. shows the two regions for the typical Cr/Si bilayer thin film.

The Urbach characteristics are from the distribution of density of localized states in the tails of the allowed bands. The Urbach energy indicates the disorder present and is related to the absorption coefficient as

$$\ln \alpha = const + \frac{h v}{E_U}$$

These optical properties are used to understand the interdiffusion-led metal-induced crystallization (MIC) mechanism of a-Si films (Naidu and Krishna 2014). Ghobadi (2013) determined the band gap of the CdSe nanostructural films from the absorption spectra by rewriting Tauc's equation as a function of the wavelength. The optical properties of thin films deposited on the opaque substrate are studied from the reflectance spectra. Shankernath et al. (2017) demonstrated the determination of the

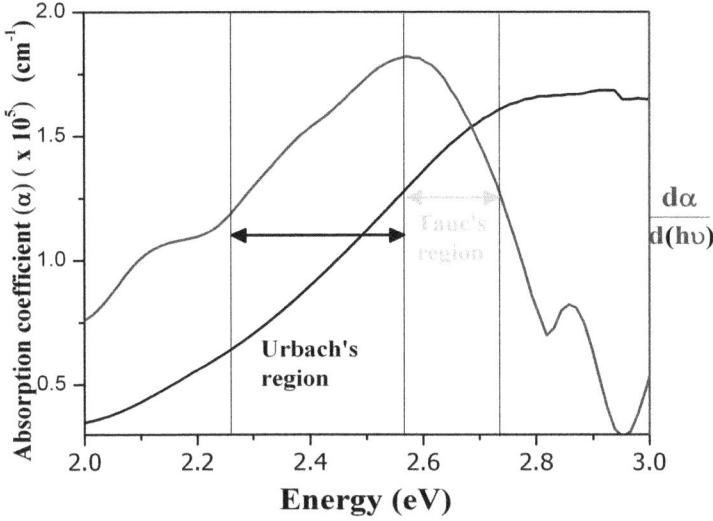

FIGURE 4.5 Typical absorption coefficient vs energy plot, depicting the Tauc's and Urbach's regions divided at the point of inflexion.

dielectric functions, refractive index and extinction coefficient of ultrathin TiN films using the Drude–Lorentz oscillator model. The reflectance R of the film is given by

$$R = \left| \frac{1 - \sqrt{\varepsilon}}{1 + \sqrt{\varepsilon}} \right|^2$$

where the dielectric function ε is given by

$$\varepsilon = \varepsilon_\infty + \sum_i \frac{f_i \omega_{oi}^2}{\omega_{oi}^2 - \omega^2 - i\gamma_i \omega}$$

where $\varepsilon\infty$ is the dielectric constant due to contributions from high energy interband transitions, $f_i = \left(\dfrac{\omega_{pi}}{\omega_{oi}} \right)^2$ is the strength of the Lorentz oscillator with values ranging between 0 and 1, ω_{pi} is the plasma frequency, ω_{oi} is the frequency of the Lorentz oscillator, ω is the frequency, and γ_i is the damping factor for both the Drude and Lorentz terms. It is shown that the reflectance minima of TiN films are the characteristic of the SPR peak and can be tuned by varying the N_2 in the sputtering system. The methods for the determination of optical properties from the reflectance spectroscopy are reported by many authors (Aly 2009; Bazkir 2007; Sreemany and Sen 2004; Filippov and Kutavichyus 2003; Filippov 2000; Stenzel, Hopfe, and Klobes 1991; Siqueiros, Regalado, and Machorro 1988; Bock 1972). Kasikov (2010) listed the dispersion formulas for theoretical fitting optical spectra to determine the optical properties of homogeneous thin films.

Certain metallic nanostructures exhibiting surface plasmon resonance show characteristic peaks in the absorption spectra (Das et al. 2010; Das and Sarkar 2015; Ngumbi, Mugo, and Ngaruiya n.d.; Ferodolin and Awang 2019; Xiong et al. 2016; Korir et al. 2019; Kumar et al. 2019). The position, shape, and shift in the peaks can be correlated to the changes in the size, shape, concentration, and refractive index of the medium near the nanomaterial. However, the interest in fabricating the nanoparticle arrays has drawn much interest with potential applications in optoelectronics, supercapacitors, sensing, and solar cells. Sim et al. (2011) reported the fabrication of 2D Au nanoarrays using nanosphere lithography and found higher LSPR sensitivity for the nanoarray structure as compared to the colloidal gold nanoparticles.

The size-dependent bandgap energy E_g of the semiconductor nanoparticles can be obtained by effective mass approximation (EMA)

$$E_g = E_{g,Bulk} + \frac{h^2}{4r^2}\left(\frac{1}{m_e^*} + \frac{1}{m_h^*}\right)$$

where $E_{g,Bulk}$ is the bulk bandgap value, h is Planck's constant, r is the radius of the nanoparticle, and m_e^* and m_h^* are the effective mass of the electron and hole, respectively. The size of the particle is calculated by substituting the bandgap E_g value obtained from the experimental UV–Vis absorption spectra. Singh, Viswanath, and Janu (2009) reported the size of the ZnO nanoparticles in the presence of capping agents of different concentrations using EMA. Memarian (2017) also demonstrated the CdS nanoparticle size from the EMA and it was found in good agreement with the size calculated from the XRD studies. However, for the case of metal nanoparticles, the size of the nanoparticles are obtained using the Mie theory (Baset et al. 2011).

The method of deconvolution of UV–Vis absorption spectra is effective in the determination of the different structures (Sasai et al. 2002) or chiralities of nanotubes (Yang, Zhang, and Li 2011), or relating the absorption peak intensity to the size distribution and surface chemistry of the carbon nanotubes (Cheng et al. 2011).

4.6   FOURIER-TRANSFORM INFRARED SPECTROSCOPY (FTIR)

Fourier-transform infrared spectroscopy (FTIR) is a necessary characterization technique for the identification of functional groups linked to the nanomaterial surface. The FTIR peaks are characteristics of the vibrational energy of different bonds present in the organic or inorganic material. A molecule is IR active if the change in dipole moment with displacement is not zero. The experimental measurement range is usually 4000–400 cm^{-1}; with absorption bands in 4000–1500 cm^{-1} attributed to different functional groups; and 1500–400 cm^{-1}, unique to a particular compound, is known as the fingerprint region. A typical list of vibrational frequencies of functional groups are discussed elsewhere (Christy, Ozaki, and Gregoriou 2001).

The uniqueness of the FTIR technique is the interaction of the modified IR beam from the interferometer with the material, allowing measurement of all frequencies simultaneously rather than the stepwise scan through the individual frequencies.

Grass, Athanassiou, and Stark (2007) reported the FTIR spectrum of chloro-func-tionalized carbon-coated cobalt magnetic nanobeads and showed the functionaliza-tion of chloro-4-ethyl-benezene instead of chlorobenzene. This observation rules out the possible physisorption of chlorobenzene on the carbon surface. The FTIR proved significant in the study of amide linkages of peptides for the study of protein structures in the medical field (Nabers et al. 2016), polymeric materials (Barrios et al. 2012), food (van de Voort 1992; Saputra, Jaswir, and Akmeliawati 2018; Jiao et al. 2019), drug delivery (Gasper and Goormaghtigh 2012), Li ion batteries (J. Li et al. 2011) and forensic exhibits (Sharma and Kumar 2019).

Many advanced techniques are developed by integrating FTIR with other techniques like Fourier-transform infrared–attenuated total reflectance (FTIR-ATR) (Kalmodia et al. 2015), Fourier-transformed infrared imaging spectroscopy (Bhargava 2012), Fourier-transform infrared–photoacoustic spectroscopy (FTIR-PAS) (Kizil and Irudayaraj 2013), Fourier-transform infrared microspectrometry (Ami et al. 2004) and nano-FTIR spectroscopy (Mester et al. 2020).

These advanced techniques assist in monitoring the degree of functionalization of the nanomaterials surfaces and in quality control in various fields (Bhowmick et al. 2015). The number of different functional groups in the nanoparticle can be calcu-lated from the size of the peaks of the FTIR spectrum (Qais, Samreen, and Ahmad 2019; Faraji, Yamini, and Rezaee 2010; Chauhan, Gupta, and Prakash 2012). Bridelli (2017) demonstrated the technique of deconvolution of the FTIR spectra of amide bands, lysozyme and collagen for the study of conformational changes induced by the hydration of the biological macromolecules.

4.7 DYNAMIC LIGHT SCATTERING (DLS) TECHNIQUE

Nanoemulsion droplets were measured by DLS in a recent study to find bioactivity against pathogenic bacteria (Ghazy et al. 2021). In DLS, the nanoparticle suspen-sion is kept in the path of a laser light and the scattered light from the suspension is recorded with a photodetector, and the Brownian motion of the particle is found by utilizing the Doppler effect. The DLS gives the particle size distribution. Similarly, electrophoretic light scattering (ELS) is used to measure the electrophoretic mobility of suspended particles. The experiment refers to the motion of suspended entities in the applied alternating electric field. This gives the electrophoretic mobility to calcu-late the zeta potential (Jain et al. 2021).

4.8 PHYSICAL PROPERTY MEASURING SYSTEM– VIBRATING SAMPLE MAGNETOMETER (PPMS-VSM)

Unpaired electrons have non-zero electron spin or orbital moments at the valency state of atoms where the magnetism originates in the material. The chemical compo-sition and the crystal structure of the nanomaterials will result in various coulombic and magnetic interactions. The structural influence is more complicated in the case of nanomaterial where highly sensitive magnetic measurements are required. The magnetic core-shell nanoparticle structures, surfactant coatings, and strong magnetic responsivity are characterized by magnetic property tests. The magnetic correlations are quite strong at low temperatures due to the formation of magnetic clusters. Also, the magnetic nature will be destroyed or lost above the transition temperature (T_c) due to the random thermal fluctuation of spins and weak magnetic interactions.

Magnetic transport properties of nanomaterials are characterized by the technique PPMS-VSM (lowest range is of the order of micro-electromagnetic units). It is possible to find the magnetic interactions and correlations in the nanoparticle system.

Quantum design, a company from North America, provides PPMS-VSM systems. The design contains four basic parts:

1. Vibrating sample holder setup
2. High magnetic field superconducting coils in a helium cryogenic environment
3. Temperature control system with liquid helium flow and a liquid nitrogen safety environment
4. Detection coil, control panel, and programming modules

The magnetic measuring design has small complications dealing with calibrations and sometimes it is a time-consuming process to obtain the accurate magnetic properties for the nanomaterials. The magnetic nature of the nanomaterials can be analyzed using the transportation curves. The magnetic interactions will lead to have various properties like spin glass, superparamagnetic, ferrimagnetic, antiferromagnetic, and paramagnetic.

The complexity can be analyzed by the magnetic transport properties and hysteresis loops of nanomaterials. The size and shape of the nanoparticles show more asymmetry with the quantum approach. The surface area will be huge for the nanoparticles and there will be high sensitivity for the small applied fields. The domain theory says that the local ferromagnetic region shows zero overall spontaneous magnetization in the ferromagnetic material. The domain is nothing but locally aligned spins acting as tiny magnets inside the materials. If the domains orient in one direction, the overall spontaneous magnetization will be high, and then the material acts as a permanent magnet. In the case of ferromagnetic nanoparticles, each particle possesses few domains. When we decrease the size of the nanoparticles up to some extent, the size reaches that of single domain particles. In this state, the nanoparticle system shows almost zero coercivity and act as a paramagnet with high magnetic susceptibility. These types of magnetic particles are used in several biomedical applications and instrumentation. There are hard ferromagnetic materials in which a high field is required to saturate the material and show large coercivity. These hard ferromagnetic nanomaterials will act as high dense storage, micro-art magnetic printing, magnetic integrated circuits, etc. Similarly, there are soft-ferromagnetic nanoparticles that are useful in switching devices, ferrofluid applications, transformer-induction devices, etc. Also, complex magnetic structured materials show enhanced properties applicable in various fields. Their nanoparticles give heterostructures and artificial multiferroic behavior where we see higher multiferroic order. This will lead to electromagnetic coupling properties.

4.9 ELECTRON SPIN RESONANCE (ESR) SPECTROSCOPY AND AC-SUSCEPTIBILITY STUDIES

Electron spin resonance (ESR) or magnetic resonance is a spectroscopic technique similar to nuclear magnetic resonance. Here the microwave absorption spectra will be with respect to the applied magnetic field strength (B) according to induced magnetic energy transitions of electrons (ΔE) as a result of magnetic interactions of unpaired spins in the atoms or compounds.

The microwave absorption can be defined as in the following equation:

$\Delta E = h\upsilon = g\,\mu_e\,B$

where g is the Landé factor, μ_e is the Bohr magneton (9.274×10^{-24} J T^{-1}), and υ is the microwave frequency.

The energy gap between the states will be increased to match microwave energy when we apply an external magnetic field. The microwave absorption is plotted as a function of the applied field. The total microwave absorption is proportional to the magnetic susceptibility or magnetization. The temperature-dependent ESR will give the idea of the magnetic transitions. The catalytic activity of nanocomposites has been recently tested by the ESR for killing bacteria (Mu et al. 2021).

Similar to the magnetic measurements in the PPMS-VSM, the magnetization will be recorded in the presence of a small AC magnetic field according to the relaxation mechanism of the nanomaterials as a function of temperature. The relaxation of nanomaterial resolves the magnetic nature in the presence of the applied field.

4.10 IMPEDANCE SPECTROSCOPY

If there are no sufficient free charges in the material, there is a phase shift in the resultant current with the application of an electric field or voltage. So, it is needed to study the behavior in the frequency domain to find the shift. The ions in the material act as bounded charges and behave like electric dipoles when we apply an electric field. Impedance is nothing but a high resistance where we see phase shift at the output (current). This refers to the electrical response of materials to correlate the properties and modeling the circuitry diagram according to the expected combinations or parameters. The impedance is similar to the dielectric spectroscopy where we measure the dielectric constant and loss of materials. Electrical contacts are important in the impedance experiment. In the case of polycrystalline materials, the density or compaction plays a role in order to minimize the loss by assuming proper grain connectivity. In nanomaterials, the connectivity between the particles will be very poor. There are two major techniques to measure the impedance: (1) dispersion of nanoparticles in a suitable base fluid with known impedance, and (2) green compaction of the nanoparticles.

The spectroscopy is used in various ways like bioimpedance, electrochemical impedance, and relaxor studies. The impedance will be measured using an impedance analyzer or LCR meter for the dipolar range. Beyond this range, it is measured by a vector network analyzer or some electromagnetic wave techniques. Figure 4.6 shows general results analysis obtained using Cole–Cole or Nyquist plots for

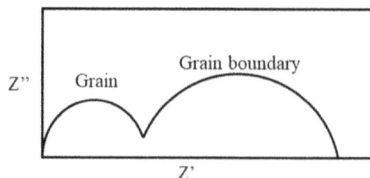

FIGURE 4.6 Cole–Cole plots.

nanomaterials. These results can be modeled as the parallel RC networks in series for the grain and grain boundary contribution of a ceramic material where the suppressed semicircles are seen. The particle connectivity will be poor, so the boundary contribution will be dominated in the plots. Further, the electrical properties will be taken into account for understanding the nature of the nanoparticles. Electrical relaxation mechanism of $MnCr_2O_4$ has been recently reported using impedance spectroscopy (Javed et al. 2021).

4.11 METHODS FOR PHYSICOCHEMICAL CHARACTERIZATION

With the wide spectrum of synthesis techniques and the characterization tools available, the physicochemical properties of the nanomaterials can be tuned to achieve the maximum efficiency for applications in biosensors, drug delivery, cellular interaction, and nanotoxicity minimization. The behavior of nanomaterials in a biological environment depends on its physicochemical properties and on the response of biomolecules toward the nanomaterials. Some of the core physicochemical properties of nanomaterials, which largely control the interactions with biomolecules, are size, shape, chemical composition, and surface properties. The size of the nanomaterial determines the interaction field in a biological environment, as a small size possesses higher surface area leading to a higher capacity for biological interaction and hence high risk of toxicity (Gatoo et al. 2014). Auffan et al. (2009) reported size-dependent properties of nanoparticles with size greater than 30 nm do not show similar properties to their bulk counterparts. Nanoparticles in human biological systems traverse through different barriers and routes and those less than 20 nm reach the kidneys and those 20 to 100 nm reach the liver (X. Wang et al. 2020). This causes inflammation, organ failure and disturbance in metabolic processes. The shape of the nanomaterial also plays a crucial role in determining the uptake into cells. Albanese, Tang, and Chan (2012) discussed the uptake into cells is higher for nanorods, followed by spheres, cylinders, and cubes. Similarly, the cell membrane proved to have slightly negative charge and nanomaterials with positive charge are favored for adhesion to the surface of the cell, leading to higher uptake value. The surface charge also affects the protein corona composition on the surface of the nanomaterial. The surface reactivity is one of the important properties, and upon interacting with biomolecules, increases reactive oxygen species (ROS) leading to cell damage and disturbance in metabolic processes (Lai 2015). The physicochemical property composition of different nanomaterials produces different degrees of damages to the biomolecules.

Titanium dioxide has two phases, rutile and anatase, but similar sizes will respond differently to the biomolecules. The anatase phase shows high cytotoxicity due to its better photocatalysts compared to the rutile phase (Lai 2015).

Understanding the physicochemical properties of nanomaterials and interaction with biomolecules will enhance progress in developing nanomedicines. The biomolecules either get adsorbed on the surface of the nanomaterials forming corona (Y. Li et al. 2017) or undergo redox reactions altering the original physicochemical properties of the nanomaterials (Arndt and Unrine 2016). The interaction of biomolecules with nanomaterials also can bring modification on the surface of nanomaterials, enhancing the biocompatibility and reducing the toxic effects from the nanomaterials. Primarily, nanomaterials interact with proteins, nucleic acids, carbohydrates and lipids. Hu et al. (2011) reviewed the coupling of nucleic acid/DNA to different groups of metal-, semiconductor-, carbon-, magnetic-, and polyacrylamide-related nanomaterials for bioimaging applications. The DNA–nanoparticles conjugate utilizes the binding characteristics of DNA with the signal-amplifying property of certain nanomaterials for the detection of several biomolecules or molecular recognitions (Xu et al. 2020; Bonyár 2020). The interaction of nanomaterials with proteins identify the protein structures, localization and functions. The XANES analysis on Ag nanoparticles during cellular uptake have shown formation of Ag-S species, followed by oxidation and formation of Ag-O species (L. Wang et al. 2015). The formation of protein corona on the surface of nanomaterials changes the biological identity, and nanomaterial–protein corona is optimized as an alternative choice in nanomedicines (Chen et al. 2020). The presence of an active functional group on nucleic acid and proteins lead to high interaction with nanomaterials, whereas there is comparatively low interaction for lipids and carbohydrates (Vigneshwaran and Jain 2011).

Although the physicochemical properties of nanomaterials and the interaction with biomolecules are largely characterized by well-known techniques, the ultracentrifugation and electrophoresis-based methods provide a novel way to quantify the interactions. Bekdemir and Stellacci (2016) reported ultracentrifugation to be the efficient method for characterizing the interaction of small nanomaterials with proteins. They studied the interaction of bovine serum albumin (BSA) with gold nanoparticles and found the shape of small nanoparticle/protein complexes to be non-spherical. Similarly, the capillary electrophoresis technique offers a path to separate the nanomaterials of varying size, shape, surface charge, and composition, and is also useful to determine the size, size distribution, charge on surface layers, composition, and zeta potential (Chetwynd et al. 2018). Integrating the capillary electrophoresis with the mass spectrometry allow the characterization of biomolecules adsorbed on the surface of nanomaterial and probing of the interactions/mechanism at the nanomaterial–biomolecule interface.

ACKNOWLEDGMENT

The author KLN acknowledges UGC-startup UGC-SRG/30-456/2018(BSR), DST-ITPAR, University of Hyderabad, University of Trento, and GITAM. GT is grateful to MITS for the support. SY acknowledges the Dr. MVVS Murthi Research fellowship, GITAM (Deemed to be University).

REFERENCES

Ajdari, Farshad Boorboor, Elaheh Kowsari, Ali Ehsani, Milan Schorowski, and Tayebeh Ameri. 2018. "New Synthesized Ionic Liquid Functionalized Graphene Oxide: Synthesis, Characterization and Its Nanocomposite with Conjugated Polymer as Effective Electrode Materials in an Energy Storage Device." *Electrochimica Acta* 292 (December): 789–804. doi:10.1016/j.electacta.2018.09.177.

Albanese, Alexandre, Peter S. Tang, and Warren C. W. Chan. 2012. "The Effect of Nanoparticle Size, Shape, and Surface Chemistry on Biological Systems." *Annual Review of Biomedical Engineering* 14 (1): 1–16. doi:10.1146/annurev-bioeng-071811-150124.

Albrecht, M. Grant, and J. Alan Creighton. 1977. "Anomalously Intense Raman Spectra of Pyridine at a Silver Electrode." *Journal of the American Chemical Society* 99 (15): 5215–17. doi:10.1021/ja00457a071.

Alexis, Frank, Eric Pridgen, Linda K. Molnar, and Omid C. Farokhzad. 2008. "Factors Affecting the Clearance and Biodistribution of Polymeric Nanoparticles." *Molecular Pharmaceutics* 5 (4): 505–15. doi:10.1021/mp800051m.

Aly, K. A. 2009. "Optical Constants of Quaternary Ge–As–Te–In Amorphous Thin Films Evaluated from Their Reflectance Spectra." *Philosophical Magazine* 89 (12): 1063–79. doi:10.1080/14786430902870542.

Ami, Diletta, Antonino Natalello, Aldo Zullini, and Silvia M. Doglia. 2004. "Fourier Transform Infrared Microspectroscopy as a New Tool for Nematode Studies." *FEBS Letters* 576 (3): 297–300. doi:10.1016/j.febslet.2004.09.022.

Ammar, M. R., and J.-N. Rouzaud. 2012. "How to Obtain a Reliable Structural Characterization of Polished Graphitized Carbons by Raman Microspectroscopy." *Journal of Raman Spectroscopy* 43 (2): 207–11. doi:10.1002/jrs.3014.

Araujo, L. L., R. Giulian, D. J. Sprouster, C. S. Schnohr, D. J. Llewellyn, P. Kluth, D. J. Cookson, G. J. Foran, and M. C. Ridgway. 2008. "Size-Dependent Characterization of Embedded Ge Nanocrystals: Structural and Thermal Properties." *Physical Review B* 78 (9): 094112. doi:10.1103/PhysRevB.78.094112.

Arndt, Devrah, and Jason Unrine. 2016. "Redox Interactions between Nanomaterials and Biological Systems." *Oxidative Stress and Biomaterials*, 187–206. doi:10.1016/B978-0-12-803269-5.00007-3.

Auffan, Mélanie, Jérôme Rose, Jean-Yves Bottero, Gregory V. Lowry, Jean-Pierre Jolivet, and Mark R. Wiesner. 2009. "Towards a Definition of Inorganic Nanoparticles from an Environmental, Health and Safety Perspective." *Nature Nanotechnology* 4 (10): 634–41. doi:10.1038/nnano.2009.242.

Ayers, J. D., V. G. Harris, J. A. Sprague, and W. T. Elam. 1994. "On the Role of Cu and Nb in the Formation of Nanocrystals in Amorphous Fe73.5Nb3Cu1Si13.5B9." *Applied Physics Letters* 64 (8): 974–76. doi:10.1063/1.110923.

Banaszak, Alexander. 2017. "Graphene-Mediated Surface Enhanced Raman Spectroscopy and Detection of Biomolecules." *Posters-at-the-Capitol*, November. https://digitalcommons.murraystate.edu/postersatthecapitol/2018/WKU/14.

Barrios, Vladimir A. Escobar, José R. Rangel Méndez, Nancy V. Pérez Aguilar, Guillermo Andrade Espinosa, and José L. Dávila Rodríguez. 2012. "FTIR - An Essential Characterization Technique for Polymeric Materials." In *Infrared Spectroscopy - Materials Science, Engineering and Technology*, edited by Theophile Theophanides. InTech. doi:10.5772/36044.

Baset, Somaye, Hossein Akbari, Hossein Zeynali, and Morteza Shafie. 2011. "Size Measurement of Metal and Semiconductor Nanoparticles via UV-Vis Absorption Spectra." *Digest Journal of Nanomaterials and Biostructures* 6 (2): 709–16.

Bazkir, Özcan. 2007. "Determination of Optical Constants of Silicon Photodiode from Reflectivity Measurements at Normal Incidence of Light." *Optics and Lasers in Engineering* 45 (1): 245–48. doi:10.1016/j.optlaseng.2006.06.003.

Bekdemir, Ahmet, and Francesco Stellacci. 2016. "A Centrifugation-Based Physicochemical Characterization Method for the Interaction between Proteins and Nanoparticles." *Nature Communications* 7 (1). Nature Publishing Group: 13121. doi:10.1038/ncomms13121.

Bhargava, Rohit. 2012. "Infrared Spectroscopic Imaging: The Next Generation." *Applied Spectroscopy*, October. SAGE Publications. doi:10.1366/12-06801.

Bhowmick, Deb Kumar, Linda Stegemann, Manfred Bartsch, Naveen Kumar Allampally, Cristian A. Strassert, and Helmut Zacharias. 2015. "Controlled 2D-Confinement of Phosphorescent Pt(II) Complexes on Quartz and 6H-SiC(0001) Surfaces." *The Journal of Physical Chemistry C* 119 (10). American Chemical Society: 5551–61. doi:10.1021/acs.jpcc.5b00377.

Bock, Jan. 1972. "Thin Film Reflection Spectroscopy. I. Determination of Optical Constants of Amorphous Selenium and As2Se3." *The Journal of Chemical Physics* 57 (4): 1464. doi:10.1063/1.1678425.

Bonyár, Attila. 2020. "Label-Free Nucleic Acid Biosensing Using Nanomaterial-Based Localized Surface Plasmon Resonance Imaging: A Review." *ACS Applied Nano Materials* 3 (9): 8506–21. doi:10.1021/acsanm.0c01457.

Bridelli, Maria Grazia. 2017. "Fourier Transform Infrared Spectroscopy in the Study of Hydrated Biological Macromolecules." *Fourier Transforms - High-Tech Application and Current Trends*, February. IntechOpen. doi:10.5772/66576.

Chauhan, Ravendra Ps, Charu Gupta, and Dhan Prakash. 2012. "Methodological Advancements in Green Nanotechnology and Their Applications in Biological Synthesis of Herbal Nanoparticles." *International Journal of Bioassays* 6.

Chen, Dongyu, Shanthi Ganesh, Weimin Wang, and Mansoor Amiji. 2020. "Protein Corona-Enabled Systemic Delivery and Targeting of Nanoparticles." *The AAPS Journal* 22 (4): 83. doi:10.1208/s12248-020-00464-x.

Cheng, Xuelian, Jun Zhong, Jie Meng, Man Yang, Fumin Jia, Zhen Xu, Hua Kong, and Haiyan Xu. 2011. "Characterization of Multiwalled Carbon Nanotubes Dispersing in Water and Association with Biological Effects." *Journal of Nanomaterials* 2011: 1–12. doi:10.1155/2011/938491.

Chetwynd, Andrew, Emily Guggenheim, Sophie Briffa, James Thorn, Iseult Lynch, and Eugenia Valsami-Jones. 2018. "Current Application of Capillary Electrophoresis in Nanomaterial Characterisation and Its Potential to Characterise the Protein and Small Molecule Corona." *Nanomaterials* 8 (2): 99. doi:10.3390/nano8020099.

Christy, Alfred A., Yukihiro Ozaki, and Vasilis G. Gregoriou. 2001. *Modern Fourier Transform Infrared Spectroscopy*. Elsevier.

Clarke, James A. 2004. *The Science and Technology of Undulators and Wigglers*. Oxforfd: OUP.

Colomban, P., and A. Slodczyk. 2009. "Raman Intensity: An Important Tool in the Study of Nanomaterials and Nanostructures." *Acta Physica Polonica A* 116 (1): 7–12. doi:10.12693/APhysPolA.116.7.

Das, Ratan, Siddarth S. Nath, Dipankar Chakdar, Gautam Gope, and Ramendhu Bhattacharjee. 2010. "Synthesis of Silver Nanoparticles and Their Optical Properties." *Journal of Experimental Nanoscience* 5 (4): 357–62. doi:10.1080/17458080903583915.

Das, Ratan, and Sumit Sarkar. 2015. "Optical Properties of Silver Nano-Cubes." *Optical Materials* 48 (October): 203–8. doi:10.1016/j.optmat.2015.07.038.

Davidson, R. Andrew, Donald S. Anderson, Laura S. Van Winkle, Kent E. Pinkerton, and T. Guo. 2015. "Evolution of Silver Nanoparticles in the Rat Lung Investigated by X-Ray Absorption Spectroscopy." *The Journal of Physical Chemistry A* 119 (2): 281–89. doi:10.1021/jp510103m.

Elleaume, P. 1988. "Undulators and Wiggler Insertion Magnets for the ESRF." *IEEE Transactions on Magnetics* 24 (2): 974–77. doi:10.1109/20.11390.

Elleaume, Pascal. 1989. "ESRF Insertion Devices, Design Considerations." *Nuclear Instruments and Methods in Physics Research Section A: Accelerators, Spectrometers, Detectors and Associated Equipment* 282 (2–3): 477–80. doi:10.1016/0168-9002(89)90025-9.

Faraji, M., Y. Yamini, and M. Rezaee. 2010. "Magnetic Nanoparticles: Synthesis, Stabilization, Functionalization, Characterization, and Applications." *Journal of the Iranian Chemical Society* 7 (1): 1–37. doi:10.1007/BF03245856.

Ferodolin, I., and A. Awang. 2019. "Tuning Surface Plasmon Resonance Peak of Glass Containing Metallic Nanoparticles." *Journal of Physics: Conference Series* 1358 (November): 012046. doi:10.1088/1742-6596/1358/1/012046.

Ferraro, John R., Kazuo Nakamoto, and Chris W. Brown. 2003. *Introductory Raman Spectroscopy.* 2nd ed. Amsterdam and Boston: Academic Press.

Filippov, V. V. 2000. "Method of the Ratio of Envelopes of the Reflection Spectrum for Measuring Optical Constants and Thickness of Thin Films." *Optics and Spectroscopy* 88 (4): 581–85. doi:10.1134/1.626854.

Filippov, V. V., and V. P. Kutavichyus. 2003. "Accuracy of Determining the Optical Parameters of Thin Films by the Method of the Reflectance-Spectrum Extrema Envelopes." *Journal of Applied Spectroscopy* 70 (1): 122–129.

Fricke., Hugo. 1920. "The K-Characteristic Absorption Frequencies for the Chemical Elements Magnesium to Chromium." *Physical Review* 16 (3): 202–15. doi:10.1103/PhysRev.16.202.

Gadipelly, T., A. Dasgupta, C. Ghosh, V. Krupa, D. Sornadurai, B. K. Sahu, and S. Dhara. 2020. "Synthesis and Structural Characterisation of Y2Ti2O7 Using Microwave Hydrothermal Route." *Journal of Alloys and Compounds* 814. doi:10.1016/j.jallcom.2019.152273.

Gasper, R., and E. Goormaghtigh. 2012. "Classification of Drug Modes of Action on Cancer Cells by FTIR Fingerprinting." *Advances in Biomedical Spectroscopy* 6: 165–94. doi:10.3233/978-1-61499-059-8-165.

Gatoo, Manzoor Ahmad, Sufia Naseem, Mir Yasir Arfat, Ayaz Mahmood Dar, Khusro Qasim, and Swaleha Zubair. 2014. "Physicochemical Properties of Nanomaterials: Implication in Associated Toxic Manifestations." *BioMed Research International* 2014: 1–8. doi:10.1155/2014/498420.

Ghazy, O. A., M. T. Fouad, H. H. Saleh, A. E. Kholif, and T. A. Morsy. 2021. "Ultrasound-Assisted Preparation of Anise Extract Nanoemulsion and Its Bioactivity against Different Pathogenic Bacteria." *Food Chemistry* 341. doi:10.1016/j.foodchem.2020.128259.

Ghobadi, Nader. 2013. "Band Gap Determination Using Absorption Spectrum Fitting Procedure." *International Nano Letters* 3 (1): 2. doi:10.1186/2228-5326-3-2.

Grass, Robert N., Evagelos K. Athanassiou, and Wendelin J. Stark. 2007. "Covalently Functionalized Cobalt Nanoparticles as a Platform for Magnetic Separations in Organic Synthesis." *Angewandte Chemie International Edition* 46 (26): 4909–12. doi:10.1002/anie.200700613.

Greaves, G. N., A. J. Dent, B. R. Dobson, S. Kalbitzer, and G. Müller. 1993. "Environments of Ion-Implanted Dopants in Amorphous Silicon at Various Stages of Annealing." *Nuclear Instruments and Methods in Physics Research Section B: Beam Interactions with Materials and Atoms* 80–81, Part 2: 966–72. doi:10.1016/0168-583X(93)90717-K.

Haensel, Ruprecht. 1989. "The European Synchrotron Radiation Facility." *Nuclear Instruments and Methods in Physics Research Section A: Accelerators, Spectrometers, Detectors and Associated Equipment* 282 (2–3): 375–79. doi:10.1016/0168-9002(89)90005-3.

Hahn, David W. 2006. "Light Scattering Theory." Department of Mechanical and Aerospace Engineering, Florida. http://expha.com/sean/UF/July2010%20Backup/Academic/Loser %20Biased%20Diagnostics/Rayleigh%20and%20Mie%20Light%20Scattering.pdf.

Hamad, Syed, G. Krishna Podagatlapalli, Md. Ahamad Mohiddon, and Venugopal Rao Soma. 2014. "Cost Effective Nanostructured Copper Substrates Prepared with Ultrafast Laser Pulses for Explosives Detection Using Surface Enhanced Raman Scattering." *Applied Physics Letters* 104 (26). American Institute of Physics: 263104. doi:10.1063/1.4885763.

Hareesh, K., B. Shateesh, R. P. Joshi, J. F. Williams, D. M. Phase, S. K. Haram, and S. D. Dhole. 2017. "Ultra High Stable Supercapacitance Performance of Conducting Polymer Coated MnO_2 Nanorods/RGO Nanocomposites." *RSC Advances* 7 (32): 20027–36. doi:10.1039/C7RA01743J.

Hayzelden, C., and J. L. Batstone. 1993. "Silicide Formation and Silicide-Mediated Crystallization of Nickel-Implanted Amorphous Silicon Thin Films." *Journal of Applied Physics* 73 (12): 8279–8289.

Hertz, G. 1920. "über die Absorptionsgrenzen in derL-Serie." *Zeitschrift für Physik* 3 (1): 19–25. doi:10.1007/BF01356225.

Hofmann, A. n.d. *The Physics of Synchrotron Radiation*. Cambridge University Press.

Holder, Cameron F., and Raymond E. Schaak. 2019. "Tutorial on Powder X-Ray Diffraction for Characterizing Nanoscale Materials." *ACS Nano* 13 (7): 7359–65. doi:10.1021/ acsnano.9b05157.

Hu, Rong, Xiao-Bing Zhang, Rong-Mei Kong, Xu-Hua Zhao, Jianhui Jiang, and Weihong Tan. 2011. "Nucleic Acid-Functionalized Nanomaterials for Bioimaging Applications." *Journal of Materials Chemistry* 21 (41): 16323. doi:10.1039/c1jm12588e.

Hwang, Ching-Shiang. 2014. "Insertion Devices." Accessed May 11. http://www.nsrrc.org.tw /OCPAschool08/lecture/2.1.pdf.

Jain, K., A. Y. Mehandzhiyski, I. Zozoulenko, and L. Wågberg. 2021. "PEDOT:PSS Nano-Particles in Aqueous Media: A Comparative Experimental and Molecular Dynamics Study of Particle Size, Morphology and Z-Potential." *Journal of Colloid and Interface Science* 584: 57–66. doi:10.1016/j.jcis.2020.09.070.

Jarvis, Roger M., and Royston Goodacre. 2004. "Discrimination of Bacteria Using Surface-Enhanced Raman Spectroscopy." *Analytical Chemistry* 76 (1): 40–47. doi:10.1021/ ac034689c.

Javed, M., A. A. Khan, J. Kazmi, M. A. Mohamed, M. S. Ahmed, and Y. Iqbal. 2021. "Impedance Spectroscopic Study of Charge Transport and Relaxation Mechanism in MnCr2O4 Ceramic Chromite." *Journal of Alloys and Compounds* 854: 156996. doi:10.1016/j.jallcom.2020.156996.

Jenkins, R., and M. Holomany. 1987. "'PC-PDF': A Search/Display System Utilizing the CD-ROM and the Complete Powder Diffraction File." *Powder Diffraction* 2 (04): 215–19. doi:10.1017/S0885715600012811.

Jiao, Leizi, Yuming Guo, Jia Chen, Xiande Zhao, and Daming Dong. 2019. "Detecting Volatile Compounds in Food by Open-Path Fourier-Transform Infrared Spectroscopy." *Food Research International* 119 (May): 968–73. doi:10.1016/j.foodres.2018.11.042.

Kalmodia, Sushma, Sowmya Parameswaran, Wenrong Yang, Colin J. Barrow, and Subramanian Krishnakumar. 2015. "Attenuated Total Reflectance Fourier Transform Infrared Spectroscopy: An Analytical Technique to Understand Therapeutic Responses at the Molecular Level." *Scientific Reports* 5 (1). Nature Publishing Group: 16649. doi:10.1038/srep16649.

Kasikov, Aarne. 2010. "Optical Characterization of Inhomogeneous Thin Films." Thesis. https://dspace.ut.ee/handle/10062/14941.

Khalil, Ibrahim, Nurhidayatullaili Muhd Julkapli, Wageeh A. Yehye, Wan Jefrey Basirun, and Suresh K. Bhargava. 2016. "Graphene–Gold Nanoparticles Hybrid—Synthesis, Functionalization, and Application in a Electrochemical and Surface-Enhanced Raman Scattering Biosensor." *Materials* 9 (6). Multidisciplinary Digital Publishing Institute: 406. doi:10.3390/ma9060406.

Kim, S. U., K. H. Kim, and Y. M. Koo. 2004. "The Crystal Fraction Determination of the Nanocrystalline Phase Transformed from the Fe-Base Amorphous Matrix Using EXAFS." *Journal of Alloys and Compounds* 368 (1–2): 357–61. doi:10.1016/j. jallcom.2003.08.085.

Kizil, Ramazan, and Joseph Irudayaraj. 2013. "Fourier Transform Infrared Photoacoustic Spectroscopy (FTIR-PAS)." In *Encyclopedia of Biophysics*, edited by Gordon C. K. Roberts, 840–44. Berlin, Heidelberg: Springer. doi:10.1007/978-3-642-16712-6_124.

Korir, Daniel K., Bharat Gwalani, Abel Joseph, Brian Kamras, Ravi K. Arvapally, Mohammad A. Omary, and Sreekar B. Marpu. 2019. "Facile Photochemical Syntheses of Conjoined Nanotwin Gold-Silver Particles within a Biologically-Benign Chitosan Polymer." *Nanomaterials* 9 (4): 596. doi:10.3390/nano9040596.

Kronig, R. de L. 1931. "Zur Theorie der Feinstruktur in den Röntgenabsorptionsspektren." *Zeitschrift für Physik* 70 (5–6): 317–23. doi:10.1007/BF01339581.

Kronig, R. de L. 1932. "Zur Theorie der Feinstruktur in den Röntgenabsorptionsspektren. III." *Zeitschrift für Physik* 75 (7–8): 468–75. doi:10.1007/BF01342238.

Krylov, A. S., S. V. Goryainov, A. N. Vtyurin, S. N. Krylova, S. N. Sofronova, N. M. Laptash, T. B. Emelina, V. N. Voronov, and S. V. Babushkin. 2012. "Raman Scattering Study of Temperature and Hydrostatic Pressure Phase Transitions in Rb2KTiOF5 Crystal." *Journal of Raman Spectroscopy* 43 (4): 577–82. doi:10.1002/jrs.3071.

Kumar, Devender, Saroj Bala, Heena Wadhwa, Geeta Kandhol, Suman Mahendia, Fakir Chand, and Shyam Kumar. 2019. "Tuning of LSPR of Gold-Silver Alloy Nanoparticles with Their Composition." In 020048. Kurukshetra, India. doi:10.1063/1.5097117.

Kumar, K. Uma Mahendra, and M. Ghanashyam Krishna. 2008. "Chromium-Induced Nanocrystallization of a-Si Thin Films into the Wurtzite Structure." *Journal of Nanomaterials* 2008: 1–6. doi:10.1155/2008/736534.

Lai, David Y. 2015. "Approach to Using Mechanism-Based Structure Activity Relationship (SAR) Analysis to Assess Human Health Hazard Potential of Nanomaterials." *Food and Chemical Toxicology* 85 (November): 120–26. doi:10.1016/j.fct.2015.06.008.

Lei, G., J. E. Anderson, M. I. Buchwald, B. C. Edwards, and R. I. Epstein. 1998. "Determination of Spectral Linewidths by Voigt Profiles in Yb^{3+}-Doped Fluorozirconate Glasses." *Physical Review B* 57 (13): 7673–78. doi:10.1103/PhysRevB.57.7673.

Lewis, Ian R., and Howell Edwards. 2001. *Handbook of Raman Spectroscopy: From the Research Laboratory to the Process Line*. London, UK: CRC Press.

Li, J., J. Fang, H. Su, and S. Sun. 2011. "Interfacial Processes of Lithium Ion Batteries by FTIR Spectroscopy." *Progress in Chemistry* 23 (2–3): 349–56.

Li, Yuancheng, Yaolin Xu, Candace C. Fleischer, Jing Huang, Run Lin, Lily Yang, and Hui Mao. 2017. "Impact of Anti-Biofouling Surface Coatings on the Properties of Nanomaterials and Their Biomedical Applications." *Journal of Materials Chemistry B* 6 (1). The Royal Society of Chemistry: 9–24. doi:10.1039/C7TB01695F.

Ling, Xi, Shengxi Huang, Jing Kong, and Mildred Dresselhaus. 2018. "Graphene-Enhanced Raman Scattering (GERS): Chemical Effect." In *Recent Developments in Plasmon-Supported Raman Spectroscopy*, edited by Katrin Kneipp, Yukihiro Ozaki, and Zhong-Qun Tian, 415–49. World Scientific (Europe). doi:10.1142/9781786344243_0015.

Liu, Hai, Luxin Yan, Yi Chang, Houzhang Fang, and Tianxu Zhang. 2013. "Spectral Deconvolution and Feature Extraction with Robust Adaptive Tikhonov Regularization." *IEEE Transactions on Instrumentation and Measurement* 62 (2): 315–27. doi:10.1109/TIM.2012.2217636.

Mangione, A., L. Torrisi, A. Picciotto, F. Caridi, D. Margarone, E. Fazio, A. La Mantia, and G. Di Marco. 2005. "Carbon Nanocrystals Produced by Pulsed Laser Ablation of Carbon." *Radiation Effects and Defects in Solids* 160 (10–12): 655–62. doi:10.1080/10420150500493188.

McKale, A. G., B. Ww Veal, A. P. Paulikas, S. K. Chan, and G. S. Knapp. 1988. "Improved Ab Initio Calculations of Amplitude and Phase Functions for Extended X-Ray Absorption Fine Structure Spectroscopy." *Journal of the American Chemical Society* 110 (12): 3763–3768.

Memarian, N. 2017. "CdS Nanoparticles: A Facile Route to Size-Controlled Synthesis of Quantum Dots on a Polymer Matrix." *Journal of Nano- and Electronic Physics* 9 (3): 03027-1-03027-5. doi:10.21272/jnep.9(3).03027.

Mester, Lars, Alexander A. Govyadinov, Shu Chen, Monika Goikoetxea, and Rainer Hillenbrand. 2020. "Subsurface Chemical Nanoidentification by Nano-FTIR Spectroscopy." *Nature Communications* 11 (1). Nature Publishing Group: 3359. doi:10.1038/s41467-020-17034-6.

Mobilio, S., and G. Vlaic. 2003. *Synchrotron Radiation: Fundamentals, Methodologies, and Applications: S. Margherita Di Pula, 17–28 September 2001.* Vol. 82. Italian Physical Society.

Mu, Q., Y. Sun, A. Guo, X. Xu, B. Qin, and A. Cai. 2021. "A Bifunctionalized NiCo2O4-Au Composite: Intrinsic Peroxidase and Oxidase Catalytic Activities for Killing Bacteria and Disinfecting Wound." *Journal of Hazardous Materials* 402. doi:10.1016/j.jhazmat.2020.123939.

Müller, J. E., and W. L. Schaich. 1983. "Single-Scattering Theory of X-Ray Absorption." *Physical Review B* 27 (10): 6489–92. doi:10.1103/PhysRevB.27.6489.

Nabers, Andreas, Julian Ollesch, Jonas Schartner, Carsten Kötting, Just Genius, Ute Haußmann, Hans Klafki, Jens Wiltfang, and Klaus Gerwert. 2016. "An Infrared Sensor Analysing Label-Free the Secondary Structure of the Abeta Peptide in Presence of Complex Fluids." *Journal of Biophotonics* 9 (3): 224–34. doi:10.1002/jbio.201400145.

Naidu, K. Lakshun, and Mamidipudi Ghanashyam Krishna. 2014. "Effect of Thermal Annealing on Disorder and Optical Properties of Cr/Si Bilayer Thin Films." *Philosophical Magazine* 94 (30): 3431–44. doi:10.1080/14786435.2014.959578.

Naidu, K. Lakshun, Md Ahamad Mohiddon, M. Ghanashyam Krishna, G. Dalba, and F. Rocca. 2013. "Metal Induced Crystallization of Amorphous Silicon Thin Films Studied by X-Ray Absorption Fine Structure Spectroscopy." *Journal of Physics: Conference Series* 430 (April): 012035. doi:10.1088/1742-6596/430/1/012035.

Newville, M., P. Līviņš, Y. Yacoby, J. J. Rehr, and E. A. Stern. 1993. "Near-Edge X-Ray-Absorption Fine Structure of Pb: A Comparison of Theory and Experiment." *Physical Review B* 47 (21): 14126–31. doi:10.1103/PhysRevB.47.14126.

Newville, M., B. Ravel, D. Haskel, J. J. Rehr, E. A. Stern, and Y. Yacoby. 1995. "Analysis of Multiple-Scattering XAFS Data Using Theoretical Standards." *Physica B: Condensed Matter*, Proceedings of the 8th International Conference on X-ray Absorption Fine Structure, 208–209 (March): 154–56. doi:10.1016/0921-4526(94)00655-F.

Newville, Matthew. 2001a. "IFEFFIT: Interactive XAFS Analysis and FEFF Fitting." *Journal of Synchrotron Radiation* 8 (2): 322–324.

Newville, Matthew. 2001b. "EXAFS Analysis Using *FEFF* and *FEFFIT*." *Journal of Synchrotron Radiation* 8 (2): 96–100. doi:10.1107/S0909049500016290.

Ngumbi, Paul K, Simon W. Mugo, and James M. Ngaruiya. n.d. "Determination of Gold Nanoparticles Sizes via Surface Plasmon Resonance," 5.

Paesler, M. A., Dale E. Sayers, Raphael Tsu, and Jesus Gonzalez-Hernandez. 1983. "Ordering of Amorphous Germanium Prior to Crystallization." *Physical Review B* 28 (8): 4550–57. doi:10.1103/PhysRevB.28.4550.

Patterson, A. L. 1939. "The Scherrer Formula for X-Ray Particle Size Determination." *Physical Review* 56 (10): 978–82. doi:10.1103/PhysRev.56.978.

Phan, Hoa T., and Amanda J. Haes. 2019. "What Does Nanoparticle Stability Mean?" *The Journal of Physical Chemistry. C, Nanomaterials and Interfaces* 123 (27): 16495–507. doi:10.1021/acs.jpcc.9b00913.

Podagatlapalli, G. Krishna, Syed Hamad, Surya P. Tewari, S. Sreedhar, Muvva D. Prasad, and S. Venugopal Rao. 2013. "Silver Nano-Entities through Ultrafast Double Ablation in Aqueous Media for Surface Enhanced Raman Scattering and Photonics Applications." *Journal of Applied Physics* 113 (7). American Institute of Physics: 073106. doi:10.1063/1.4792483.

Podagatlapalli, G. Krishna, Syed Hamad, Md. Ahamad Mohiddon, and S. Venugopal Rao. 2014. "Effect of Oblique Incidence on Silver Nanomaterials Fabricated in Water via Ultrafast Laser Ablation for Photonics and Explosives Detection." *Applied Surface Science* 303 (June): 217–32. doi:10.1016/j.apsusc.2014.02.152.

Podagatlapalli, G. Krishna, Syed Hamad, Ahamad Mohiddon, and S. Venugopal Rao. 2015. "Fabrication of Nanoparticles and Nanostructures Using Ultrafast Laser Ablation of Silver with Bessel Beams." *Laser Physics Letters* 10.

Podagatlapalli, G. Krishna, Syed Hamad, and S. Venugopal Rao. 2015. "Trace-Level Detection of Secondary Explosives Using Hybrid Silver–Gold Nanoparticles and Nanostructures Achieved with Femtosecond Laser Ablation." *Journal of Physical Chemistry C* 12.

Qais, Faizan Abul, Samreen, and Iqbal Ahmad. 2019. "Green Synthesis of Metal Nanoparticles: Characterization and Their Antibacterial Efficacy." In *Antibacterial Drug Discovery to Combat MDR: Natural Compounds, Nanotechnology and Novel Synthetic Sources*, edited by Iqbal Ahmad, Shamim Ahmad, and Kendra P. Rumbaugh, 635–80. Singapore: Springer. doi:10.1007/978-981-13-9871-1_28.

Ravel, B., and M. Newville. 2005. "ATHENA and ARTEMIS: Interactive Graphical Data Analysis Using IFEFFIT." *Physica Scripta* 2005 (T115): 1007. doi:10.1238/Physica. Topical.115a01007.

Ravel, Bruce. 2001. "ATOMS: Crystallography for the X-Ray Absorption Spectroscopist." *Journal of Synchrotron Radiation* 8 (2): 314–16. doi:10.1107/S090904950001493X.

Ravel, Bruce. 2008. "ATHENA User's Guide." *Document Version 1.* http://cars9.uchicago .edu/~ravel/software/doc/Athena/html/athena.pdf.

Rehr, J. J., and R. C. Albers. 1990. "Scattering-Matrix Formulation of Curved-Wave Multiple-Scattering Theory: Application to X-Ray-Absorption Fine Structure." *Physical Review B* 41 (12): 8139–49. doi:10.1103/PhysRevB.41.8139.

Rehr, J. J., R. C. Albers, C. R. Natoli, and E. A. Stern. 1986. "New High-Energy Approximation for X-Ray-Absorption Near-Edge Structure." *Physical Review B* 34 (6): 4350–53. doi:10.1103/PhysRevB.34.4350.

Rehr, J. J., and A. L. Ankudinov. 2001. "Progress and Challenges in the Theory and Interpretation of X-Ray Spectra." *Journal of Synchrotron Radiation* 8 (2): 61–65.

Rehr, J. J., J. Mustre de Leon, S. I. Zabinsky, and R. C. Albers. 1991. "Theoretical X-Ray Absorption Fine Structure Standards." *Journal of the American Chemical Society* 113 (14): 5135–40. doi:10.1021/ja00014a001.

Rehr, John J., and R. C. Albers. 2000. "Theoretical Approaches to X-Ray Absorption Fine Structure." *Reviews of Modern Physics* 72 (3): 621.

Revol, J.-L., J.-C. Biasci, J.-F. Bouteille, F. Ewald, L. Farvacque, A. Franchi, G. Gautier, et al. 2013. "ESRF Operation and Upgrade Status." In *IPAC 2013: Proceedings of the 4th International Particle Accelerator Conference*, 82–84.

Rusi, and S. R. Majid. 2015. "Green Synthesis of in Situ Electrodeposited RGO/MnO2 Nanocomposite for High Energy Density Supercapacitors." *Scientific Reports* 5 (1): 16195. doi:10.1038/srep16195.

Saputra, Irwan, Irwandi Jaswir, and Rini Akmeliawati. 2018. "Identification of Pig Adulterant in Mixture of Fat Samples and Selected Foods Based on FTIR-PCA Wavelength Biomarker Profile." *International Journal on Advanced Science, Engineering and Information Technology* 8 (6). INSIGHT – Indonesian Society for Knowledge and Human Development: 2341–48.

Sasai, Ryo, Taketoshi Fujita, Nobuo Iyi, Hideaki Itoh, and Katsuhiko Takagi. 2002. "Aggregated Structures of Rhodamine 6G Intercalated in a Fluor-Taeniolite Thin Film." *Langmuir* 18 (17): 6578–83. doi:10.1021/la020183y.

Sayers, Dale E., Edward A. Stern, and Farrel W. Lytle. 1971. "New Technique for Investigating Noncrystalline Structures: Fourier Analysis of the Extended X-Ray—Absorption Fine Structure." *Physical Review Letters* 27 (18): 1204–7. doi:10.1103/PhysRevLett.27.1204.

Schuster, Julian J., Stefan Will, Alfred Leipertz, and Andreas Braeuer. 2014. "Deconvolution of Raman Spectra for the Quantification of Ternary High-Pressure Phase Equilibria Composed of Carbon Dioxide, Water and Organic Solvent." *Journal of Raman Spectroscopy* 45 (3): 246–52. doi:10.1002/jrs.4451.

Shaik, Ummar Pasha, Syed Hamad, Md. Ahamad Mohiddon, Venugopal Rao Soma, and M. Ghanashyam Krishna. 2016. "Morphologically Manipulated Ag/ZnO Nanostructures as Surface Enhanced Raman Scattering Probes for Explosives Detection." *Journal of Applied Physics* 119 (9). American Institute of Physics: 093103. doi:10.1063/1.4943034.

Shankernath, V., K. Lakshun Naidu, M. Ghanashyam Krishna, and K. A. Padmanabhan. 2017. "Optical Response of Ultra-Thin Titanium Nitride Films on Brass and Gold Plated Brass Surfaces." *Materials Research Bulletin* 85 (January): 121–30. doi:10.1016/j.materresbull.2016.09.006.

Sharma, Vishal, and Raj Kumar. 2019. "CHAPTER 8:FTIR and NIRS in Forensic Chemical Sensing." In *Forensic Analytical Methods*, 164–97. doi:10.1039/9781788016117-00164.

Shenoy, Gopal. 2003. "Basic Characteristics of Synchrotron Radiation." *Structural Chemistry* 14 (1): 3–14.

Sim, Brandon, Fernando Monjaraz, Yong-Joong Lee, and So-Yeun Park. 2011. "Two-Dimensional Arrays of Gold Nanoparticles for Plasmonic Nanosensor." *Korean Journal of Materials Research* 21 (10): 525–31. doi:10.3740/MRSK.2011.21.10.525.

Singh, A. K., V. Viswanath, and V. C. Janu. 2009. "Synthesis, Effect of Capping Agents, Structural, Optical and Photoluminescence Properties of ZnO Nanoparticles." *Journal of Luminescence* 129 (8): 874–78. doi:10.1016/j.jlumin.2009.03.027.

Siqueiros, Jesus M., Luis E. Regalado, and Roberto Machorro. 1988. "Determination of (n, k) for Absorbing Thin Films Using Reflectance Measurements." *Applied Optics* 27 (20): 4260–4264.

Siyar, Muhammad, Asghari Maqsood, and Sadaf Khan. 2014. "Synthesis of Mono Layer Graphene Oxide from Sonicated Graphite Flakes and Their Hall Effect Measurements." *Materials Science-Poland* 32 (2): 292–96. doi:10.2478/s13536-013-0189-2.

Skoog, Douglas A., F. James Holler, and Stanley R. Crouch. 2007. *Principles of Instrumental Analysis*. 6th ed. Belmont, CA: Thomson Brooks/Cole.

Smith, Ewen, and Geoffrey Dent. 2005. *Modern Raman Spectroscopy: A Practical Approach*. Hoboken, NJ: John Wiley & Sons.

Smith, Ewen, and Geoffrey Dent. 2019. *Modern Raman Spectroscopy: A Practical Approach*. John Wiley & Sons.

Sreemany, Monjoy, and Suchitra Sen. 2004. "A Simple Spectrophotometric Method for Determination of the Optical Constants and Band Gap Energy of Multiple Layer TiO2 Thin Films." *Materials Chemistry and Physics* 83 (1): 169–77. doi:10.1016/j.matchemphys.2003.09.030.

Stenzel, O., V. Hopfe, and P. Klobes. 1991. "Determination of Optical Parameters for Amorphous Thin Film Materials on Semitransparent Substrates from Transmittance and Reflectance Measurements." *Journal of Physics. D. Applied Physics* 24 (11): 2088–2094.

Strutt, J. W. 1871. "LVIII. On the Scattering of Light by Small Particles." *Philosophical Magazine Series 4* 41 (275): 447–54. doi:10.1080/14786447108640507.

Tan, Zhengquan, S. M. Heald, M. Rapposch, C. E. Bouldin, and J. C. Woicik. 1992. "Gold-Induced Germanium Crystallization." *Physical Review B* 46 (15): 9505–10. doi:10.1103/PhysRevB.46.9505.

Tauc, J., R. Grigorovici, and A. Vancu. 1966. "Optical Properties and Electronic Structure of Amorphous Germanium." *Physica Status Solidi (b)* 15 (2): 627–37. doi:10.1002/pssb.19660150224.

Vigneshwaran, N., and P. Jain. 2011. "Biomolecules–Nanoparticles: Interaction in Nanoscale." In *Metal Nanoparticles in Microbiology*, edited by Mahendra Rai and Nelson Duran, 135–50. Berlin, Heidelberg: Springer Berlin Heidelberg. doi:10.1007/978-3-642-18312-6_6.

Voort, F. R. van de. 1992. "Fourier Transform Infrared Spectroscopy Applied to Food Analysis." *Food Research International* 25 (5): 397–403. doi:10.1016/0963-9969(92)90115-L.

Wakagi, M., and Y. Maeda. 1994. "Structural Study of Crystallization of A-Ge Using Extended X-Ray-Absorption Fine Structure." *Physical Review B* 50 (19): 14090–95. doi:10.1103/PhysRevB.50.14090.

Wakagi, Masatoshi, Toshiki Kaneko, Kiyoshi Ogata, and Asao Nakano. 1992. "Crystallinity Analysis of Amorphous-Crystalline Mixed Phase Silicon Films Using Exafs Method." In *Symposium F – Microcrystalline Semiconductors–Materials Science and Devices*. Vol. 283. MRS Online Proceedings Library. doi:10.1557/PROC-283-555.

Wall, Mark. 2011. "The Raman Spectroscopy of Graphene and the Determination of Layer Thickness." Thermo Fisher Scientific, Madison, 5. http://www.thermoscientific.com/content.

Wang, Liming, Tianlu Zhang, Panyun Li, Wanxia Huang, Jinglong Tang, Pengyang Wang, Jing Liu, et al. 2015. "Use of Synchrotron Radiation-Analytical Techniques to Reveal Chemical Origin of Silver-Nanoparticle Cytotoxicity." *ACS Nano* 9 (6): 6532–47. doi:10.1021/acsnano.5b02483.

Wang, Xiaoyu, Xuejing Cui, Yuliang Zhao, and Chunying Chen. 2020. "Nano-Bio Interactions: The Implication of Size-Dependent Biological Effects of Nanomaterials." *Science China Life Sciences* 63 (8): 1168–82. doi:10.1007/s11427-020-1725-0.

Wiley, Benjamin J, Sang Hyuk Im, Zhi-Yuan Li, Joeseph McLellan, Andrew Siekkinen, and Younan Xia. n.d. "Maneuvering the Surface Plasmon Resonance of Silver Nanostructures through Shape-Controlled Synthesis." 10.

Willets, Katherine A., and Richard P. Van Duyne. 2007. "Localized Surface Plasmon Resonance Spectroscopy and Sensing." *Annual Review of Physical Chemistry* 58 (1): 267–97. doi:10.1146/annurev.physchem.58.032806.104607.

Williams, David B., and C. Barry Carter. 2009. *Transmission Electron Microscopy: A Textbook for Materials Science*. 2nd ed. Springer US. doi:10.1007/978-0-387-76501-3.

Xiong, Ziye, Fen Qin, Po-Shun Huang, Ian Nettleship, and Jung-Kun Lee. 2016. "Effect of Synthesis Techniques on Crystallization and Optical Properties of Ag-Cu Bimetallic Nanoparticles." *JOM* 68 (4): 1163–68. doi:10.1007/s11837-015-1757-1.

Xu, Wentao, Wanchong He, Zaihui Du, Liye Zhu, Kunlun Huang, Yi Lu, and Yunbo Luo. 2020. "Functional Nucleic Acid Nanomaterials: Development, Properties, and Applications." *Angewandte Chemie International Edition*, August, anie.201909927. doi:10.1002/anie.201909927.

Yadav, L. D. S. 2005. "Ultraviolet (UV) and Visible Spectroscopy." In *Organic Spectroscopy*, edited by L. D. S. Yadav, 7–51. Dordrecht: Springer Netherlands. doi:10.1007/978-1-4020-2575-4_2.

Yang, Juan, Daqi Zhang, and Yan Li. 2011. "How to Remove the Influence of Trace Water from the Absorption Spectra of SWNTs Dispersed in Ionic Liquids." *Beilstein Journal of Nanotechnology* 2 (September): 653–58. doi:10.3762/bjnano.2.69.

Zabinsky, S. I., J. J. Rehr, A. Ankudinov, R. C. Albers, and M. J. Eller. 1995. "Multiple-Scattering Calculations of X-Ray-Absorption Spectra." *Physical Review B* 52 (4): 2995–3009. doi:10.1103/PhysRevB.52.2995.

Zanatta, A. R., and I. Chambouleyron. 2005. "Low-Temperature Al-Induced Crystallization of Amorphous Ge." *Journal of Applied Physics* 97 (9): 094914. doi:10.1063/1.1889227.

Zhang, B., S. Shrestha, M. A. Green, and G. Conibeer. 2010. "Size Controlled Synthesis of Ge Nanocrystals in SiO2 at Temperatures below 400°C Using Magnetron Sputtering." *Applied Physics Letters* 96 (26): 261901. doi:10.1063/1.3457864.

5 Metal Nanoparticles against Bacteria

*Amjad Islam Aqib, Iqra Muzammil, Saad Ahmad,
Muhammad Luqman Sohail, Ahmad Ali,
Kashif Prince, Amna Ahmad, and Hina Afzal Sajid*

CONTENTS

Acronyms 120
5.1 How Antimicrobial Resistance Invites Nanoparticles 120
5.2 Types of Metal Nanoparticles and Their Mechanisms 122
    5.2.1 Silver Nanoparticles (Ag NPs) 122
    5.2.2 Copper and Copper Oxide Nanoparticles (Cu NPs, Cu_2O NPs,
            and CuO NPs) 122
    5.2.3 Magnesium Nanoparticles (Mg NPs) 123
    5.2.4 Bismuth Nanoparticles (Bi NPs) 123
    5.2.5 Gold Nanoparticles (Au NPs) 123
    5.2.6 Zinc Nanoparticles (Zn NPs) 124
5.3 Metal Oxide Nanoparticles 124
5.4 Characterization Effects on Antibacterial Activity of
     Metal Nanoparticles 129
    5.4.1 Shape of Nanoparticles 129
    5.4.2 Size of Nanoparticles 133
    5.4.3 Roughness of Nanoparticles 133
    5.4.4 Zeta Potential 133
    5.4.5 Doping Modification 134
5.5 In Vivo Studies of Metal Nanoparticles as Antibacterials 134
5.6 Nanoparticle- and Nanomaterial-Based Vaccines against Bacterial
     Infections 134
    5.6.1 Mechanisms of Action 136
5.7 Types of Nanovaccines 137
    5.7.1 Nanoparticle-Conjugated Subunit Vaccines 137
          5.7.1.1 Outer Membrane Vesicles (OMVs) 137
          5.7.1.2 Self-Assembled Nanomaterials 138
          5.7.1.3 Polymeric Nanoparticles 138
    5.7.2 Toxoid Vaccine Nanoparticles 139
          5.7.2.1 Single-Toxin Nanotoxoids 139
          5.7.2.2 Multi-toxin Nanotoxoids 140
5.8 Metal Nanoparticles and Antibiotic Products 140

DOI: 10.1201/9781003126256-5

5.9   Metal Nanoparticles and the Industry 142
      5.9.1   Role of Nanotechnology in Medical Industry 142
      5.9.2   Global Economic Impact of Nanotechnology 142
      5.9.3   Global Nanotechnology Market and Forecast to 2024 – By
               Component.. 144
      5.9.4   Global Nanotechnology Market and Forecast to 2024 – By
               Application.. 144
5.10   Challenges to Using Metal Nanoparticles as Antibacterials 144
5.11   Conclusion 146
References.. 146

ACRONYMS

APCs:	antigen-presenting cells
DCs:	dendritic cells
DDA:	dimethyl-dioctadecyl-ammonium bromide
DOTAP:	1,2-dioleoyl-3-trimethyl-ammonium-propane chloride
IL:	interleukin
LPS:	lipopolysaccharides
MDR:	multidrug resistant
MIC:	minimum inhibitory concentration
MO:	metal oxide
MPLA:	monophosphoryl lipid A
MRSA:	methicillin-resistant *Staphylococcus aureus*
NPs:	nanoparticles
OD:	optical density
OMVs:	outer membrane vesicles
PAMPs:	pathogen-associated molecule patterns
PLGA:	poly(d, l-lactic-co-glycolic acid)
PRRs:	pattern recognition receptors
PS:	polystyrene
ROS:	reactive oxygen species
TNF:	tumor necrosis factor
WHO:	World Health Organization

5.1   HOW ANTIMICROBIAL RESISTANCE INVITES NANOPARTICLES

Infectious diseases are the most important cause of deaths throughout the world, and it is necessary to search for antimicrobials without resistance. A public health concern is the prevalence of both gram-positive and gram-negative microbes of antibiotic-resistant and multidrug-resistant (MDR) strains (Raghunath and Perumal, 2017). About 70% of bacterial infections are resistant to one or more of the antibiotics commonly used to eliminate the infection (Dizaj et al., 2014). Such proliferation of resistance is because of the haphazard use of antibiotics, which results in the rapid development of new antimicrobial-resistant strains (Raghunath and Perumal, 2017), and, consequently, diseases caused by these strains become difficult to treat (Gharpure et al., 2020). Moreover, several organic antimicrobial agents are toxic to human and other species, and might cause multiple allergic reactions in them (Hossain et al., 2015; Rajawat and Qureshi, 2012). According to a 2016 WHO report, microbes, including *Klebsiella pneumoniae*, *Escherichia coli*, MRSA (methicillin-resistant *Staphylococcus aureus*), *Enterobacteriaceae*, *Plasmodiumfalciparum*, HIV, and influenza, are resistant to advanced antimicrobials, including carbapenem antibiotics, fluoroquinolones, and third-generation cephalosporins. When microbes are exposed to an antimicrobial environment, they develop several mechanisms to adapt to the environment and then develop antimicrobial resistance (AMR) (Alekshun and Levy, 2007).

Bacteria having antibiotic resistance are mainly responsible for deficient efficacy in antimicrobial agents. Microorganisms are capable of showing resistance because of modifications so that they oppose antibacterial agents by two means: either to make them inactive or by decreasing their therapeutic efficacy. Eventually, with the passage of time, microorganisms show resistance impulsively because of modifications in their genetic makeup (Aslam et al., 2018). This causes the extension of infectious periods, higher death rates, and high health care costs (IACG, 2016). Along with the genetic variations in microscopic organisms, the exchange of genetic material is also responsible for bacterial resistance between bacteria or phages by means of (i) DNA transformation, or the ability of bacteria to integrate a DNA fragment from the surrounding environment; or (ii) via the transduction process in which transmission of bacterial genes takes place using a virus and through the conjugation method, comprising the transmission of hereditary material between both donor and receptor microorganisms (Munita and Arias, 2016). There are diverse mechanisms for antibiotic resistance, such as enzymatic means by using β-lactamases, amino glycoside, or acetyltransferases altering enzymes (Munita and Arias, 2016). Modifications in the permeability of membranes to prevent the diffusion of the antimicrobial mediator are a mechanism for resistance, in cooperation with alteration in the antimicrobial target (e.g., mutations in DNA gyrase and topoisomerase IV or penicillin-binding proteins) (Dugassa and Shukuri, 2017). Ever since the "golden age," there are three new groups of antibiotics that are vigorous in opposing gramp-positive bacteria, for instance, oxazolidinones and daptomycin. Conversely, high numerals of analogues of accessible groups and antibiotic alternatives have arrived at the market (Wang et al., 2017).

Nanoparticles exhibit antibacterial activity because of the surface area touching the microorganisms (Chiriac et al., 2016; Tang et al., 2013). The quantity of metal ions or other non-essential ions (such as Ag) can be fatal to cells, often due to oxidative stress, protein degradation, or membrane damage (Lemire et al., 2013). The morphological and physicochemical characteristics of nanometals have also been shown to affect their antimicrobial activity (Mohammadi et al., 2010) and the highest bactericidal effect (Besinis et al., 2014; Fellahi et al., 2013). The positive surface charge of the metal nanoparticles facilitates their binding to the negatively charged surface of the bacteria, which can lead to an increased bactericidal effect (Seil and Webster, 2012). The type of metal, physical properties, and composition of the particles influence the effectiveness of the agents tested against bacterial and fungal stress. Efficient agents for the quick spreading and killing of microorganisms are of the smallest particle size and the more aggregated particles (El-Refai et al., 2018). Metal-based compounds and their ions are highly toxic to microbes and have demonstrated substantial biocidal activity due to the existence of large-area reactive species (Ghosh et al., 2012).

5.2 TYPES OF METAL NANOPARTICLES AND THEIR MECHANISMS

Metal nanoparticles, such as silver, gold, and magnesium, exhibit promising antimicrobial properties (Syafiuddin et al., 2017). Metallic nanoparticles have been widely studied for a broad range of biomedical-related applications. The World Health Organization (WHO) has explained the effectiveness of metallic nanoparticles together with their diminutive size and selection of bacteria for the pathogens that are of top priority. These metal nanoparticles are recognized to contain an imprecise bacterial toxicity system (i.e., they lack the capability of binding to a particular receptor within the cells of bacteria), which makes it difficult for bacteria to develop resistance, as well as widen the range of antibacterial activity that gives an effective and potential outcome in both gram-positive and gram-negative bacteria. In recent times, nanomaterial applications have gained great interest from researchers for their antimicrobial works. Studies on nanoparticles demonstrate the toxicity potential of a few metals and metal oxide nanoparticles (NPs), such as gold, silver, copper, zinc, nickel, bismuth, aluminum, and magnesium, toward successful killing of bacteria.

5.2.1 SILVER NANOPARTICLES (AG NPS)

Silver as an antimicrobial agent has been extensively used in two physical states (solids and solution forms) of salts for the healing and disinfection of wounds. At the present time, dressings infused with silver nitrate ($AgNO_3$) can be created (Chen and Schluesener, 2008). Silver displays fascinating properties because of its good conductivity and chemical stability as well as antibacterial and catalytic activity. Furthermore, silver nanoparticles (Ag NPs) are being extensively studied these days (Aderibigbe, 2017). In biomedical areas, these particles are acquiring influence particularly because of their large implementation as coatings for medical approaches, antimicrobial agents and for chemotherapeutic drugs as a carrier (Zhang et al., 2016). There are four antimicrobial mechanisms of action of silver nanoparticles: (i) desirability to bacterial surface, (ii) deterioration of bacterial cell membrane and cell wall with variation in its permeability, (iii) modulation of signal transduction pathways, and (iv) instigation of oxidative stress and toxicity by production of reactive oxygen species (Dakal et al., 2016). Bactericidal characteristics of Ag NPs are largely studied; however, an extensive application of further biomedical properties – for instance, antiviral, antifungal, antitumor, anti-inflammatory, and antiangiogenic activities – are being discussed (Zhang et al., 2016).

5.2.2 COPPER AND COPPER OXIDE NANOPARTICLES (CU NPS, CU$_2$O NPS, AND CUO NPS)

Copper is regarded as an outstanding antimicrobial candidate because it is a good semiconductor material for the preparation of metallic nanoparticles. It is highly temperature resistant as well as stable, economical, and effortlessly synthesized

(Abbaszadegan et al., 2015). Cu NPs have been suggested greatly to restrain growth of bacteria because they get in touch with bacterial cells directly. The mode of action for antibacterial activity of CuO nanoparticles is still not exposed completely, but it is supposed that it entails cell wall adhesion of bacteria activated by means of electrostatic interactions. The antibacterial activity of these particles has been considered predominantly against microorganisms like *Pseudomonas aeruginosa*, *Escherichia coli*, *Staphylococcus aureus*, *Bacillus subtilis*, *Streptococcus faecalis*, and *Enterococcus faecalis* (Rajendran et al., 2017). Results explained that the 0.0113, 0.00113 and 0.000113 mol L^{-1} concentrations of elemental copper with CuO NPs eradicated all the assessed microorganisms (Kumar et al., 2019).

5.2.3 Magnesium Nanoparticles (Mg NPs)

Magnesium nanoparticles (Mg NPs) inhibit microbial enzymes by damaging cell membranes destroying the pathogens through induction of reactive oxygen species (ROS). They also inhibit the formation of biofilm (Ravishankar, 2011). Their resistance appears unviable because they act concurrently on numerous sites of microbes. These nanoparticles have highly reported activity against *Bacillus megaterium* and *E. coli* and have modest activity against *Bacillus subtilis* spores (Gharpure et al., 2020).

5.2.4 Bismuth Nanoparticles (Bi NPs)

Nanoparticles made up of bismuth in combination with X-rays can be meritoriously exploited to treat MDR pathogens. X-rays have high ability of penetration that is used to control profound skin infection, but it cannot be utilized because of its toxicity. However, their toxicity can be reduced when used in combination with Bi NPs by decreasing the radiation quantity. On the other hand, complete eradication of the biofilm formation was caused by the mutants of *Streptococcus* by preparing the stable colloidal bismuth nanoparticles (Gharpure et al., 2020).

5.2.5 Gold Nanoparticles (Au NPs)

Gold nanoparticles (Au NPs) are clustered or colloidal particles made of gold core, a reactive and biocompatible composite (Gerber et al., 2013). These particles are valued for their synthetic flexibility, which controls their shape, size, and surface characteristics. Moreover, the particles' stability, solubility, and contact with the environment can be controlled by modifying their coating. In addition, the particle surface has the binding ability to amines and thiols, giving functional groups to the gold nanoparticles for the purpose of targeting, labeling, and conjugating the molecules of pharmacological interest (Her et al., 2017). These gold particles have unique characteristics of optical and electronic properties. Gold nanoparticles are applicable in several fields including bioimaging and biosensors, drug delivery systems, and cancer treatment (Paciotti et al., 2004; Podsiadlo et al., 2008). Gold nanoparticles are highly effective against various bacteria, including *E. coli*, *S. typhi*, *K. pneumoniae*,

E. faecalis, *Serratia* sp., *B. subtilis*, and *S. aureus* (Ahmed et al., 2015). Recently, the pharmaceutical characteristics of gold nanoparticles have been investigated. These nanoparticles have effective anticancer and antioxidant activities in addition to their antimicrobial activity.

5.2.6 Zinc Nanoparticles (Zn NPs)

Zinc is such a vital mineral extensively dispersed all over the body tissues, and also involved in the catalyst activities of various enzymes available in living organisms (Krol et al., 2017). Zinc oxide (ZnO) has multifunctional properties, and is a biocompatible material (semiconductor) being used for the manufacturing of several products like ceramics, plastics, paints, and batteries due to antibacterial activity (Darvishi et al., 2019). Due to the substantial optical properties of ZnO NPs, they are often used as anticancer, anti-diabetic, and antibacterial drugs. ZnO nanoparticles hold antimicrobial action against gram-positive bacteria (*B. subtilis*, *S. epidermis*, *S. aureus*, *L. monocytogenes*, and *B. cereus*) and gram-negative bacteria (such as *P. aeruginosa*, *Salmonella*, *E. coli*, and *K. pneumonia* sp.) (Kaliamurthi et al., 2019) as shown in Figure 5.1 and Table 5.1.

Metallic nanoparticles are broadly used in various biomedical systems and engineering fields. Researchers have modified the most important features of Au NPs, Ag NPs, Zn NPs, and Cu NPs, frequently being exploited for their applications in medical areas and pharmaceutical industries (antibacterial, antiviral, antifungal, antiangiogenic, anticancer, and anti-inflammatory) (Zhang et al., 20106). These nanoparticles have been considered as substitute to customary antibiotics to prevail over bacterial resistance because of their well-explained antimicrobial action against gram-positive and gram-negative bacteria. The mode of action that is used by nanoparticles diverges from the traditional treatments, by means of being dynamic against those bacteria that have previously evolved resistance for antibiotics. When nanoparticles are used in humans, any risk of toxicity is ascribed to their physicochemical characteristics, route of administration, and dosage, which control their pharmacodynamics and pharmacokinetics. While these particles could possibly have a small therapeutic window, a comprehensive physicochemical categorization of their pharmaceutical progress is suggested. Moreover, during their preclinical and clinical testing, the consideration of in vivo behavior of nanoparticles is a fundamental basis of information designed for pharmaceutical progress, so that disappointments in delayed stages of research and advancements are avoided (Table 5.2).

5.3 METAL OXIDE NANOPARTICLES

Metal oxide nanoparticles (MO-NPs) are a new class of nanomaterials varying in size from 1 to 100 nm. MO-NPs exhibit distinct physical and chemical properties related to the size of their nanometers and thus have versatility (Corr, 2012; Seabra and Duran, 2015). MO-NPs have been used as antimicrobial agents (Mahapatra et al., 2008; Jones et al., 2008) and have recently gained significant attention. The antibacterial mode of action of metal and metal oxide nanoparticles is mainly due to the formation of ROS as well as damage to cell walls or cell membranes. The smaller

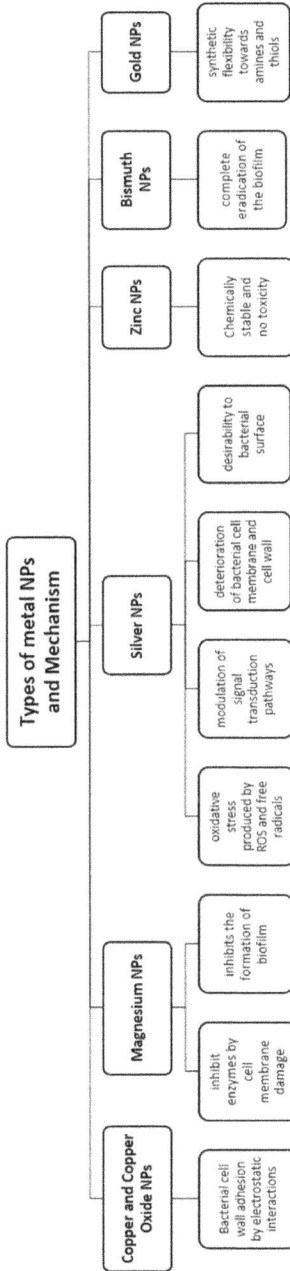

FIGURE 5.1 Description of nanoparticles along with their specific mechanism of actions.

TABLE 5.1

Efficacy of Metal Nanoparticles against Bacteria

Nanoparticle	Antibacterial Effect of Nanoparticle	Physicochemical Characteristics of the Nanoparticle	Method of Production	Remedial Action	MIC Value	Reference
Silver Based	Coliform bacteria	Monodispersed spherical Ag NPs	Green method from *Olea europaea* leaves extracts	Possess antimicrobial activity	Data not shown	Matteis et al., 2019
	Staphylococcus aureus, Bacillus subtili	Spherical shape, size 13.2 ± 2.9 nm	Bacterial-mediated Synthesis By using acidophilic actinobacterial SH11	Show antibacterial activity	*S. aureus, B. subtilis* MIC 40 μg/mL	Wypij et al., 2017
	Vibrio natriegens	Average size 10 ± 5 nm	Green method used to synthesize it using casein hydrolysate as a reducing reagent and NaOH as a catalyst	Antimicrobial activity	MIC 1.0–11.5 μg/mL	Dong et al., 2019
Copper Based	*Escherichia coli, Staphylococcus aureus*	Spherical shape and sized 7–14 nm	Mechano-chemical method using $CuSO_4.5H_2O$ and $CuCl_2.2H_2O$ as precursors	Exhibit antibacterial activity	*E. coli* MIC 3.75 mg/mL, *S. aureus* MIC 2.50 mg/mL	Moniri et al., 2019
Gold Based	*Pseudomonas aeruginosa, Staphylococcus aureus*	Size 18.32 nm	By using extract of A. comosus	Disc diffusion	MIC 4 μg/mL	Shamaila et al., 2016
	Klebsiella pneumoniae	Average size 77.13 nm	By using extract of G. elongate	Agar well diffusion	MIC 3.3 μg/mL	Shamaila et al., 2016

(Continued)

TABLE 5.1 (CONTINUED)
Efficacy of Metal Nanoparticles against Bacteria

Nanoparticle	Antibacterial Effect of Nanoparticle	Physicochemical Characteristics of the Nanoparticle	Method of Production	Remedial Action	MIC Value	Reference
Zinc Based	Escherichia coli	Spherical and hexagonal shape	By using Punica granatum fruit peels extract	Antimicrobial sensitivity testing	MIC E. coli 64.53 µg/mL	Nur et al., 2019
	Pseudomonas otitidis, Pseudomonas oleovorans, Enterococcus faecalis	Spherical shape, average size 25–45 nm	By using P.seudomonas putida broth Culture	Antibacterial activity	MIC 10 µg/mL in all bacteria	Jayabalan et al., 2019
	Staphylococcus aureus	Hexagonal shape	By using Cinnamomum tamala leaf extract	Antimicrobial susceptibility testing	MIC 40 µg/mL	Agarwal et al., 2019
	Bacillus subtilis, Escherichia coli, Klebsiella pneumoniae, Serratia marcescens	Needle-like shape, size 90–110 nm	By using Berberis aristata leaf extract	Antibacterial activity assay	MICB. subtilis, S. marcescens, 64 µg/mL.E. coli, K. pneumoniae, 256 µg/mL	Chandra et al., 2019

TABLE 5.2

Antimicrobial Effects of Metal Nanoparticles against Pathogenic Bacteria

Metal Nanoparticle	Size Range	Microorganism Tested	Mode of Action	Reference
Ag	7.1 ± 1.2 nm	E. coli and P. aeruginosa	Mechanical damage to the cell membrane and ROS production	Ramalingam et al., 201673
	2–15 nm	K. pneumoniae, A.hydrophila, Acinetobacter sp., and S. aureus	Disruption of cell membrane, DNA damage, protein degradation, and ROS generation	Aziz et al., 201578
	10 ± 5.4 nm	E. coli, S. aureus, and C. albicans	Cell membrane damage	Biao et al., 2017
	~2 nm	E. coli, S. aureus, P. aeruginosa, E. faecalis, E. cloacae, and A. azurea	Membrane damage and release of Ag+ ions	Huma et al., 2018
	5–10 nm	E. coli and S. aureus	Cell membrane damage	Zhao et al., 2017
	2–5 nm	E. coli	Perforations caused on the cell wall	Gogoi et al., 2006
	16–20 nm	E. coli and A. tumefaciens	Attachment on cell surface followed by membrane perforations	Radzig et al., 2013
	20–30 nm	B. thuringiensis, S. aureus, E. coli, and S. typhimurium	Membrane damage and ROS production	Verma et al., 2017
	10–70 nm	E. coli, S. aureus, and P. aeruginosa	ROS generation	Korshed et al., 2018
Au	7–30 nm	E. coli, B. subtilis, S. aureus, and K. pneumonia	Disrupted their respiratory systems and induced ROS generation	Shamaila et al., 2016
	(Capped with vancomycin)	E. coli, S. aureus, and A. baumannii	Binding of vancomycin to peptide layer of the bacterial cell wall	Hernández-Sierra et al., 2008
		E. coli, S. aureus, K. pneumonia, and S. typhimurium	Subtle capping of the gold nanoparticle with protein, enzymes as well as citric acid present in the cell extract plays vital role its antibacterial activity	Ahmad et al., 2013 Mishra et al., 2011

the size of the nanoparticles, the greater their surface-to-volume ratio, which would increase their interaction with components of microbial cytoplasm cells and the cell wall or cell membrane (Sotiriou and Pratsinis, 2010; Slavin et al., 2017). The antimicrobial effectiveness is determined by particle size, the aqueous medium used, and light intensity (Table 5.3). The key mechanisms are modifications or deterioration of the cell wall, and disruption of the enzyme and nucleic acid pathway (Zhu et al., 2013). Metal oxide nanoparticles cause cell damage in prokaryotic cells by changing cellular functions (Kumar et al., 2011).

5.4 CHARACTERIZATION EFFECTS ON ANTIBACTERIAL ACTIVITY OF METAL NANOPARTICLES

Shape, size, zeta potential, and core structure are important factors that determine biological effects of nanoparticles, such as cellular uptake, activation, and distribution (Helmlinger et al., 2016). It is important to consider these intrinsic and controllable properties of nanomaterials to achieve high precision downstream biological applications. In the following, we will discuss the effect of shape and characterization on antibacterial properties of nanoparticles.

5.4.1 SHAPE OF NANOPARTICLES

Shapes of certain nanoparticles play a vital role in antimicrobial activity. Different shapes of nanoparticles cause varying degrees of damage to bacterial cells by interacting with the enzymes present in the space between the inner and outer membrane of bacteria (Cha et al., 2015). A study comprising the comparison of plate-, pyramid-, and sphere-shaped zinc oxide nanoparticles reported that the combination of β-galactosidase and shape-specific zinc oxide nanoparticles produced photocatalytic activity by obstructing and restructuring the enzymes. This study showed that pyramid-shaped nanoparticles prevented the dissolution of enzymes, causing fatal damage to bacterial cells (Kumar et al., 2015). In a recent study, the highest antimicrobial activity of cube-shaped nanoparticles was attributed to their three-dimensional shape, surface charge, and components attached on the surface (Hameed et al., 2020). As three-dimensional nanoparticles have the largest contact area, it causes deformation in bacterial cells. On the other hand, gold nanospheres having zero-dimensional shape provide less contact area to interact with bacterial cells, causing lesser damage to bacterial structure. Cube-shaped silver nanoparticles displayed antimicrobial activity due to the facet reactivity and specific surface area (Actis et al., 2015). Similarly, nanocubes showed better antibacterial activity against E. faecium and E. coli when compared to nanospheres (Alshareef et al., 2017). Truncated triangular silver nanoplates displayed the strongest bactericidal potential against E. coli (Dong et al. 2012). The facets on the surface of nanoparticles changes with a change in their shape. Moreover, the density of atoms varies with variation in facets, which changes the reactivity of facets. Reactive facets amplify the affinity of nanoparticles for the cell membranes, which enhance their antibacterial activity (Rojas-Andrade et al., 2015).

TABLE 5.3

Activity of the Metal Oxide Nanoparticles (MO-NPs) against Various Pathogenic Bacteria

MO-NP	Size Range	Pathogenic Microorganism	Result	Reference
Aluminum oxide (Al$_2$O$_3$) NPs	<50 nm	Escherichia coli	Growth of E. coli reduced	Ansari et al., 2014
Antimony trioxide NPs	90–210 nm	E. coli, Bacillus subtilis, and Staphylococcus aureus	Toxic to all 3 pathogenic microorganisms	Baek and An et al., 2011
Silver-decorated titanium dioxide NPs	112 nm (TiO$_2$:Ag) and <10 nm (Ag)	MRSA and Candida spp.	Effective against both MRSA and Candida sp.	Carre et al., 2014
α-Fe$_{2-x}$Ag$_x$O$_3$ nanocrystals	37.13–41.87 nm	B. subtilis, S. aureus, P. fluorescens, and E. coli	Stopped growth of all these bacteria	Bhushan et al., 2015
Fe$_3$O$_4$-encapsulated silica sulphonic acid NPs	22 nm	S. aureus and E. coli	Antibacterial activity	Naeimi et al., 2015
Carvone-functionalized iron oxide NPs	12 nm	S. aureus and E. coli	Reduced colonization and biofilm formation	Holban et al., 2015
Silver-deposited titanium dioxide	Ag-TiO$_2$ crystallites (22–30 nm)	E. coli	Antibacterial activity	Gomathi et al., 2014
Silver oxide and ZnO NP composite	AgO (20 nm) and ZnO (50 nm)	Streptococcus mutans and Lactobacillus sp.	Antibacterial activity	Kasraei et al., 2014
Zinc-doped CuO nanocomposite	30 nm	S. aureus, MRSA and MDR E. coli	Outstanding bactericidal activity	Malka et al., 2013
Superparamagnetic iron oxide nanoparticles (SPION) with fructose metabolite	9.92 ± 3.14 nm	MRSA E. coli and P. aeruginosa	Effective against both gram-positive and gram-negative biofilms	Durmus et al., 2013

(Continued)

TABLE 5.3 (CONTINUED)
Activity of the Metal Oxide Nanoparticles (MO-NPs) against Various Pathogenic Bacteria

MO-NP	Size Range	Pathogenic Microorganism	Result	Reference
Vanadium pentoxide nanowires	300 nm long and 20 nm wide	S. aureus and E. coli	Possess strong antibacterial activity	Natalio et al., 2012
Nanosilver-decorated TiO₂ nanofibers	200–300 nm	S. aureus and E. coli	Higher bactericidal effect	Srisitthiratkul et al., 2011
Multifunctional Fe3O4@Au nanoeggs	<100 nm	S. pyogenes, A. baumannii, MRSA	Useful for killing pathogenic bacteria	Huang et al., 2009
Zinc oxide (ZnO) NPs	<10–100 nm	MRSA, methicillin-susceptible Staphylococcus aureus, P. aeruginosa, S. aureus, Streptococcus agalactiae, S. epidermidis, K. pneumonia, Salmonella choleraesuis, E. coli, B. subtilis, Salmonella paratyphi, and B. cereus	Showed good antimicrobial activity against MRSAShowed protective effects against E. papillata and S. Aureus	Jones et al., 2008Jesline et al., 2015Reddy et al., 2007Ansari et al., 2012Yousef and Danial, 2012Espitia et al., 2012Shoeb et al., 2013Shateri-Khalilabad and Yazdanshenas., 2013Reddy et al., 2014Watson et al., 2015Dkhil et al., 2015
Titanium dioxide (TiO2) NPs	<50–60 nm	MRSA, K. pneumonia	Bactericidal effect	Rezaei et al., 2010Sundaresan et al., 2012
Nickel oxide (NiO) NPs	10–30 nm	P. aeruginosa, B. subtilis, and S. aureus	Antimicrobial property	Baek and An et al., 2011Rakshit et al., 2013
Magnesium oxide (MgO)	45–70 nm	E. coli (O157:H7) and Salmonella	Killed all the tested microbes	Tang et al., 2013
Maghemite (Fe₂O₃) NPs	25–30 nm	MRSA, MRSE, E. coli, K. pneumoniae, and P. mirabilis	Antibacterial activity	Agarwala et al., 2014
Magnetite (Fe3O4) NPs	8 nm	E. coli	Possess bacteriostatic property	Chatterjee et al., 2011

(Continued)

TABLE 5.3 (CONTINUED)

Activity of the Metal Oxide Nanoparticles (MO-NPs) against Various Pathogenic Bacteria

MO-NP	Size Range	Pathogenic Microorganism	Result	Reference
Copper oxide NPs	<40–100 nm	*E. coli, P. vulgaris, S. epidermis, P. aeruginosa*, MRSA, *B. subtilis, Pseudomonas* spp., *Shigella dysenteriae, Vibrio cholera*, and *Proteus mirabilis*	Killed all bacteria tested	Baek and An et al., 2011Ren et al., 2009Sutradhar et al., 2014Kumar et al., 2015
Cobalt oxide NPs	10–25 nm	*S. aureus* and *E. coli*	Antibacterial activity	Ghosh et al., 2014
Chromium oxide NPs	65 nm 79 nm 41 nm	*E. coli* and *P. aeruginosa*	Stopped growth of tested bacteria	Rakesh et al., 2013
Cadmium oxide NPs	60 nm	Pathogenic strain of *E. coli*	Bacteriostatic effect	Rezaei et al., 2010
Calcium oxide NPs	15–180 nm	*Staphylococcus epidermidis* and *Lactobacillus plantarum*	Antimicrobial effect	Tang et al., 2013Roy et al., 2013

5.4.2 Size of Nanoparticles

The size of a nanoparticle also immensely affects its antimicrobial activity (Pan et al., 2013). Smaller nanoparticles have larger surface area, which results in higher odds of interaction with bacterial cells as compared to the large-sized nanoparticles (Gurunathan et al., 2014), as nanoparticles with a size of 200 nm showed better antibacterial activity compared to large-sized (400 nm) nanoparticles against *E. coli*. The antibacterial activity of silver nanoparticles against *Streptococcus mutants* and *E. coli* is also found to be size dependent, signifying that silver nanoparticles of 5 nm showed higher antibacterial activity as compared to 15 nm and 55 nm particles (Lu et al., 2013). However, the size of the nanomaterials may not be the only goal. Various other aspects of characterization in terms of physicochemical properties should also be considered for efficient antimicrobial activity (Pal et al., 2007).

5.4.3 Roughness of Nanoparticles

As the roughness of nanoparticles increases, the surface area-to-mass ratio and size also increase, which leads to the adsorption of bacterial proteins reducing the bacterial adhesion potential, thus is preventing their potential to cause cellular damage (Sukhorukova et al., 2015). Antibacterial activity of rough and smooth nanoparticles toward *E. coli* demonstrated by evaluating the optical density (OD) at 600 nm of a bacterial suspension concluded that the rough particles displayed higher antimicrobial activity compared to their smooth counterparts (Ahmed et al., 2013). The main reason for rough nanoparticles is to prevent adhesion of bacteria and biofilm formation. Microscale roughness has been reported as an undesirable characteristic because bacteria can form biofilm in the grooves of the biomaterial surface. However, the interaction of bacteria and nanoscale rough surfaces may reduce the adhesion of bacteria (Seil and Webster, 2012).

5.4.4 Zeta Potential

Nanoparticles have positive, negative, or neutral charge, depending on the procedure of their synthesis. As the bacterial surface is negatively charged, positively charged nanoparticles are attracted toward the surface of the bacteria, which enhance the antimicrobial potential (Sanchez-Lopez et al., 2020). Net charge on the bacterial surface is a significant parameter in its survival and alteration in it can have fatal outcomes. Bacterial adhesion is affected by the zeta potential, so its neutralization is said to be a tool used by various antimicrobial agents (Ong et al., 2019). The zeta potential indicates the extent of stability of nanostructures. Values closer to –30 mV indicate the nanoparticles are stable (Skladanowski et al., 2017). However, higher concentrations of negatively charged metal nanoparticles show antibacterial activity because of molecular crowding, which results in interactions between the nanoparticles and the bacterial surface (Arakha et al., 2015).

5.4.5 Doping Modification

Various studies have experimented with doping modifications to prevent nanoparticles from aggregating; preventing this allows nanoparticles to efficiently diffuse in an aqueous environment. Nanocomposites of ZnO and gold showed increased ROS generation. Zinc oxide nanoparticles doped with fluorine also generated more ROS than ZnO alone, causing greater damage to bacteria (Guo et al., 2015; Podporska-Carroll et al., 2017). In a study, cobalt-doped ZnO was used against bacterial strains of *Klebsiella pneumonia, Escherichia coli, Salmonella typhi, Pseudomonas aeruginosa, Shigella dysenteriae, Staphylococcus aureus*, and *Bacillus subtilis* and exhibited significant antibacterial activity (Nirmala and Anukaliani, 2011).

In conclusion, by altering the shape, size, and characterization of nanoparticles, we can achieve better and efficient antibacterial activity (Figure 5.2).

5.5 IN VIVO STUDIES OF METAL NANOPARTICLES AS ANTIBACTERIALS

Gold and silver were used to make coins and jewelry in the past due to their ability to survive the harmful effects of the microorganisms. Even before the advent of antibiotics, metals like copper and silver were used as chemotherapeutic agents against bacterial infections (Li et al., 2011; Umer et al., 2012). Nanoparticles are also used in the treatment of water to make it clear of all sorts of infectious agents (Theron et al., 2008).

Though the use of metallic nanoparticles in vivo is debatable, many studies are still ongoing. There are a few studies that covered the effects of metallic nanoparticles in the animal model. Khanal et al. (2018) conducted a study to see the in vivo effect of nanoparticles on *Klebsiella pneumoniae*, in which it was concluded that silver nanoparticles have a positive effect in controlling the pathogenic effect of bacteria in mice. The effect of green synthesized silver nanoparticles on *Acacia rigidula* was observed by Escárcega-González et al. (2018). The results confirmed that silver nanoparticles can eradicate the pathogenic ability. To confirm the toxicity of nanoparticles on vital organs, including skin, liver, and kidney, damage profiles were also checked, and it was concluded that there is no significant damage to these vital organs (Morones et al., 2005). Metallic nanoparticles have been studied as antimicrobial agents and also in combination with antibiotics as conjugates in various in vitro studies. Results are very promising and could be implemented in in vivo studies.

5.6 NANOPARTICLE- AND NANOMATERIAL-BASED VACCINES AGAINST BACTERIAL INFECTIONS

For the cure and deterrence of contagious diseases, several nanoparticle-based drug carriage devices and vaccines have been found (Fox et al., 2011; Roldão et al., 2010). The practice of nanotechnology in vaccine exploration has gradually increased (Zhao et al., 2014), as a result of nanoparticles exclusive benefits (Irvine et al., 2015;

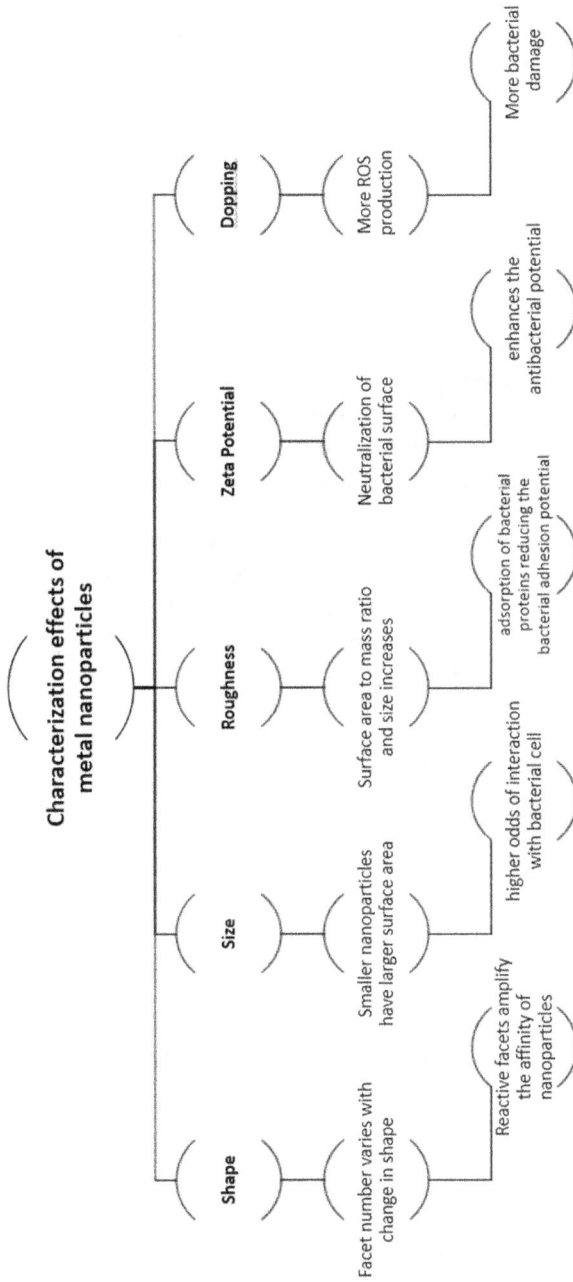

FIGURE 5.2   Summary of the effects of changing characterization on vaccine potential of nanoparticles.

Smith et al., 2015; Fang and Zhang, 2016; Fang et al., 2015). Nanoparticulates are generally susceptible to cellular uptake, and this characteristic allows antigen-presenting cells (APCs) to competently adopt and manage the nanoparticle-adjuvant antigens. Nanoparticles can be designed with capricious materials and physical characteristics as carriage approaches for antigens and/or adjuvants, all of which can be well organized to have positive impacts on the effectiveness of vaccines (Gregory et al., 2013; Mamo and Poland, 2012). Nanocarriers can protect encapsulated antigens from intimidating environments, and can target cells and tissues of concern. Due to these properties, both subunit and toxoid bacteria vaccines have been amplified by nanoparticulate preparation (Angsantikul et al., 2017).

5.6.1 Mechanisms of Action

In order to mimic natural bacteria to cause greater immune reactions, nanoparticles will simultaneously bear antigens and immune-stimulating adjuvants, such as proteins, nucleic acids, and synthetic peptides (Yoo et al., 2011). Release of supplementary permits for coordinated immune triggering of these immune determining factors, resulting in improved vaccination efficiency (Demento et al., 2012; Silva et al., 2013). Targeted nanoparticles can lead to increased aggregation in a particular type of cell, tissue, or organ, making it easier for antigen-presenting cells to accept them to stimulate more effective T cell reactions (Demento et al., 2011; Nochi et al., 2010). In addition, when exposed to external stimuli, environmental-responsive nanoparticles can alter the physical assembly and surface physiognomy in a way that helps site-specific deposition of vaccines for immune stimulation (Yoo and Mitragotri, 2010; Wilson et al., 2013). Recently, some nanoparticle organizations have been developed with advanced intraparticle structural designs that convert nanoparticles from simple payloads to complex antigen-presenting bodies that cause robust humoral responses (Moon et al., 2011; Li et al., 2013). Nanopreparations can protect payloads from enzymatic obstacles and promote upregulation. In this way, as opposed to conventional antigen–adjuvant combinations, vaccine distribution can be enormously developed by nanopreparations (Lin et al., 2018).

Conjugation of antigens to nanoparticles surfaces imitates the natural presentation of antigen bacteria and viruses (Little, 2012). Nanoparticles can also be suggested for elongated discharge of encapsulated antigens (Demento et al., 2012; Silva et al., 2013). In this way, using a variety of nanoformulations, the repetitive presentation of immunogens has been progressively applied (Nam et al., 2017; Dowling et al., 2017), as this facilitates enhanced immune system recognition, leading to increased immune activation.

Despite these attractive features, the stabilization of the exact replication of bacterial immune physiognomies and the conservation of output controllability and versatility for immune modulations is a key challenge in constructing bacterium-mimicking nanoparticles (Gregory et al., 2013; Li et al., 2014). An exceptional bacterial membrane-coated nanoparticle organization was developed and investigated as a novel platform for antibacterial vaccines in order to face this challenge (Gao et al., 2015). Since they comprise many immunogenic antigens with inherent adjuvant features, bacterial membranes are attractive vaccination resources (Poetsch and Wolters, 2008). Numerous pathogen-related molecular configurations are also shown by bacterial membranes that play an important role in initiating innate immunity (Lee et al., 2007). Due to coating of bacterial membranes on the surfaces of synthetic nanoparticles, real pathogenic bacteria are mimicked by these entities. In the meantime, synthetic nanoparticle nuclei provide the immune cells with a great variety of tunable physicochemical characteristics such as particle shape and size for successful antigen display (Hu et al., 2013; Luk et al., 2014). The bacterial membrane-coated nanoparticles are likely to produce long-lasting immune antibacterial defenses (Angsantikul et al., 2015).

5.7 TYPES OF NANOVACCINES

Collectively, antibacterial nanovaccines can be divided into two main categories: subunit vaccines and toxoid vaccines.

5.7.1 NANOPARTICLE-CONJUGATED SUBUNIT VACCINES

Only fragments of a pathogen comprise subunit vaccines. Critical antigens are isolated and purified from the pathogens rather than whole cells, to provoke critical protective events against infection. However, this often arises at the cost of weak immunogenicity, as the development of different immune-stimulating bacterial constituents and pathogen-associated molecule patterns (PAMPs) is omitted. Bacterial immune stimulators such as mono-phosphoryl lipid A (MPLA), lipopolysaccharides (LPS), and flagellin are therefore commonly required to be reinstated as subunit vaccine adjuvants in order to encourage an effective immune response to the target pathogen. Because of the deviation of the required protective immune response against different bacteria and antigenic targets, it is important to prudently observe antigen and adjuvant amalgamations to draw an effective response via the proper immune-stimulating pathway (Lin et al., 2018). In addition, based on their structure and functionality, subunit vaccines are divided into subsequent classes.

5.7.1.1 Outer Membrane Vesicles (OMVs)

Bacterial outer membrane vesicles (OMVs) are a category of materials that are increasingly regarded as a combined source of bacterial surface antigens other than polysaccharides and single peptides (Angsantikul et al., 2017). Initially recognized in gram-negative bacteria, but later also reported in mycobacteria and many gram-positive bacteria, OMVs are naturally formed proteo-liposomes (Fauci and Morens, 2012). Typically, the vesicles are 50–250 nm in size, holding LPS, phospholipids, proteins, and periplasmic components. The solid core of nanoparticles provides homogeneous tunability in size and increases the vesicle's firmness (Angsantikul et al., 2017), as they can contain additional virulence elements such as antigens, toxins, and even immune regulators that inhibit the immune reactions of the host. However, because OMVs have an exceptional similarity to bacterial surface antigens, they have been studied as a tool for vaccination. They offer the ideal small size for acceptance of APCs and effective drainage of lymph nodes. OMVs are proposed as a suitable medium for vaccine production, accompanied by possible intrinsic immune-stimulating PAMPs. Due to the existence of bacterial PAMPs, the strong immune-stimulating action of OMVs is attributed to their intrinsic pro-inflammatory feature. OMVs are responsible for activating the TLR4 signaling pathway, except for the TLR5 ligand flagellin (Koeberling et al., 2014; Fantappiè et al., 2014). For strong immune reactions, the ability to designate distinct pattern recognition receptors (PRRs) upon immunization is essential. In growth, the stimulation of pro-inflammatory cytokines and dendritic cells (DCs) offers crucial indications for the provoking of CD4+ helper T cells, a major facilitator in the coordination of both immunity weapons. Harmonized with increased tumor necrosis factor (TNF) emission-alpha, interleukin (IL)-12, and pro-inflammatory type I interferons along with upregulated costimulatory substances in DCs, multiple studies have shown evidence of OMV-triggered CD4+ T cell reactions (Rosenthal et al., 2014). The ability of OMVs to stimulate CD8+ T cell reactions to promote the death of infected cells, however, is not as much investigated and remains blurred (Pal et al., 2007).

5.7.1.2   Self-Assembled Nanomaterials

Amphipathic molecules can be self-produced in aqueous solutions into nano assemblies. Phospholipids, such as 1,2-dioleoyl-3-trimethyl-ammonium-propane chloride (DOTAP), dimethyl-dioctadecyl-ammonium bromide (DDA), and DC-cholesterol (3β-[N-(N′,N′-dimethyl-amino-ethane)-carbamyl] cholesterol hydrochloride), are commonly used to construct self-assembled liposomes (Bozzuto and Molinari, 2015). Liposomes were comprehensively analyzed for drug delivery and vaccine production with the loading ability, adjuvancy, and biocompatibility of cargo (Hume and Lua, 2017; Watson et al., 2012; Marasini et al., 2017).

5.7.1.3   Polymeric Nanoparticles

Biodegradable nontoxic and biocompatible polymers like chitosan and poly(d, l-lactic-co-glycolic acid) (PLGA) are widely favored as the cargo medium, while other resources like polyethyleneimine and polystyrene (PS) have also been analyzed. Polymers are assembled into spherical nanoparticles by practicing nanoprecipitation or emulsion-evaporation techniques, which provide the source of interest for encapsulating or conjugating payloads (Danhier et al., 2012). The nanoparticle size can vary from 10 nm to 1 μm in diameter, depending on the polymer and the surface adaptation, and the zeta potential can be negative, positive, or neutral (Gaumet et al., 2008). Since these physicochemical properties are primarily studied by immune cell recognition, polymeric nanoparticles can thus be modified to target cross-presenting phagocytic cells to primary immune comebacks (Hans and Lowman, 2002).

Either by surface association or internal encapsulation, polymeric nanoparticles can load and transport antigens. The method of antigen pairing has been shown to influence the effects of immune reactions, which is interesting. Antigen-coated nanoparticles have the potential to activate a stronger cytotoxic CD8+ T cell reaction when compared to encapsulation that favorably triggers CD4+ T cells (Lin et al., 2018).

5.7.2 TOXOID VACCINE NANOPARTICLES

Among the most general types of antivirulence vaccines are deactivated types of live bacterial toxins or toxoids. The body generates defensive measures against bacterial attacks by vaccinating against bacterial pathogenic elements, thus reducing their intrusiveness (Angsantikul et al., 2017). To remove the cytotoxic and lethal factors comprising bacterial contaminants, cell membrane-coated nanoparticles were revealed. Non-denatured and intact toxins become incapable of their free movement after interaction with nanoparticles and are arrested by cell membrane-coated nanoparticles. In order to initiate their standard virulence activities, these captive toxins are prohibited and can therefore be administered securely in vivo for effective immune procedures (Hu et al., 2013). In the manufacture of toxoid vaccines, the nanoparticle–toxin composite, called "nanotoxoid," has intense hints that can be used for the cure and prevention of many types of bacterial infestations (Hu and Zhang, 2014). It covers all the weaknesses in the production of toxoid vaccines since particle loads are administered in the nanoparticle-detainment method to seize the toxin's virulence, thus enabling intact toxins to be applied for immune response (Angsantikul et al., 2015).

The nanoparticles have been affected by their bio-interfacing abilities, which allow them to prevent bacteria from pore-forming toxins (Fang et al., 2015). Red blood cell (RBC) membrane-coated nanoparticles can be constructed by combining biological membrane vesicles of red blood cells and poly(lactic-co-glycolic acid) polymeric nanoparticle cores. By familiarizing ultrasonic energy or mechanical interference to assist membrane fusion, all these components are separately produced and then jointly assembled (Hu et al., 2011; Copp et al., 2014). Nanoparticles can also be referred to as "nanosponges" that can draw and deliver an anchoring substrate to membrane-directed toxins (Hu et al., 2013), and nanosponges can theoretically counteract the membrane-destructive action of pore-forming toxins and inhibit RBC breakdown. This emphasizes the part of the inner polymeric core that does work to relieve the covering of the erythrocyte membrane and inhibit its fusion with healthy cells (Hu et al., 2013).

Broadly, nanotoxoids have been classified into main categories: single toxin and multi-toxin.

5.7.2.1 Single-Toxin Nanotoxoids

As nanoparticle-based vaccines, nanotoxoids show numerous appealing characteristics. Nanotoxoids are constructed, unlike conventionally deactivated toxins, by loading toxins on their natural substrates in their natural condition, a development that nullifies virulence by maintaining antigenic determining factors. This nanoparticle-based hold back approach can probably be widespread to a wide variety of toxins, given the great responses produced against alpha-hemolysin. With the biocompatible existence of PLGA polymers and RBC membranes, despite the powerful antitoxin reactions caused, no noteworthy long-term or acute immune reactions are seen against the nanoparticulate vector during vaccination with nanotoxoids (Luk et al., 2016). Finally, this allows elasticity in the treatment pattern scheme, allowing repeated administration in order to further increase immunity (Angsantikul et al., 2017).

5.7.2.2 Multi-toxin Nanotoxoids

Furthermore, the strong ability of cell membrane-coated nanosponges to promote high immunogenicity and prevent toxins in a non-adjuvant manner has prompted a new approach to the production of multi-antigen vaccine preparations (Wei et al., 2017). Multiple infectious proteins, including Panton-Valentine leucocidin, alpha-hemolysin, and γ-hemolysin, have been shown to be ingested simultaneously. In addition, the nanoparticles are capable of fully removing the cytotoxic and hemolytic features of MRSA proteins through their ability to arrest and deactivate toxins. The argument that even heat-impervious toxins can be deactivated by nanoparticle-established toxin impedance indicates the methodology's versatility. Antibody titers could be produced against all three of the aforementioned toxins by consuming the arrested toxins as a multi-antigenic nanotoxoid preparation (Barnett et al., 2015). A desirable source of multi-antigenic material for the manufacture of toxoid vaccines (Shewen and Wilkie, 1988) has been shown to be the supernatant from bacterial cultures, which is possibly due to the inherent toxicity of such formulations. In addition, the occurrence of extraneous proteins may weaken the attention of the immune system (Figure 5.3).

5.8   METAL NANOPARTICLES AND ANTIBIOTIC PRODUCTS

Presently, a variety of pharmaceuticals based on nanoparticles have effectively come into the market and are being utilized by many patients on daily basis. These pharmaceutical products come from assorted companies from all over the globe and signify the current and (possibly) future accomplishment of nanomaterials as therapeutic agents. These pharmaceutical groups include nanocrystals, polymeric (with pegylated biologics, emulsions, and gels), liposome, lipid-based, and protein-based metallic NPs. But metallic nano products are the focus here. NPs based on metals have gained massive attention in several biomedical areas because of their theranostic applications, like ease of manufacturing and functionalization, along with the development of plasmonic features for their use in therapeutics and diagnostics (Khlebtsov et al., 2013).

Over the years, the number of publications based on metal nanoparticles has increased. Ag, Au, Pd, Pt, Cd, Zn, Fe, and Cu are the most important metals being studied for nanotechnology. Versatile properties of these metallic nanoparticles, i.e., large surface energies, large surface-area-to-volume ratio, transition between metallic and molecular states providing specific electronic structure, quantum confinement, plasmon excitation, short-range ordering, ability to store excess electrons, and increased number of kinks and dangling bonds, make them exhibit a variety of characteristics (Sanchez-Lopez et al., 2020). Industrialization is considered a sign of economic progress. It produces more employment opportunities, encourages innovative technological changes, increases the income of people, and aids in improving living standards of society. The utilization of material properties at the nanoscale

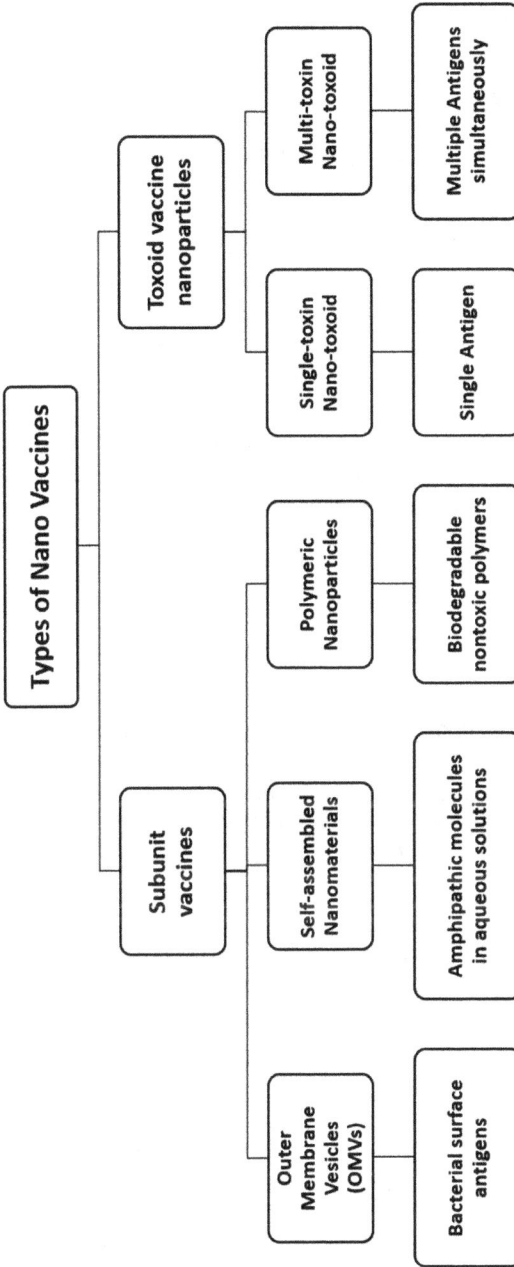

FIGURE 5.3 Types of nanovaccines and their constituents.

level has important applications in all aspects of life. Using nanotechnology in the manufacturing sector can help place atoms in a specific reaction, which can lead to more precise, improved, and refined products as compared to those produced the traditional way and thus enhancing the quality of products (Adam et al., 2019).

5.9 METAL NANOPARTICLES AND THE INDUSTRY

Nanotechnology, a multidisciplinary scientific field, is undergoing drastic development and has covered almost every aspect of industrial areas from drugs to machinery. Due to its widespread applicability, its benefit-to-risk ratio has become a topic of debate worldwide. In biomedical science and engineering, metallic nanoparticles are being extensively used. There has been a heavy increase in the market of metal-based nanoparticles over the last few years, and this is not very likely to decrease. Medical and pharmaceutical applications have exploited the important antibacterial, antifungal, antiamoebic, anti-inflammatory, antiangiogenic, and antiviral features of these metallic nanoparticles.

The nanotechnology industry plays a crucial role in strengthening the economy, as it has strong connection channels with other industries. The nanotechnology industry has promising potential, because it offers a wide range of applications in all sectors of life and for scientists, researchers, investors, and policy makers all over the world (Adam et al. 2019).

5.9.1 ROLE OF NANOTECHNOLOGY IN MEDICAL INDUSTRY

The nanotechnology pharmaceutical industry not only uses a completely different mechanism of action as compared to that of traditional antibiotics that have already failed in being effective against bacteria owing to developed resistance, but also these metallic nanoparticles target numerous biomolecules that compromise the improvement of resilient strains of bacteria (Table 5.4) (Slavin et al., 2017).

5.9.2 GLOBAL ECONOMIC IMPACT OF NANOTECHNOLOGY

In most parts of the world, the potential power of nanotechnology to improve economics is gauged in terms of job opportunities, education, national security, research activity, environmental protection, commercialization of products and processes, adoption of new technologies, improved capacity and quality of industrial sectors, and significant positive change in human life. It is very challenging to explain the measuring scales of nanotechnology, as they vary widely. We can assess the significance of research by calculating published papers and articles; however, research differs over a broad range in its importance and worth. The reserves can be assessed in terms of new employment opportunities, formation of innovative companies, decline of built-up costs, advanced technology for health, involvement of export industries, formation of improved yields and innovative facilities, lasting social change and sustainability benefits, and improved strategic areas, but very rarely is there a straight-line metric that reliably interprets public support of nanotechnology initiatives into

TABLE 5.4
Metallic Nanoparticles Conjugated with Antibiotics

Particles	Antibiotics	Bacteria	Reference
Au NPs	Gentamicin	*E. coli* K12	Burygin et al., 2009
	Vancomycin	*E. coli, VRSA, VSSA*	Fayaz et al., 2011
	Vancomycin	*E. coli, VRE*	Gu et al., 2003
	Cefaclor	*S. aureus, E. coli*	Rai et al., 2010
	Gentamicin	*S. aureus*	Ahangari et al., 2013
	Penicillin G, vancomycin, amoxicillin, methicillin, cephalexin, cefixime, amikacin, tetracycline, clindamycin, nitrofurantoin, nalidixic acid, erythromycin, gentamicin, ciprofloxacin	*P. aeruginosa, E. coli, S. aureus*	Nazari et al., 2012
	Streptomycin, gentamycin, neomycin	*S. aureus, M. luteus, E. coli, P. aeruginosa*	Grace and Pandian, 2007
	Ampicillin, streptomycin, kanamycin	*E. coli* DH5α, *M. luteus, S. aureus*	Bhattacharya et al., 2012
	Ciprofloxacin	–	Tom et al., 2004
Ag NPs	Ampicillin, penicillin G, isoniazid	*E. coli, P. aeruginosa, S. aureus, B. subtilis, C. albicans*	Tudose et al., 2015
	Ampicillin, erythromycin, tetracycline, vancomycin, chloramphenicol, gentamicin	*P. aeruginosa, S. flexneri, S. aureus, K. pneumoniae*	Gurunathan et al., 2014
	Amoxicillin	*E. coli*	Li et al., 2005
	Streptomycin, tetracycline, ampicillin,	*E. coli, S. aureus*	Kora and Rastogi, 2013

commercial commitments. Taking into consideration the preceding points, it must be figured how business models are going to evolve because of nanotechnology plans, based on nanomaterials (nanotubes, nanoparticles, nanostructured materials, and nanocomposites), nanotools (tools related to nanolithography and scanning probe microscopes), and nanodevices (nanosensors and nanoelectronics) (Santos et al., 2015). The global nanotechnology market is anticipated to surpass the US$125 billion mark by 2024. Nanotechnology continues to provide new opportunities for venture capital, and it will produce lasting social changes and economic benefits. Its impact is not confined to one sector, but it has revolutionized multiple regions of the international economy, including biomedical, energy, electronics, defense, cosmetics, agriculture, and automotive sectors, among others. Increased support from government bodies, sophisticated investors, private sector funding for corporate research, and development across wide range of industries are opening more commercialization avenues for nanotechnology industrialization. Advancement in technology, increasing call for reduction of devices, and alliances among countries are fueling growth in the inclusive nanotechnology market. However, the ecological and health implications, and potential security risks and concerns may hamper the market growth of nanotechnology commercialization (Di, 2017).

5.9.3 GLOBAL NANOTECHNOLOGY MARKET AND FORECAST TO 2024 – BY COMPONENT

Nanomaterials, by means of component, account for the maximum share of the nanotechnology worldwide market. Nanoparticles hold over 85% of the nanomaterial market globally. Nanotools have the second highest share of the nanotechnology market. Nanolithography apparatuses dominate the international nanotools component market. Nanodevices account for a small share of the nanotechnology market globally.

5.9.4 GLOBAL NANOTECHNOLOGY MARKET AND FORECAST TO 2024 – BY APPLICATION

Biomedical, electronics, and energy are the topmost three applications of nanotechnology. Together, they have an over 70% share of the international nanotechnology market. Electronics is the leading application for nanotechnology, followed by energy and then biomedical applications. Among the most enthusiastic early adopters of nanotechnology was the cosmetic industry. Automotive applications only have an approximately 5% share of the nanotechnology market globally.

Nanotechnology will continue to revolutionize many different industries. It is considered an essential enabling technology for the evolution of a wide range of applications, including healthcare, electronics, cosmetics, chemicals, energy, and composites. Nanotechnology is becoming an international footrace. It is bringing innovation and providing innovative openings for the development of day-to-day products with improved performance. It is proficient for modernizing our approaches to common challenges, reducing manufacturing costs and using fewer raw materials. Regardless of its development and direct beneficial applications, there are certain hindrances to a superior influence of nanotechnology in the industry. The deficiency of data, the probability of unintended and unfavorable impacts on surroundings, together with the human health risks and protection and sustainability of the natural ecosystem, are static challenges.

5.10 CHALLENGES TO USING METAL NANOPARTICLES AS ANTIBACTERIALS

The ascendance of nanoantibiotics has been well researched and documented, but less work has been done on their pitfalls. This is due to several hurdles, like production problems, economic issues, and negative health impacts. The advantages of nanoantibiotics are widely studied, but the disadvantages have been a neglected topic and need detailed research. A standard method of dosage of nanoparticles administration in different organs needs to be developed. Administration of nanoantibiotics may cause multiorgan nanotoxicity. Nanoparticles and their degraded products can obstruct the coagulation pathways by implicating hemolysis (Laloy et al., 2014). Other examples are silver NPs, which may cause skin and eye pigmentation, and fullerenes and nanotubes, which may cause toxicity if not administered at desired concentration. NPs, when administered through the IV route, may cause direct damage to the spleen, lung, liver, and bone marrow (Buzea et al., 2007). Similarly, due to their small size and more cell vitality, NPs may cause damage to the brain, heart, lungs, and liver when inhaled. More studies on stability, dispersion, biodistribution, the final product, and possible side effects of nanoparticles need to be carried out before putting them in the clinical trials on animal models (Manzano and Vallet, 2020). The safety profiles of the available nanoparticles from green nanotechnology need to be reviewed before using them in humans (Foko et al., 2019).

Temperature, pressure, time, and pH have an effect on shape and morphology of NPs materials to achieve/optimize specific product characteristics. A study of specific characterization techniques need to be used for better product properties and reproduction. Use of nanoparticles, especially heavy metals, cause some serious health hazards and ecological risks. These issues need to be addressed before using nanoparticles (Khan et al., 2019). Detailed preclinical safety of products and side effects, either long-term or short-term, regarding nanoparticles use need to be studied. Nanomaterials have specific toxicity databases. Nanoparticles work as nanodevices and have unique characteristics that affect cell vitality and allow better drug delivery at the cellular level unlike conventional medicines. For this purpose, NPs should be well-dispersed, but on the contrary, they form aggregation and agglomeration when they come in contact with body fluids or in cell culture media, negatively affecting their pharmacokinetics. The size and surface properties of NPs affect their biodistribution and therapeutic abilities. By working on this we can optimize drug stability by lowering their clearance and potential side effects and improving biodistribution and pharmacokinetic parameters (Limongi et al., 2019). Another area that needs further research and trials is the production of non-antibiotics at the commercial level. Different methods are used to artificially design or engineer the specific type of nanoparticles and each method has its own advantages and limitations. Here are the few methods currently being used at the commercial scale for nanomaterial formation. The top-down method is used at a larger scale and works by reducing the size of large materials into nanoparticles by chemical, mechanical, or physical methods. The most used commercial method is the bottom-up method and works based on compilation of the small atomic-sized particles to generate nanoparticles by chemical or physical processes. Another method is bottom-to-bottom, which is based on the principles of molecular and mechano-synthetic chemistry and is being developed by researchers but requires much work (Dolez, 2015). These are toxic, biocompatible, and non-ecofriendly ways due to contamination of precursor chemicals, use of toxic solvents, and production of hazardous by-products. Another method to produce NPs is the emulsion system, but it is also limited to the production of NPs soluble in organic solvents. Further studies and researches are needed to design and use cost-effective and eco-friendly production methods. Biological methods, such as green synthesis, can be an alternative approach regarding their production. Green synthesis is the production of nanoparticles from plant or herbal extracts. They are produced by the polysaccharide method, irradiation method, biological method, polyoxometalate method, and Tollens method. But these production methods have some limitations, like change in the physicochemical properties of the nanoparticles produced due to reduction in their size (Thangadurai et al., 2020). Production protocols of nanoparticles need to be standardized to get reproducible synthesis of nanoparticles like mesoporous silica nanoparticles (MSNs).

Many economic issues are also attached to the production of the nanobiotics because of lack of economical production plans. Most companies and global stakeholders are not ready to invest in the new technology, which is not yet well accepted by consumers. Furthermore, there are no national plans formed in many underdeveloped countries to combat antimicrobial resistance, and the public lacks proper awareness and guidance about it. This lowers the interest of private companies and they are more attracted toward incentives as public–private partnerships.

5.11 CONCLUSION

Indiscriminate use of antibiotics is causing serious issues in the medical field, like multidrug resistance (MDR), and due to drug residues, new diseases are emerging so it is getting tough to treat many diseases. In this vulnerable condition, the morphological and physicochemical characteristics of nanometals have also been shown to affect their antimicrobial activity. Keeping in mind their lethal potential, they have also been used as antibacterial vaccines that not only resolved the problems generated by antibiotics but are also helpful in prevention of fatal diseases. In the medical field, this approach has been trending. Research and development sectors all over the world are promoting it and surveys show its positive impact on the global economy. Although a lack of information about their toxicity and harmful effects are hurdles, they still gain attention toward resolving all the current issues in the field of medicine.

REFERENCES

Abbaszadegan, A., Ghahramani, Y., Gholami, A., Hemmateenejad, B., Dorostkar, S., Nabavizadeh, M. and Sharghi, H., 2015. The effect of charge at the surface of silver nanoparticles on antimicrobial activity against Gram-positive and Gram-negative bacteria: A preliminary study. *Journal of Nanomaterials*, 2015, p.720654.

Actis, L., Srinivasan, A., Lopez-Ribot, J.L., Ramasubramanian, A.K. and Ong, J.L., 2015. Effect of silver nanoparticle geometry on methicillin susceptible and resistant Staphylococcus aureus, and osteoblast viability. *Journal of Materials Science: Materials in Medicine*, 26(7), p.215.

Adam, H. and Youssef, A., 2019. Economic impacts of nanotechnology industry: Case study on Egypt. In: *Proceedings of 10th International Conference on Digital Strategies for Organizational Success*. (January 6, 2019).

Aderibigbe, B.A., 2017. Metal-based nanoparticles for the treatment of infectious diseases. *Molecules*, 22(8), p.1370.

Agarwal, H., Nakara, A., Menon, S. and Shanmugam, V., 2019. Eco-friendly synthesis of zinc oxide nanoparticles using Cinnamomum Tamala leaf extract and its promising effect towards the antibacterial activity. *Journal of Drug Delivery Science and Technology*, 53, pp.1773–2247.

Agarwala, M., Choudhury, B. and Yadav, R.N.S., 2014. Comparative study of antibiofilm activity of copper oxide and iron oxide nanoparticles against multidrug resistant biofilm forming uropathogens. *Indian Journal of Microbiology*, 54(3), pp.365–368.

Ahangari, A., Salouti, M., Heidari, Z., Kazemizadeh, A.R. and Safari, A.A., 2013. Development of gentamicin-gold nanospheres for antimicrobial drug delivery to Staphylococcal infected foci. *Drug Delivery*, 20(1), pp.34–39.

Ahmad, T., Wani, I.A., Manzoor, N., Ahmed, J. and Asiri, A.M., 2013. Biosynthesis, structural characterization and antimicrobial activity of gold and silver nanoparticles. *Colloids and Surfaces, Part B: Biointerfaces*, 107, pp.227–234.

Ahmed, H.A., Bhattacharyya, D.K. and Kalita, J.K., 2015. Strew index. *Network Modeling Analysis in Health Informatics and Bioinformatics*, 4(1), p.24.

Alekshun, M.N. and Levy, S.B., 2007. Molecular mechanisms of antibacterial multidrug resistance. *Cell*, 128(6), pp.1037–1050.

Alshareef, A., Laird, K. and Cross, R.B.M., 2017. Shape-dependent antibacterial activity of silver nanoparticles on Escherichia coli and Enterococcus faecium bacterium. *Applied Surface Science*, 424, pp.310–315.

Anghel, A.G., Grumezescu, A.M., Chirea, M., Grumezescu, V., Socol, G., Iordache, F., Oprea, A.E., Anghel, I. and Holban, A.M., 2014. MAPLE fabricated Fe3O4@ Cinnamomum verum antimicrobial surfaces for improved gastrostomy tubes. *Molecules*, 19(7), pp.8981–8994.

Angsantikul, P., Fang, R.H. and Zhang, L., 2017. Toxoid vaccination against bacterial infection using cell membrane-coated nanoparticles. *Bioconjugate Chemistry*, 29(3), pp.604–612.

Angsantikul, P., Thamphiwatana, S., Gao, W. and Zhang, L., 2015. Cell membrane-coated nanoparticles as an emerging antibacterial vaccine platform. *Vaccines*, 3(4), pp.814–828.

Ansari, M.A., Khan, H.M., Khan, A.A., Ahmad, M.K., Mahdi, A.A., Pal, R. and Cameotra, S.S., 2014. Interaction of silver nanoparticles with *Escherichia coli* and their cell envelope biomolecules. *Journal of Basic Microbiology*, 54(9), pp.905–915.

Ansari, M.A., Khan, H.M., Khan, A.A., Sultan, A. and Azam, A., 2012. Characterization of clinical strains of MSSA, MRSA and MRSE isolated from skin and soft tissue infections and the antibacterial activity of ZnO nanoparticles. *World Journal of Microbiology and Biotechnology*, 28(4), pp.1605–1613.

Arakha, M., Pal, S., Samantarrai, D., Panigrahi, T.K., Mallick, B.C., Pramanik, K., Mallick, B. and Jha, S., 2015. Antimicrobial activity of iron oxide nanoparticle upon modulation of nanoparticle-bacteria interface. *Scientific Reports*, 5(1), pp.1–12.

Aslam, B., Wang, W., Arshad, M. I., Khurshid, M., Muzammil, S., Rasool, M. H. and Salamat, M. K. F. Antibiotic resistance: A rundown of a global crisis. *Infection and Drug Resistance*, 11, pp.1645–1658.

Aslam, M.A., Qamar, M.U., et al, 2018. Antibiotic resistance: A rundown of a global crisis. *Infection and Drug Resistance*, 11, pp.1645–1658.

Aziz, N., Faraz, M., Pandey, R., Shakir, M., Fatma, T., Varma, A., Barman, I. and Prasad, R., 2015. Facile algae-derived route to biogenic silver nanoparticles: Synthesis, antibacterial, and photocatalytic properties. *Langmuir*, 31(42), pp.11605–11612.

Baek, Y.W. and An, Y.J., 2011. Microbial toxicity of metal oxide nanoparticles (CuO, NiO, ZnO, and Sb2O3) to *Escherichia coli*, *Bacillus subtilis*, and *Streptococcus aureus*. *Science of the Total Environment*, 409(8), pp.1603–1608.

Barnett, T.C., Cole, J.N., Rivera-Hernandez, T., Henningham, A., Paton, J.C., Nizet, V. and Walker, M.J., 2015. Streptococcal toxins: Role in pathogenesis and disease. *Cellular Microbiology*, 17(12), pp.1721–1741.

Besinis, A., De Peralta, T. and Handy, R.D., 2014. The antibacterial effects of silver, titanium dioxide and silica dioxide nanoparticles compared to the dental disinfectant chlorhexidine on *Streptococcus mutans* using a suite of bioassays. *Nanotoxicology*, 8(1), pp.1–16.

Bhattacharya, D., Saha, B., Mukherjee, A., Santra, C.R. and Karmakar, P., 2012. Gold nanoparticles conjugated antibiotics: Stability and functional evaluation. *Nanoscience and Nanotechnology*, 2(2), pp.14–21.

Bhushan, M., Muthukamalam, S., Sudharani, S. and Viswanath, A.K., 2015. Synthesis of α-Fe 2– x Ag x O 3 nanocrystals and study of their optical, magnetic and antibacterial properties. *RSC Advances*, 5(40), pp.32006–32014.

Biao, L., Tan, S., Wang, Y., Guo, X., Fu, Y., Xu, F., Zu, Y. and Liu, Z., 2017. Synthesis, characterization and antibacterial study on the chitosan-functionalized Ag nanoparticles. *Materials Science and Engineering: Part C*, 76, pp.73–80.

Bozzuto, G. and Molinari, A., 2015. Liposomes as nanomedical devices. *International Journal of Nanomedicine*, 10, p.975.

Burygin, G.L., Khlebtsov, B.N., Shantrokha, A.N., Dykman, L.A., Bogatyrev, V.A. and Khlebtsov, N.G., 2009. On the enhanced antibacterial activity of antibiotics mixed with gold nanoparticles. *Nanoscale Research Letters*, 4(8), pp.794–801.

Buzea, C., Pacheco, I.I. and Robbie, K., 2007. Nanomaterials and nanoparticles: Sources and toxicity. *Biointerphases*, 2(4), pp.17–71.

Carré, G., Hamon, E., Ennahar, S., Estner, M., Lett, M.C., Horvatovich, P., Gies, J.P., Keller, V., Keller, N. and Andre, P., 2014. TiO2 photocatalysis damages lipids and proteins in Escherichia coli. *Applied and Environmental Microbiology*, 80(8), pp.2573–2581.

Cha, S.H., Hong, J., McGuffie, M., Yeom, B., VanEpps, J.S. and Kotov, N.A., 2015. Shape-dependent biomimetic inhibition of enzyme by nanoparticles and their antibacterial activity. *ACS Nano*, 9(9), pp.9097–9105.

Chandra, H., Patel, D., Kumari, P., Jangwan, J.S. and Yadav, S., 2019. Phyto-mediated synthesis of zinc oxide nanoparticles of *Berberis aristata*: Characterization, antioxidant activity and antibacterial activity with special reference to urinary tract pathogens. *Materials Science and Engineering. Part C*, 102, pp.212–220.

Chatterjee, S., Bandyopadhyay, A. and Sarkar, K., 2011. Effect of iron oxide and gold nanoparticles on bacterial growth leading towards biological application. *Journal of Nanobiotechnology*, 9(1), p.34.

Chen, H.W., Huang, C.Y., Lin, S.Y., Fang, Z.S., Hsu, C.H., Lin, J.C., Chen, Y.I., Yao, B.Y. and Hu, C.M.J., 2016. Synthetic virus-like particles prepared via protein corona formation enable effective vaccination in an avian model of coronavirus infection. *Biomaterials*, 106, pp.111–118.

Chen, X. and Schluesener, H.J., 2008. Nanosilver: A nanoproduct in medical application. *Toxicology Letters*, 176(1), pp.1–12.

Chen, Y.S., Hung, Y.C., Lin, W.H. and Huang, G.S., 2010. Assessment of gold nanoparticles as a size-dependent vaccine carrier for enhancing the antibody response against synthetic foot-and-mouth disease virus peptide. *Nanotechnology*, 21(19), p.195101.

Chiriac, V., Stratulat, D.N., Calin, G., Nichitus, S., Burlui, V., Stadoleanu, C., Popa, M. and Popa, I.M., 2016. Antimicrobial property of zinc based nanoparticles. In *IOP Conference Series: Materials Science and Engineering* (Vol. 133, No. 1, p. 012055). IOP Publishing.

Coates, A.R., Halls, G. and Hu, Y., 2011. Novel classes of antibiotics or more of the same? *British Journal of Pharmacology*, 163(1), pp.184–194.

Copp, J.A., Fang, R.H., Luk, B.T., Hu, C.M.J., Gao, W., Zhang, K. and Zhang, L., 2014. Clearance of pathological antibodies using biomimetic nanoparticles. *Proceedings of the National Academy of Sciences of the United States of America*, 111(37), pp.13481–13486.

Corr, S.A., 2012. Metal oxide nanoparticles. *Nanoscience*, 1, pp.180–207.

Dakal, T.C., Kumar, A., Majumdar, R.S. and Yadav, V., 2016. *Mechanistic Basis of Antimicrobial Actions of Silver. Frontiers in Microbiology*, 7, pp.1831.

Danhier, F., Ansorena, E., Silva, J.M., Coco, R., Le Breton, A. and Préat, V., 2012. PLGA-based nanoparticles: An overview of biomedical applications. *Journal of Controlled Release*, 161(2), pp.505–522.

Darvishi, E., Kahrizi, D. and Arkan, E., 2019. Comparison of different properties of zinc oxide nanoparticles synthesized by the green (using Juglans regia L. leaf extract) and chemical methods. *Journal of Molecular Liquids*, 286, p.110831.

De Matteis, V., Rizzello, L., Ingrosso, C., Liatsi-Douvitsa, E., De Giorgi, M.L., De Matteis, G. and Rinaldi, R., 2019. Cultivar-dependent anticancer and antibacterial properties of silver nanoparticles synthesized using leaves of different Olea europaea trees. *Nanomaterials*, 9(11), p.1544.

Demento, S.L., Cui, W., Criscione, J.M., Stern, E., Tulipan, J., Kaech, S.M. and Fahmy, T.M., 2012. Role of sustained antigen release from nanoparticle vaccines in shaping the T cell memory phenotype. *Biomaterials*, 33(19), pp.4957–4964.

Demento, S.L., Siefert, A.L., Bandyopadhyay, A., Sharp, F.A. and Fahmy, T.M., 2011. Pathogen-associated molecular patterns on biomaterials: A paradigm for engineering new vaccines. *Trends in Biotechnology*, 29(6), pp.294–306.

Di Sia, P., 2017. Nanotechnology among innovation, health and risks. *Procedia-Social and Behavioral Sciences*, 237, pp. 1076–1080.

Dizaj, S.M., Lotfipour, F., Barzegar-Jalali, M., Zarrintan, M.H. and Adibkia, K., 2014. Antimicrobial activity of the metals and metal oxide nanoparticles. *Materials Science and Engineering: Part C*, 44, pp.278–284.

Dkhil, M.A., Al-Quraishy, S. and Wahab, R., 2015. Anticoccidial and antioxidant activities of zinc oxide nanoparticles on *Eimeria papillata*-induced infection in the jejunum. *International Journal of Nanomedicine*, 10, p.1961.

Dolez, P.I. ed., 2015. *Nanoengineering: Global Approaches to Health and Safety Issues*. Elsevier.

Dong, Y., Zhu, H., Shen, Y., Zhang, W. and Zhang, L., 2019. Antibacterial activity of silver nanoparticles of different particle size against Vibrio natriegens. *PLOS ONE*, 14(9), p.e0222322.

Dowling, D.J., Scott, E.A., Scheid, A., Bergelson, I., Joshi, S., Pietrasanta, C., Brightman, S., Sanchez-Schmitz, G., Van Haren, S.D., Ninković, J. and Kats, D., 2017. Toll-like receptor 8 agonist nanoparticles mimic immunomodulating effects of the live BCG vaccine and enhance neonatal innate and adaptive immune responses. *Journal of Allergy and Clinical Immunology*, 140(5), pp.1339–1350.

Dugassa, J. and Shukuri, N., 2017. Antibiotic resistance and its mechanism of Development. *Journal of Health, Medicine and Nursing*, 1, pp.1–17.

Durmus, N.G., Taylor, E.N., Kummer, K.M. and Webster, T.J., 2013. Enhanced efficacy of superparamagnetic iron oxide nanoparticles against antibiotic-resistant biofilms in the presence of metabolites. *Advanced Materials*, 25(40), pp.5706–5713.

Dykman, L.A. and Khlebtsov, N.G., 2017. Immunological properties of gold nanoparticles. *Chemical Science*, 8(3), pp.1719–1735.

El-Refai, A.A., Ghoniem, G.A., El-Khateeb, A.Y. and Hassaan, M.M., 2018. Eco-friendly synthesis of metal nanoparticles using ginger and garlic extracts as biocompatible novel antioxidant and antimicrobial agents. *Journal of Nanostructure in Chemistry*, 8(1), pp.71–81.

Escárcega-González, C.E., Garza-Cervantes, J.A., Vazquez-Rodríguez, A., Montelongo-Peralta, L.Z., Treviño-Gonzalez, M.T., Castro, E.D.B., Saucedo-Salazar, E.M., Morales, R.C., Soto, D.R., and González, F.T., 2018. In vivo antimicrobial activity of silver nanoparticles produced via a green chemistry synthesis using *Acacia rigidula* as a reducing and capping agent. *International Journal of Nanomedicine*, 13, p.2349.

Espitia, P.J.P., Soares, N.D.F.F., dos Reis Coimbra, J.S., de Andrade, N.J., Cruz, R.S. and Medeiros, E.A.A., 2012. Zinc oxide nanoparticles: Synthesis, antimicrobial activity and food packaging applications. *Food and Bioprocess Technology*, 5(5), pp.1447–1464.

Fang, R.H. and Zhang, L., 2016. Nanoparticle-based modulation of the immune system. *Annual Review of Chemical and Biomolecular Engineering*, 7, pp.305–326.

Fang, R.H., Kroll, A.V. and Zhang, L., 2015. Nanoparticle-based manipulation of antigen-presenting cells for cancer immunotherapy. *Small*, 11(41), pp.5483–5496.

Fang, R.H., Luk, B.T., Hu, C.M.J. and Zhang, L., 2015. Engineered nanoparticles mimicking cell membranes for toxin neutralization. *Advanced Drug Delivery Reviews*, 90, pp.69–80.

Fantappiè, L., De Santis, M., Chiarot, E., Carboni, F., Bensi, G., Jousson, O., Margarit, I. and Grandi, G., 2014. Antibody-mediated immunity induced by engineered *Escherichia coli* OMVs carrying heterologous antigens in their lumen. *Journal of Extracellular Vesicles*, 3(1), p.24015.

Fauci, A.S. and Morens, D.M., 2012. The perpetual challenge of infectious diseases. *New England Journal of Medicine*, 366(5), pp.454–461.

Fayaz, A.M., Girilal, M., Mahdy, S.A., Somsundar, S.S., Venkatesan, R. and Kalaichelvan, P.T., 2011. Vancomycin bound biogenic gold nanoparticles: A different perspective for development of anti VRSA agents. *Process Biochemistry*, 46(3), pp.636–641.

Fellahi, O., Sarma, R.K., Das, M.R., Saikia, R., Marcon, L., Coffinier, Y., Hadjersi, T., Maamache, M. and Boukherroub, R., 2013. The antimicrobial effect of silicon nanowires decorated with silver and copper nanoparticles. *Nanotechnology*, 24(49), p.495101.

Foko, L.P.K., Meva, F.E.A., Moukoko, C.E.E., Ntoumba, A.A., Njila, M.I.N., Kedi, P.B.E., Ayong, L. and Lehman, L.G., 2019. A systematic review on anti-malarial drug discovery and antiplasmodial potential of green synthesis mediated metal nanoparticles: Overview, challenges and future perspectives. *Malaria Journal*, 18(1), p.337.

Fox, C.B., Baldwin, S.L., Duthie, M.S., Reed, S.G. and Vedvick, T.S., 2011. Immunomodulatory and physical effects of oil composition in vaccine adjuvant emulsions. *Vaccine*, 29(51), pp.9563–9572.

Gao, W., Fang, R.H., Thamphiwatana, S., Luk, B.T., Li, J., Angsantikul, P., Zhang, Q., Hu, C.M.J. and Zhang, L., 2015. Modulating antibacterial immunity via bacterial membrane-coated nanoparticles. *Nano Letters*, 15(2), pp.1403–1409.

Gaumet, M., Vargas, A., Gurny, R. and Delie, F., 2008. Nanoparticles for drug delivery: The need for precision in reporting particle size parameters. *European Journal of Pharmaceutics and Biopharmaceutics*, 69(1), pp.1–9.

Gerber, A., Bundschuh, M., Klingelhofer, D. and Groneberg, D.A., 2013. Gold nanoparticles: Recent aspects for human toxicology. *Journal of Occupational Medicine and Toxicology*, 8(1), p.32.

Gharpure, S., Akash, A. and Ankamwar, B., 2020. A review on antimicrobial properties of metal nanoparticles. *Journal of Nanoscience and Nanotechnology*, 20(6), pp.3303–3339.

Ghosh, S., Patil, S., Ahire, M., Kitture, R., Kale, S., Pardesi, K., Cameotra, S.S., Bellare, J., Dhavale, D.D., Jabgunde, A. and Chopade, B.A., 2012. Synthesis of silver nanoparticles using *Dioscorea bulbifera* tuber extract and evaluation of its synergistic potential in combination with antimicrobial agents. *International Journal of Nanomedicine*, 7, p.483.

Ghosh, T., Dash, S.K., Chakraborty, P., Guha, A., Kawaguchi, K., Roy, S., Chattopadhyay, T. and Das, D., 2014. *Preparation of Antiferromagnetic Co₃O₄ Nanoparticles from Two Different Precursors by Pyrolytic Method: In Vitro Antimicrobial Activity.* RSC advances 2014 v.4 no.29 pp. 15022-15029.

Gogoi, S.K., Gopinath, P., Paul, A., Ramesh, A., Ghosh, S.S. and Chattopadhyay, A., 2006. Green fluorescent protein-expressing *Escherichia coli* as a model system for investigating the antimicrobial activities of silver nanoparticles. *Langmuir*, 22(22), pp.9322–9328.

Gomathi Devi, L. and Nagaraj, B., 2014. Disinfection of *Escherichia coli* Gram negative Bacteria Using Surface Modified TiO2: Optimization of Ag Metallization and Depiction of Charge Transfer Mechanism. *Photochemistry and Photobiology*, 90(5), pp.1089–1098.

Gordon, T., Perlstein, B., Houbara, O., Felner, I., Banin, E. and Margel, S., 2011. Synthesis and characterization of zinc/iron oxide composite nanoparticles and their antibacterial properties. *Colloids and Surfaces A: Physicochemical and Engineering Aspects*, 374(1–3), pp.1–8.

Grace, A.N. and Pandian, K., 2007. Antibacterial efficacy of aminoglycosidic antibiotics protected gold nanoparticles—A brief study. *Colloids and Surfaces A: Physicochemical and Engineering Aspects*, 297(1–3), pp.63–70.

Gregory, A.E., Williamson, D. and Titball, R., 2013. Vaccine delivery using nanoparticles. *Frontiers in Cellular and Infection Microbiology*, 3, p.13.

Gu, H., Ho, P.L., Tong, E., Wang, L. and Xu, B., 2003. Presenting vancomycin on nanoparticles to enhance antimicrobial activities. *Nano Letters*, 3(9), pp.1261–1263.

Guo, B.L., Han, P., Guo, L.C., Cao, Y.Q., Li, A.D., Kong, J.Z., Zhai, H.F. and Wu, D., 2015. The antibacterial activity of Ta-doped ZnO nanoparticles. *Nanoscale Research Letters*, 10(1), p.336.

Gurunathan, S., Han, J.W., Kwon, D.N. and Kim, J.H., 2014. Enhanced antibacterial and anti-biofilm activities of silver nanoparticles against Gram-negative and Gram-positive bacteria. *Nanoscale Research Letters*, 9(1), pp.1–17.

Hamal, D.B., Haggstrom, J.A., Marchin, G.L., Ikenberry, M.A., Hohn, K. and Klabunde, K.J., 2010. A multifunctional biocide/sporocide and photocatalyst based on titanium dioxide (TiO2) codoped with silver, carbon, and sulfur. *Langmuir*, 26(4), pp.2805–2810.

Hameed, A.S.H., Karthikeyan, C., Sasikumar, S., Kumar, V.S., Kumaresan, S. and Ravi, G., 2013. Impact of alkaline metal ions Mg 2+, Ca 2+, Sr 2+ and Ba 2+ on the structural, optical, thermal and antibacterial properties of ZnO nanoparticles prepared by the co-precipitation method. *Journal of Materials Chemistry B*, 1(43), pp.5950–5962.

Hameed, S., Wang, Y., Zhao, L., Xie, L., and Ying, Y. 2020. Shape-dependent significant physical mutilation and antibacterial mechanisms of gold nanoparticles against foodborne bacterial pathogens (Escherichia coli, Pseudomonas aeruginosa and Staphylococcus aureus) at lower concentrations. *Materials Science and Engineering: C*, 108, 110338.

Hans, M.L. and Lowman, A.M., 2002. Biodegradable nanoparticles for drug delivery and targeting. *Current Opinion in Solid State and Materials Science*, 6(4), pp.319–327.

Helmlinger, J., Sengstock, C., Groß-Heitfeld, C., Mayer, C., Schildhauer, T.A., Köller, M. and Epple, M., 2016. Silver nanoparticles with different size and shape: Equal cytotoxicity, but different antibacterial effects. *RSC Advances*, 6(22), pp.18490–18501.

Her, S., Jaffray, D.A. and Allen, C., 2017. Gold nanoparticles for applications in cancer radiotherapy: Mechanisms and recent advancements. *Advanced Drug Delivery Reviews*, 109, pp.84–101.

Hernández-Sierra, J.F., Ruiz, F., Pena, D.C.C., Martínez-Gutiérrez, F., Martínez, A.E., Guillén, A.D.J.P., Tapia-Pérez, H. and Castañón, G.M., 2008. The antimicrobial sensitivity of *Streptococcus mutans* to nanoparticles of silver, zinc oxide, and gold. *Nanomedicine: Nanotechnology, Biology and Medicine*, 4(3), pp.237–240.

Holban, A.M., Andronescu, E., Grumezescu, V., Oprea, A.E., Grumezescu, A.M., Socol, G., Chifiriuc, M.C., Lazar, V. and Iordache, F., 2015. Carvone functionalized iron oxide nanostructures thin films prepared by MAPLE for improved resistance to microbial colonization. *Journal of Sol-Gel Science and Technology*, 73(3), pp.605–611.

Hossain, K.M.Z., Patel, U. and Ahmed, I., 2015. Development of microspheres for biomedical applications: A review. *Progress in Biomaterials*, 4(1), pp.1–19.

Hsueh, Y.-H., Lin, K.-S., Ke, W.-J., Hsieh, C.-T., Chiang, C.-L., Tzou, D.-Y. and Liu, S.-T., 2015. The antimicrobial properties of silver nanoparticles in Bacillus subtilis are mediated by released Ag+ ions. *PLOS ONE*, 10(12), p.e0144306.

Hu, C.M.J. and Zhang, L., 2014. Nanotoxoid vaccines. *Nano Today*, 9(4), pp.401–404.

Hu, C.M.J., Fang, R.H., Copp, J., Luk, B.T. and Zhang, L., 2013. A biomimetic nanosponge that absorbs pore-forming toxins. *Nature Nanotechnology*, 8(5), pp.336–340.

Hu, C.M.J., Fang, R.H., Luk, B.T. and Zhang, L., 2013. Nanoparticle-detained toxins for safe and effective vaccination. *Nature Nanotechnology*, 8(12), pp.933–938.

Hu, C.M.J., Fang, R.H., Luk, B.T., Chen, K.N., Carpenter, C., Gao, W., Zhang, K. and Zhang, L., 2013. 'Marker-of-self' functionalization of nanoscale particles through a top-down cellular membrane coating approach. *Nanoscale*, 5(7), pp.2664–2668.

Hu, C.M.J., Zhang, L., Aryal, S., Cheung, C., Fang, R.H. and Zhang, L., 2011. Erythrocyte membrane-camouflaged polymeric nanoparticles as a biomimetic delivery platform. *Proceedings of the National Academy of Sciences of the United States of America*, 108(27), pp.10980–10985.

Huang, W.C., Tsai, P.J. and Chen, Y.C., 2009. Multifunctional Fe3O4@ Au nanoeggs as photothermal agents for selective killing of nosocomial and antibiotic-resistant bacteria. *Small*, 5(1), pp.51–56.

Huma, Z.E., Gupta, A., Javed, I., Das, R., Hussain, S.Z., Mumtaz, S., Hussain, I. and Rotello, V.M., 2018. Cationic silver nanoclusters as potent antimicrobials against multidrug-resistant bacteria. *ACS Omega*, 3(12), pp.16721–16727.

Hume, H.K.C. and Lua, L.H., 2017. Platform technologies for modern vaccine manufacturing. *Vaccine*, 35(35), pp.4480–4485.

IACG, 2016. No time to wait: Infections from drug-resistant securing the future. *Artforum International*, 54, pp.113–114.

Irvine, D.J., Hanson, M.C., Rakhra, K. and Tokatlian, T., 2015. Synthetic nanoparticles for vaccines and immunotherapy. *Chemical Reviews*, 115(19), pp.11109–11146.

Jayabalan, J., Mani, G., Krishnan, N., Pernabas, J., Milton, J. and Tae, H., 2019. Green biogenic synthesis of zinc oxide nanoparticles using *Pseudomonas putida* culture and its in vitro antibacterial and anti-biofilm activity. *Biocatalysis and Agricultural Biotechnology*, 21, pp.1–9.

Jesline, A., John, N.P., Narayanan, P.M., Vani, C. and Murugan, S., 2015. Antimicrobial activity of zinc and titanium dioxide nanoparticles against biofilm-producing methicillin-resistant *Staphylococcus aureus*. *Applied Nanoscience*, 5(2), pp.157–162.

Jones, N., Ray, B., Ranjit, K.T. and Manna, A.C., 2008. Antibacterial activity of ZnO nanoparticle suspensions on a broad spectrum of microorganisms. *FEMS Microbiology Letters*, 279(1), pp.71–76.

Kaliamurthi, S., Selvaraj, G., Elibol, Z., Demir, A. and Cakmak, T., 2019. The relationship between *Chlorella* sp. and zinc oxide nanoparticles: Changes in biochemical, oxygen evolution, and lipid production ability. *Process Biochemistry*, 85, pp.43–50.

Kasraei, S., Sami, L., Hendi, S., AliKhani, M.Y., Rezaei-Soufi, L. and Khamverdi, Z., 2014. Antibacterial properties of composite resins incorporating silver and zinc oxide nanoparticles on *Streptococcus mutans* and *Lactobacillus*. *Restorative Dentistry and Endodontics*, 39(2), pp.109–114.

Khan, I., Saeed, K. and Khan, I., 2019. Nanoparticles: Properties, applications and toxicities. *Arabian Journal of Chemistry*, 12(7), pp.908–931.

Khanal, R., Dahal, S., Aryal, A., Tamang, M., Ranjit, S. and Sudeep, K., 2018. Study of in vitro and in vivo antibacterial effects of silver nanoparticles. *International Research Journal of Engineering and Technology*, 05, pp.1143–1147.

Khlebtsov, N., Bogatyrev, V., Dykman, L., Khlebtsov, B., Staroverov, S., Shirokov, A., Matora, L., Khanadeev, V., Pylaev, T., Tsyganova, N. and Terentyuk, G., 2013. Analytical and theranostic applications of gold nanoparticles and multifunctional nanocomposites. *Theranostics*, 3(3), pp.167–180.

Koeberling, O., Ispasanie, E., Hauser, J., Rossi, O., Pluschke, G., Caugant, D.A., Saul, A. and MacLennan, C.A., 2014. A broadly-protective vaccine against meningococcal disease in sub-Saharan Africa based on generalized modules for membrane antigens (GMMA). *Vaccine*, 32(23), pp.2688–2695.

Kora, A.J. and Rastogi, L., 2013. Enhancement of antibacterial activity of capped silver nanoparticles in combination with antibiotics, on model gram-negative and gram-positive bacteria. *Bioinorganic Chemistry and Applications*, 2013: 7.

Korshed, P., Li, L., Liu, Z. and Wang, T., 2018. Correction: The molecular mechanisms of the antibacterial effect of picosecond laser generated silver nanoparticles and their toxicity to human cells. *PLOS ONE*, 13(8), p.e0203636.

Krol, A., Pomastowski, P., Rafińska, K., Railean-Plugaru, V. and Buszewski, B., 2017. Zinc oxide nanoparticles: Synthesis, antiseptic activity and toxicity mechanism. *Advances in Colloid and Interface Science*, 249, pp.37–52.

Kumar, A., Pandey, A.K., Singh, S.S., Shanker, R. and Dhawan, A., 2011. Cellular response to metal oxide nanoparticles in bacteria. *Journal of Biomedical Nanotechnology*, 7(1), pp.102–103.

Kumar, A., Pandey, A.K., Singh, S.S., Shanker, R. and Dhawan, A., 2011. Engineered ZnO and TiO2 nanoparticles induce oxidative stress and DNA damage leading to reduced viability of *Escherichia coli*. *Free Radical Biology and Medicine*, 51(10), pp.1872–1881.

Kumar, P.V., Shameem, U., Kollu, P., Kalyani, R.L. and Pammi, S.V.N., 2015. Green synthesis of copper oxide nanoparticles using Aloe vera leaf extract and its antibacterial activity against fish bacterial pathogens. *BioNanoScience*, 5(3), pp.135–139.

Kumar, S.V., Bafana, A.P., Pawar, P., Faltane, M., Rahman, A., Dahoumane, S.A., Kucknoor, A. and Jeffreys, C.S. 2019. Optimized production of antibacterial copper oxide nanoparticles in a microwave-assisted synthesis reaction using response surface methodology. *Colloids and Surfaces A: Physicochemical and Engineering Aspects*, 573, pp.170–178.

Laloy, J., Minet, V., Alpan, L., Mullier, F., Beken, S., Toussaint, O., Lucas, S. and Dogné, J.M., 2014. Impact of silver nanoparticles on haemolysis, platelet function and coagulation. *Nanobiomedicine*, 1, p.4.

Lee, E.Y., Bang, J.Y., Park, G.W., Choi, D.S., Kang, J.S., Kim, H.J., Park, K.S., Lee, J.O., Kim, Y.K., Kwon, K.H. and Kim, K.P., 2007. Global proteomic profiling of native outer membrane vesicles derived from *Escherichia coli*. *Proteomics*, 7(17), pp.3143–3153.

Lemire, J.A., Harrison, J.J. and Turner, R.J., 2013. Antimicrobial activity of metals: Mechanisms, molecular targets and applications. *Nature Reviews in Microbiology*, 11(6), pp.371–384.

Li, G., Li, X. and Zhang, Z., 2011. Preparation methods of copper nanomaterials. *Progress in Chemistry*, 23, p.1644.

Li, A.V., Moon, J.J., Abraham, W., Suh, H., Elkhader, J., Seidman, M.A., Yen, M., Im, E.J., Foley, M.H., Barouch, D.H. and Irvine, D.J., 2013. Generation of effector memory T cell–based mucosal and systemic immunity with pulmonary nanoparticle vaccination. *Science Translational Medicine*, 5(204), pp.204ra130–204ra130.

Li, L.L., Xu, J.H., Qi, G.B., Zhao, X., Yu, F. and Wang, H., 2014. Core–shell supramolecular gelatin nanoparticles for adaptive and "on-demand" antibiotic delivery. *ACS Nano*, 8(5), pp.4975–4983.

Li, P., Li, J., Wu, C., Wu, Q. and Li, J., 2005. Synergistic antibacterial effects of β-lactam antibiotic combined with silver nanoparticles. *Nanotechnology*, 16(9), p.1912.

Limongi, T., Canta, M., Racca, L., Ancona, A., Tritta, S., Vighetto, V. and Cauda, V., 2019. Improving dispersal of therapeutic nanoparticles in the human body. *Nanomedicine*, 14(7): 797–801.

Lin, L.C.W., Chattopadhyay, S., Lin, J.C. and Hu, C.M.J., 2018. Advances and opportunities in nanoparticle-and nanomaterial-based vaccines against bacterial infections. *Advanced Healthcare Materials*, 7(13), p.1701395.

Little, S.R., 2012. Reorienting our view of particle-based adjuvants for subunit vaccines. *Proceedings of the National Academy of Sciences of the United States of America*, 109(4), pp.999–1000.

Lu, Z., Rong, K., Li, J., Yang, H. and Chen, R., 2013. Size-dependent antibacterial activities of silver nanoparticles against oral anaerobic pathogenic bacteria. *Journal of Materials Science: Materials in Medicine*, 24(6), pp.1465–1471.

Luk, B.T., Fang, R.H., Hu, C.M.J., Copp, J.A., Thamphiwatana, S., Dehaini, D., Gao, W., Zhang, K., Li, S. and Zhang, L., 2016. Safe and immunocompatible nanocarriers cloaked in RBC membranes for drug delivery to treat solid tumors. *Theranostics*, 6(7), p.1004.

Luk, B.T., Hu, C.M.J., Fang, R.H., Dehaini, D., Carpenter, C., Gao, W. and Zhang, L., 2014. Interfacial interactions between natural RBC membranes and synthetic polymeric nanoparticles. *Nanoscale*, 6(5), pp.2730–2737.

Mahapatra, O., Bhagat, M., Gopalakrishnan, C. and Arunachalam, K.D., 2008. Ultrafine dispersed CuO nanoparticles and their antibacterial activity. *Journal of Experimental Nanoscience*, 3(3), pp.185–193.

Malka, E., Perelshtein, I., Lipovsky, A., Shalom, Y., Naparstek, L., Perkas, N., Patick, T., Lubart, R., Nitzan, Y., Banin, E. and Gedanken, A., 2013. Eradication of multi-drug resistant bacteria by a novel Zn-doped CuO nanocomposite. *Small*, 9(23), pp.4069–4076.

Mamo, T. and Poland, G.A., 2012. Nanovaccinology: The next generation of vaccines meets 21st century materials science and engineering. *Vaccine*, 30(47), p.6609.

Manzano, M. and Vallet-Regí, M., 2020. Mesoporous silica nanoparticles for drug delivery. *Advanced Functional Materials*, 30(2), p.1902634.

Marasini, N., Ghaffar, K.A., Skwarczynski, M. and Toth, I., 2017. Liposomes as a vaccine delivery system. In *Micro and Nanotechnology in Vaccine Development* (pp.221–239). William Andrew Publishing.

Market, G.N., 2018. *Funding & Investment, Patent Analysis and 27 Companies Profile & Recent Developments–Forecast to 2024. Global.* Available at:——Economics and Management.

Mirhosseini, M., 2015. Synergistic antibacterial effect of metal oxide nanoparticles and ultrasound stimulation. *Journal of Biology and Today's World*, 4, pp.138–144.

Mishra, A., Tripathy, S.K. and Yun, S.I., 2011. Bio-Synthesis of gold and silver nanoparticles from *Candida guilliermondii* and their antimicrobial effect against pathogenic bacteria. *Journal of Nanoscience and Nanotechnology*, 11(1), pp.243–248.

Mohammadi, G., Valizadeh, H., Barzegar-Jalali, M., Lotfipour, F., Adibkia, K., Milani, M., Azhdarzadeh, M., Kiafar, F. and Nokhodchi, A., 2010. Development of azithromycin–PLGA nanoparticles: Physicochemical characterization and antibacterial effect against Salmonella typhi. *Colloids and Surfaces, Part B: Biointerfaces*, 80(1), pp.34–39.

Moniri Javadhesari, S., Alipour, S., Mohammadnejad, S. and Akbarpour, M.R., 2019. Antibacterial activity of ultra-small copper oxide (II) nanoparticles synthesized by mechanochemical processing against *S. aureus* and *E. coli*. *Materials Science and Engineering. Part C*, 105, p.110011.

Moon, J.J., Suh, H., Bershteyn, A., Stephan, M.T., Liu, H., Huang, B., Sohail, M., Luo, S., Um, S.H., Khant, H. and Goodwin, J.T., 2011. Interbilayer-crosslinked multilamellar vesicles as synthetic vaccines for potent humoral and cellular immune responses. *Nature Materials*, 10(3), pp.243–251.

Moon, J.J., Suh, H., Li, A.V., Ockenhouse, C.F., Yadava, A. and Irvine, D.J., 2012. Enhancing humoral responses to a malaria antigen with nanoparticle vaccines that expand Tfh cells and promote germinal center induction. *Proceedings of the National Academy of Sciences of the United States of America*, 109(4), pp.1080–1085.

Morones, J.R., Elechiguerra, J.L., Camacho, A., Holt, K., Kouri, J.B., Ramírez, J.T. and Yacaman, M.J., 2005. The bactericidal effect of silver nanoparticles. *Nanotechnology*, 16(10), pp.2346–2353.

Munita, J.M. and Arias, C.A., 2016. Mechanisms of antibiotic resistance. *Microbiology Spectrum*, 2(2), pp.1–37.

Naeimi, H., Nazifi, Z.S. and Amininezhad, S.M., 2015. Preparation of Fe3O4 encapsulated-silica sulfonic acid nanoparticles and study of their in vitro antimicrobial activity. *Journal of Photochemistry and Photobiology, Part B: Biology*, 149, pp.180–188.

Nam, J., Son, S. and Moon, J.J., 2017. Adjuvant-loaded spiky gold nanoparticles for activation of innate immune cells. *Cellular and Molecular Bioengineering*, 10(5), pp.341–355.

Natalio, F., André, R., Hartog, A.F., Stoll, B., Jochum, K.P., Wever, R. and Tremel, W., 2012. Vanadium pentoxide nanoparticles mimic vanadium haloperoxidases and thwart biofilm formation. *Nature Nanotechnology*, 7(8), pp.530–535.

Nazari, Z.E., Banoee, M., Sepahi, A.A., Rafii, F. and Shahverdi, A.R., 2012. The combination effects of trivalent gold ions and gold nanoparticles with different antibiotics against resistant Pseudomonas aeruginosa. *Gold Bulletin*, 45(2), pp.53–59.

Nirmala, M. and Anukaliani, A., 2011. Synthesis and characterization of undoped and TM (Co, Mn) doped ZnO nanoparticles. *Materials Letters*, 65(17–18), pp.2645–2648.

Nochi, T., Yuki, Y., Takahashi, H., Sawada, S.I., Mejima, M., Kohda, T., Harada, N., Kong, I.G., Sato, A., Kataoka, N. and Tokuhara, D., 2010. Nanogel antigenic protein-delivery system for adjuvant-free intranasal vaccines. *Nature Materials*, 9(7), pp.572–578.

Nur, S., Shameli, K., Mei-Theng, W.M., Teow, S.-Y., Chew, J. and Ismail, N.A., 2019. Cytotoxicity and antibacterial activities of plant-mediated synthesized zinc oxide (ZnO) nanoparticles using Punica granatum (pomegranate) fruit peels extract. *Journal of Molecular Structure* J, 1189, pp.57–65.

Ong, T.H., Chitra, E., Ramamurthy, S., Ling, C.C.S., Ambu, S.P. and Davamani, F., 2019. Cationic chitosan-propolis nanoparticles alter the zeta potential of S. epidermidis, inhibit biofilm formation by modulating gene expression and exhibit synergism with antibiotics. *PloS one*, 14(2), p.e0213079

Paciotti, G.F., Myer, L., Weinreich, D., Goia, D., Pavel, N., McLaughlin, R.E. and Tamarkin, L., 2004. Colloidal gold: A novel nanoparticle vector for tumor directed drug delivery. *Drug Delivery*, 11(3), pp.169–183.

Pal, S., Tak, Y.K. and Song, J.M., 2007. Does the antibacterial activity of silver nanoparticles depend on the shape of the nanoparticle? A study of the gram-negative bacterium Escherichia coli. *Applied and Environmental Microbiology*, 73(6), pp.1712–1720.

Pan, H., Wang, X., Xiao, S., Yu, L. and Zhang, Z., 2013. *Preparation and Characterization of TiO 2 Nanoparticles Surface-Modified by Octadecyltrimethoxysilane*. NISCAIR-CSIR, India.

Patil, S.P. and Kumbhar, S.T., 2017. Antioxidant, antibacterial and cytotoxic potential of silver nanoparticles synthesized using terpenes rich extract of Lantana camara L. leaves. *Biochemistry and Biophysics Reports*, 10, pp.76–81.

Podporska-Carroll, J., Myles, A., Quilty, B., McCormack, D.E., Fagan, R., Hinder, S.J., Dionysiou, D.D. and Pillai, S.C., 2017. Antibacterial properties of F-doped ZnO visible light photocatalyst. *Journal of Hazardous Materials*, 324(A), pp.39–47.

Podsiadlo, P., Sinani, V.A., Bahng, J.H., Kam, N.W.S., Lee, J. and Kotov, N.A., 2008. Gold nanoparticles enhance the anti-leukemia action of a 6-mercaptopurine chemotherapeutic agent. *Langmuir*, 24(2), pp.568–574.

Poetsch, A. and Wolters, D., 2008. Bacterial membrane proteomics. *Proteomics*, 8(19), pp.4100–4122.

Poole Jr, C.P. and Owens, F.J., 2003. *Introduction to Nanotechnology*. John Wiley & Sons.

Radzig, M.A., Nadtochenko, V.A., Koksharova, O.A., Kiwi, J., Lipasova, V.A. and Khmel, I.A., 2013. Antibacterial effects of silver nanoparticles on gram-negative bacteria: Influence on the growth and biofilms formation, mechanisms of action. *Colloids and Surfaces, Part B: Biointerfaces*, 102, pp.300–306.

Raghunath, A. and Perumal, E., 2017. Metal oxide nanoparticles as antimicrobial agents: A promise for the future. *International Journal of Antimicrobial Agents*, 49(2), pp.137–152.

Rai, A., Prabhune, A. and Perry, C.C., 2010. Antibiotic mediated synthesis of gold nanoparticles with potent antimicrobial activity and their application in antimicrobial coatings. *Journal of Materials Chemistry*, 20(32), pp.6789–6798.

Rajawat, S. and Qureshi, M.S., 2012. Comparative study on bactericidal effect of silver nanoparticles, synthesized using green technology, in combination with antibiotics on Salmonella typhi. *Journal of Biomaterials and Nanobiotechnology*, 3(4), p.480.

Rajendran, K., Anwar, A., Khan, N.A. and Siddiqui, R., 2017. Brain-eating amoebae: Silver nanoparticle conjugation. Enhanced efficacy of anti-amoebic drugs against *Naegleria fowleri*. *ACS Chemical Neuroscience*, 8(12), pp.2626–2630.

Rakesh, S., Ananda, S. and Gowda, N.M., 2013. Synthesis of chromium (III) oxide nanoparticles by electrochemical method and *Mukia maderaspatana* plant extract, characterization, KMnO4 decomposition and antibacterial study. *Modern Research in Catalysis*, 2, pp.127–135.

Ramalingam, B., Parandhaman, T. and Das, S.K., 2016. Antibacterial effects of biosynthesized silver nanoparticles on surface ultrastructure and nanomechanical properties of gram-negative bacteria viz. *Escherichia coli* and *Pseudomonas aeruginosa*. *ACS Applied Materials and Interfaces*, 8(7), pp.4963–4976.

Ravishankar Rai, V., 2011. *Nanoparticles and Their Potential Application as Antimicrobials*.

Reddy, L.S., Nisha, M.M., Joice, M. and Shilpa, P.N., 2014. Antimicrobial activity of zinc oxide (ZnO) nanoparticle against *Klebsiella pneumoniae*. *Pharmaceutical Biology*, 52(11), pp.1388–1397.

Reddy, M.P., Venugopal, A. and Subrahmanyam, M., 2007. Hydroxyapatite-supported Ag–TiO2 as Escherichia coli disinfection photocatalyst. *Water Research*, 41(2), pp.379–386.

Ren, G., Hu, D., Cheng, E.W., Vargas-Reus, M.A., Reip, P. and Allaker, R.P., 2009. Characterisation of copper oxide nanoparticles for antimicrobial applications. *International Journal of Antimicrobial Agents*, 33(6), pp.587–590.

Rezaei, Z.S., Javed, A., Ghani, M.J., Soufian, S., Barzegari, F.F., Bayandori, M.A. and Mirjalili, S.H., 2010. *Comparative Study of Antimicrobial Activities of TiO2 and CdO Nanoparticles against the Pathogenic Strain of Escherichia coli*. *Iranian Journal of Pathology*, 5(2):83–89.

Rojas-Andrade, M., Cho, A.T., Hu, P., Lee, S.J., Deming, C.P., Sweeney, S.W., Saltikov, C. and Chen, S., 2015. Enhanced antimicrobial activity with faceted silver nanostructures. *Journal of Materials Science*, 50(7), pp.2849–2858.

Roldão, A., Mellado, M.C.M., Castilho, L.R., Carrondo, M.J. and Alves, P.M., 2010. Virus-like particles in vaccine development. *Expert Review of Vaccines*, 9(10), pp.1149–1176.

Rosenthal, J.A., Chen, L., Baker, J.L., Putnam, D. and DeLisa, M.P., 2014. Pathogen-like particles: Biomimetic vaccine carriers engineered at the nanoscale. *Current Opinion in Biotechnology*, 28, pp.51–58.

Roy, A., Gauri, S.S., Bhattacharya, M. and Bhattacharya, J., 2013. Antimicrobial activity of CaO nanoparticles. *Journal of Biomedical Nanotechnology*, 9(9), pp.1570–1578.

Sanchez-Lopez, E., Gomes, D., Esteruelas, G., Bonilla, L., Lopez-Machado, A.L., Galindo, R., Cano, A., Espina, M., Ettcheto, M., Camins, A. and Silva, A.M., 2020. Metal-based nanoparticles as antimicrobial agents: An overview. *Nanomaterials*, 10(2), p.292.

Santos, C.S., Gabriel, B., Blanchy, M., Menes, O., García, D., Blanco, M., Arconada, N. and Neto, V., 2015. Industrial applications of nanoparticles–a prospective overview. *Materials Today: Proceedings*, 2(1), pp.456–465.

Seabra, A.B. and Duran, N., 2015. Nanotoxicology of metal oxide nanoparticles. *Metals*, 5(2), pp.934–975.

Seil, J.T. and Webster, T.J., 2012. Antimicrobial applications of nanotechnology: Methods and literature. *International Journal of Nanomedicine*, 7, p.2767.

Shamaila, S., Zafar, N., Riaz, S., Sharif, R., Nazir, J. and Naseem, S., 2016. Gold nanoparticles: An efficient antimicrobial agent against enteric bacterial human pathogen. *Nanomaterials*, 6(4), p.71.

Shamaila, S., Zafar, N., Riaz, S., Sharif, R., Nazir, J. and Naseem, S., 2016. Gold nanoparticles: An efficient antimicrobial agent against enteric bacterial human pathogen. *Nanomaterials*, 6(4), p.71.

Shewen, P.E. and Wilkie, B.N., 1988. Vaccination of calves with leukotoxic culture supernatant from *Pasteurella haemolytica*. *Canadian Journal of Veterinary Research*, 52(1), p.30.

Shoeb, M., Singh, B.R., Khan, J.A., Khan, W., Singh, B.N., Singh, H.B. and Naqvi, A.H., 2013. ROS-dependent anticandidal activity of zinc oxide nanoparticles synthesized by using egg albumen as a biotemplate. *Advances in Natural Sciences: Nanoscience and Nanotechnology*, 4(3), p.035015.

Silva, A.L., Rosalia, R.A., Sazak, A., Carstens, M.G., Ossendorp, F., Oostendorp, J. and Jiskoot, W., 2013. Optimization of encapsulation of a synthetic long peptide in PLGA nanoparticles: Low-burst release is crucial for efficient CD8+ T cell activation. *European Journal of Pharmaceutics and Biopharmaceutics*, 83(3), pp.338–345.

Składanowski, M., Wypij, M., Laskowski, D., Golińska, P., Dahm, H. and Rai, M., 2017. Silver and gold nanoparticles synthesized from Streptomyces sp. isolated from acid forest soil with special reference to its antibacterial activity against pathogens. *Journal of Cluster Science*, 28(1), pp.59–79.

Slavin, Y.N., Asnis, J., Häfeli, U.O. and Bach, H., 2017. Metal nanoparticles: Understanding the mechanisms behind antibacterial activity. *Journal of Nanobiotechnology*, 15(1), pp.1–20.

Smith, J.D., Morton, L.D. and Ulery, B.D., 2015. Nanoparticles as synthetic vaccines. *Current Opinion in Biotechnology*, 34, pp.217–224.

Sotiriou, G.A. and Pratsinis, S.E., 2010. Antibacterial activity of nanosilver ions and particles. *Environmental Science and Technology*, 44(14), pp.5649–5654.

Srisitthiratkul, C., Pongsorrarith, V. and Intasanta, N., 2011. The potential use of nanosilver-decorated titanium dioxide nanofibers for toxin decomposition with antimicrobial and self-cleaning properties. *Applied Surface Science*, 257(21), pp.8850–8856.

Sukhorukova, I.V., Zhitnyak, I.Y., Kovalskii, A.M., Matveev, A.T., Lebedev, O.I., Li, X., Gloushankova, N.A., Golberg, D. and Shtansky, D.V., 2015. Boron nitride nanoparticles with a petal-like surface as anticancer drug-delivery systems. *ACS Applied Materials and Interfaces*, 7(31), pp.17217–17225.

Sundaresan, K., Sivakumar, A., Vigneswaran, C. and Ramachandran, T., 2012. Influence of nano titanium dioxide finish, prepared by sol-gel technique, on the ultraviolet protection, antimicrobial, and self-cleaning characteristics of cotton fabrics. *Journal of Industrial Textiles*, 41(3), pp.259–277.

Sutradhar, P., Saha, M. and Maiti, D., 2014. Microwave synthesis of copper oxide nanoparticles using tea leaf and coffee powder extracts and its antibacterial activity. *Journal of Nanostructure in Chemistry*, 4(1), p.86.

Syafiuddin, A., Salim, M.R., Beng Hong Kueh, A., Hadibarata, T. and Nur, H., 2017. A review of silver nanoparticles: Research trends, global consumption, synthesis, properties, and future challenges. *Journal of the Chinese Chemical Society*, 64(7), pp.732–756.

Tang, Z.X., Yu, Z., Zhang, Z.L., Zhang, X.Y., Pan, Q.Q. and Shi, L.E., 2013. Sonication-assisted preparation of CaO nanoparticles for antibacterial agents. *Química Nova*, 36(7), pp.933–936.

Thangadurai, D., Sangeetha, J. and Prasad, R., 2020. *Nanotechnology for Food, Agriculture, and Environment*. Springer.

Theron, J., Walker, J. and Cloete, T., 2008. Nanotechnology and water treatment: Applications and emerging opportunities. *Critical Reviews in Microbiology*, 34(1), pp.43–69.

Tom, R.T., Suryanarayanan, V., Reddy, P.G., Baskaran, S. and Pradeep, T., 2004. Ciprofloxacin-protected gold nanoparticles. *Langmuir*, 20(5), pp.1909–1914.

Tudose, M., Culita, D.C., Munteanu, C., Pandele, J., Hristea, E., Ionita, P., Zarafu, I. and Chifiriuc, M.C., 2015. Antibacterial activity evaluation of silver nanoparticles entrapped in silica matrix functionalized with antibiotics. *Journal of Inorganic and Organometallic Polymers and Materials*, 25(4), pp.869–878.

Umer, A., Naveed, S., Ramzan, N. and Rafique, M.S., 2012. Selection of a suitable method for the synthesis of copper nanoparticles. *Nano*, 7(5), 1230005.

Verma, S.K., Jha, E., Sahoo, B., Panda, P.K., Thirumurugan, A., Parashar, S.K.S. and Suar, M., 2017. Mechanistic insight into the rapid one-step facile biofabrication of antibacterial silver nanoparticles from bacterial release and their biogenicity and concentration-dependent in vitro cytotoxicity to colon cells. *RSC Advances*, 7(64), pp.40034–40045.

Van Dong, P., Ha, C.H. and Kasbohm, J., 2012. Chemical synthesis and antibacterial activity of novel-shaped silver nanoparticles. *International Nano Letters*, 2(1), pp.1–9.

Wang, L., Hu, C. and Shao, L., 2017. The antimicrobial activity of nanoparticles: Present situation and prospects for the future. *International Journal of Nanomedicine*, 12, pp.1227–1249.

Watson, C.Y., Molina, R.M., Louzada, A., Murdaugh, K.M., Donaghey, T.C. and Brain, J.D., 2015. Effects of zinc oxide nanoparticles on Kupffer cell phagosomal motility, bacterial clearance, and liver function. *International Journal of Nanomedicine*, 10, p.4173.

Watson, D.S., Endsley, A.N. and Huang, L., 2012. Design considerations for liposomal vaccines: Influence of formulation parameters on antibody and cell-mediated immune responses to liposome associated antigens. *Vaccine*, 30(13), pp.2256–2272.

Wei, X., Gao, J., Wang, F., Ying, M., Angsantikul, P., Kroll, A.V., Zhou, J., Gao, W., Lu, W., Fang, R.H. and Zhang, L., 2017. In situ capture of bacterial toxins for antivirulence vaccination. *Advanced Materials*, 29(33), p.1701644.

Wilson, J.T., Keller, S., Manganiello, M.J., Cheng, C., Lee, C.C., Opara, C., Convertine, A. and Stayton, P.S., 2013. pH-Responsive nanoparticle vaccines for dual-delivery of antigens and immunostimulatory oligonucleotides. *ACS Nano*, 7(5), pp.3912–3925.

Wypij, M., Golinska, P., Dahm, H. and Rai, M., 2017. Actinobacterial-mediated synthesis of silver nanoparticles and their activity against pathogenic bacteria. *IET Nanobiotechnology*, 11(3), pp.336–342.

Yoo, J.W., Irvine, D.J., Discher, D.E. and Mitragotri, S., 2011. Bio-inspired, bioengineered and biomimetic drug delivery carriers. *Nature Reviews. Drug Discovery*, 10(7), pp.521–535.

Yoo, J.W. and Mitragotri, S., 2010. Polymer particles that switch shape in response to a stimulus. *Proceedings of the National Academy of Sciences of the United States of America*, 107(25), pp.11205–11210.

Yousef, J.M. and Danial, E.N., 2012. In vitro antibacterial activity and minimum inhibitory concentration of zinc oxide and nano-particle zinc oxide against pathogenic strains. *Journal of Health Sciences*, 2(4), pp.38–42.

Yu-sen, E.L., Vidic, R.D., Stout, J.E., McCartney, C.A. and Victor, L.Y., 1998. Inactivation of *Mycobacterium avium* by copper and silver ions. *Water Research*, 32(7), pp.1997–2000.

Zhang, X.F., Liu, Z.G., Shen, W. and Gurunathan, S., 2016. Silver nanoparticles: Synthesis, characterization, properties, applications, and therapeutic approaches. *International Journal of Molecular Sciences*, 17(9), p.9.

Zhao, L., Seth, A., Wibowo, N., Zhao, C.X., Mitter, N., Yu, C. and Middelberg, A.P., 2014. Nanoparticle vaccines. *Vaccine*, 32(3), pp.327–337.

Zhao, R., Lv, M., Li, Y., Sun, M., Kong, W., Wang, L., Song, S., Fan, C., Jia, L., Qiu, S. and Sun, Y., 2017. Stable nanocomposite based on PEGylated and silver nanoparticles loaded graphene oxide for long-term antibacterial activity. *ACS Applied Materials and Interfaces*, 9(18), pp.15328–15341.

Zhu, X., Hondroulis, E., L+iu, W. and Li, C.Z., 2013. Biosensing approaches for rapid genotoxicity and cytotoxicity assays upon nanomaterial exposure. *Small*, 9(9–10), pp.1821–1830.

6 Non-metallic Nanoparticles Eliminating Bacteria

Muhammad Ijaz, Amjad Islam Aqib,
Muhammad Muddassir Ali, and Yung-Fu Chang

CONTENTS

Abbreviations 162
6.1 Need for Alternatives 162
6.2 Types of Non-metal Nanoparticles and Their Antibacterial Mechanisms ... 163
    6.2.1 Fullerenes 163
    6.2.2 Carbon Nanotubes (CNTs) 165
    6.2.3 Graphene 165
    6.2.4 Carbon Nanofiber 165
    6.2.5 Carbon Black 165
    6.2.6 Sulfur Nanoparticles 166
    6.2.7 Silica Nanoparticles 166
6.3 Antibacterial Mechanism of Nanoparticles 166
6.4 Methods of Preparation of Non-metallic Nanoparticles 167
    6.4.1 Green Synthesis 168
    6.4.2 Chemical Vapor Deposition (CVD) Method 168
6.5 Shapes and Characterization Effects on Antibacterial Activity
    of Non-metal Nanoparticles 169
6.6 In Vitro and In Vivo Evaluation of Non-metallic Nanoparticles 169
6.7 Non-metal Nanoparticles and Bacterial Vaccines 170
    6.7.1 Mesoporous Silica Nanoparticles As Antigen Carriers 170
    6.7.2 Carbon-Based Nanodelivery Systems 172
    6.7.3 Carbon Nanotubes Stimulate Specific Immune Responses 172
6.8 Non-metal Nanoparticles and Antibiotic Products 173
    6.8.1 Nitric Oxide-Releasing Silica Nanoparticles 174
    6.8.2 Metal-Modified Silica Nanoparticles 174
        6.8.2.1 Silver–Silica Nanoparticles 174
        6.8.2.2 Copper–Silica Nanoparticles 175
6.9 Biomedical Applications of Non-metallic Nanoparticles 177
6.10 Challenges to Using Non-metal Nanoparticles as Antibacterials 177
6.11 Conclusions 178
References 178

DOI: 10.1201/9781003126256-6

ABBREVIATIONS

AHAP3   N-(6-inohexyl) aminopropyltrimethoxysilane
BCG   bacille Calmette–Guerin
CCSSNPs chitosan-covered stabilized SNPs
CNPs   carbon nanoparticles
CNTs   carbon nanotubes
CVD   chemical vapor deposition
DWCNTs   double-walled carbon nanotubes
FTIR   Fourier-transform infrared
GO   graphene oxide
HMSNs   hollow mesoporous silica nanoparticles
IRIVs   immunopotentiating reconstituted influenza virosomes
MAP3   N-methylaminopropyltrimethoxysilane
MRSA   methicillin-resistant *Staphylococcus aureus*
MSNs   mesoporous silica nanoparticles
MWCNTs multiwalled carbon nanotubes
                   NM
                   nanomaterials
NRS   nitrogen reactive species
PPD   pure protein derivative
QY   quantum yield
rGO   reduced graphene oxide
ROS   reactive oxygen species
SEM   scanning electron microscopy
SiNPs   silica nanoparticles
SLN   sentinel lymph node
SWCNTs single-walled carbon nanotubes
TEM   transmission electron microscopy

6.1   NEED FOR ALTERNATIVES

An antibacterial drug is used in medication for treatment of bacterial infections. It is of great interest to formulate new antimicrobial agents from natural sources, which are cost-effective and resistance-free (Dong et al., 2012). Now scientists and some pharmaceutical companies are trying to synthesize antimicrobial agents because of pathogenic microbes that cause contagious ailments. Most are highly toxic, and cause drug resistance and allergies to humans. In the present situation, non-metallic nanoscale materials, i.e., carbon nanoparticles (CNPs), sulfur nanoparticles, silica nanoparticles, and nitric oxide nanoparticles, have been considered as innovative antimicrobial agents having good physiochemical characteristics (Mocan et al., 2017).

Nanomaterials (NMs) are gaining great importance because they are used as antibacterials complementary to antibiotics. They consist of mutants that are drug resistant and biofilm producing (Zhang et al., 2010). They may be metal or non-metal having different properties. Bacteria are different from one another other in genetics, structure of cell wall, and metabolic pathways. The sensitivity of bacteria to nanomaterials depends on the plankton, biofilm, growing phase, immobile phase, and starving state of bacteria (Nath and Banerjee, 2013). The toxic level between the bacteria and nanomaterials is high (Lee et al., 2019). Environmental elements such as aeration, pH, and heat affect the lethality of NMs toward bacteria. Physicochemical characteristics such as mass, shape, and chemical changes disturb antibacterial activity of NPs during preparation (Gatoo et al., 2014). As per studies, the mechanism of action, antibacterial properties, and hazard level of these particles is not clear, leading to their practical field use as questionable (Hajipour et al., 2012).

There are many properties of NPs. Firstly, they have a large surface area, which increases contact with target organisms. NPs can interact with bacterial cells, regulate penetration of cell sheath, or interfere with indifferent molecular paths (Hemeg, 2017). Secondly, inhibitory effects of antibiotics are increased by them. Saha et al. (2007) showed that a combination of gold nanoparticles with kanamycin, streptomycin, or ampicillin can reduce the minimum inhibitory concentration (MIC) of the antibiotic in contradiction of gram-positive and gram-negative bacteria. Similarly, multidrug-resistant *Escherichia coli* infections can be treated with both gold NPs and fluoroquinolone antibiotics (Lee et al., 2019). The mutual effect of antibiotics and NPs provide great antimicrobial properties (Huh and Kwon, 2011). Silica nanoparticles are considered as therapeutic agents against methicillin-resistant *Staphylococcus aureus* (MRSA) (Zaidi et al., 2017). The combination of nanotechnology and microbiology involve advanced preparation of new types of antibacterial agents for accurate treatment of bacterial infections and to avoid resistance.

6.2 TYPES OF NON-METAL NANOPARTICLES AND THEIR ANTIBACTERIAL MECHANISMS

Nanomaterials of carbon have high significance. Due to special physiochemical characteristics, carbon nanotubes, fullerenes, nanofibers, and nanodiamonds have been manufactured (Figure 6.1). They help in decreasing the weight of materials and improve strength by increasing the surface area per unit volume. They act as good electrical conductors, and thermal and dimensional stabilizers, due to their optical properties and flame resistant characteristics (Han et al., 2009b).

6.2.1 FULLERENES

Fullerenes are made up of nanomaterials in a hollow cage structure with diameter of 0.71 nm. They may be carbon 60 or carbon 70 in pentagonal and hexagonal shape with sp2 hybridization. Fullerenes may be single layer or multilayer (Astefanei et al., 2015). They have great importance due to their structure, strength, versatility, electrical conductivity, and electron affinity. In microorganisms, they cause disruption of cell membranes and DNA breakage. By manipulating their energy uptake, microorganisms are deactivated. On a light microscope, they yield higher values of reactive oxygen species (ROS), which increases their antibacterial activity (Lyon et al., 2006).

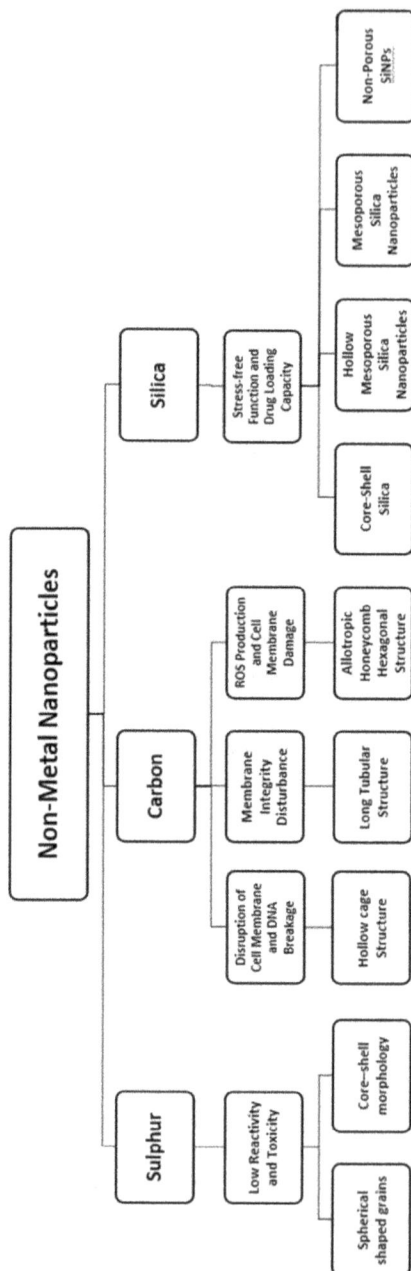

FIGURE 6.1 Types of salient non-metallic nanoparticles, their modes of action and their morphology.

6.2.2 CARBON NANOTUBES (CNTs)

Carbon nanotubes (CNTs) are long tubular structures with diameters of 1–2 nm. They act as semiconductors. Their structure is just like a rolled-up graphite sheet. Based on rolling, there are various types: single-walled, double-walled, and multi-walled nanotubes (Khan et al., 2019). Single-walled carbon nanotubes (SWCNTs) are made of a single-rolled sheet of nanotubes with 0.7 nm diameter. Double-walled carbon nanotubes (DWCNTs) are made of a double-rolled sheet. Multiwalled carbon nanotubes (MWCNTs) are made up of multiple rolled sheets with a diameter of 100 nm. CNTs have a honeycomb carbon frame structure, which is twisted into a hollow cylinder. They can be bent, but regain their strong structure. They have different structures, shapes, and thickness levels. The properties of CNTs are based on sheets of graphene (Winkin et al., 2016).

CNTs show a antibacterial property by membrane integrity disturbance due to electrostatic forces between CNTs and the microbial outer surface, which cause deterioration of the membrane. Production of reactive oxygen damages biological molecules of bacteria and incidentally damage deoxyribonucleic acid. During fabrication processes, many impurities incorporate (i.e., metallic nanoparticles, catalysts, suspension) into carbon nanotubes, which contribute to antibacterial activities (Lawrence et al., 2016). A combination of C60 with an antibacterial mechanism can apply to SWCNTs, as they both are pure carbon and their diameters are similar (Liu et al., 2009).

6.2.3 GRAPHENE

Graphene possesses a carbon allotropic form of a honeycomb carbon having a hexagonal structure with 2D planar surface of 1 nm thickness. Graphene-related materials have antibacterial mechanisms such as damage of cell membrane, harmful removal of phospholipids from lipid membranes, reactive oxygen species causing oxidative stress, and microorganism separation from their microenvironment (Mejías Carpio et al., 2012).

6.2.4 CARBON NANOFIBER

Graphene nanofoils are moved to carbon nanofiber, which are coiled into a cup shape.

6.2.5 CARBON BLACK

Carbon black is an amorphous carbon that shows antibacterial characteristics by incorporating into the bacterial cell. It has a spherical shape with diameter of 20 to 70 nm. It is in aggregate form. The size of the carbon particles is inversely proportional to the antimicrobial activity. Carbon black shows a large surface area and huge quantity of atoms present within the surface and near to the layers of surface (Dong et al., 2012). Carbon nanomaterials are used against human pathogens. Recent research on nanofibers enable their use for drug delivery systems and controlled drug release, tissue-engineered scaffolds, dressings for wound healing, biosensors, biomedical devices, medical implants, and skin care, as well as air, water, and blood purification systems (Rasouli et al., 2019).

6.2.6 SULFUR NANOPARTICLES

Sulfur is a biologically active element that is easily available and helpful in dermatological treatment, for example, in acne ointments and dandruff shampoos (Abuyeva et al., 2018), and has antimicrobial, anticancer, antioxidant, radical-scavenging, and antifungal properties (Teng et al., 2019; Saedi et al., 2020). Sulfur is a non-metallic element and inorganic in nature, abundantly present on earth's surface and more than 60 million tons of elemental sulfur is made in the petroleum refining process annually as a by-product (Chung and Kim, 2018). Its nanoparticles also present in lithium-sulfur batteries and in photocatalysts. But sulfur nanoparticles are unusable as drug delivery agents due to their low reactivity and toxicity. To solve such issues, sulfur NPs should be altered (Li et al., 2020).

6.2.7 SILICA NANOPARTICLES

Silica nanoparticles (SiNPs) are used to treat gram-positive and gram-negative bacterial infections (Şen Karaman et al., 2018), including lead-resistant bacteria and water contaminants (Fatemi et al., 2020). SiNPs are of different types, i.e., coreshell silica, non-porous SiNPs, mesoporous silica nanoparticles (MSNs), and hollow mesoporous silica nanoparticles (HMSNs). MSNs are used in drug delivery due to their stress-free function and drug loading capacity.

6.3 ANTIBACTERIAL MECHANISM OF NANOPARTICLES

Nanomaterials behave in two lethal ways: (1) membrane disruption and (2) production of ROS (Pelgrift and Friedman, 2013). When nanomaterials statically bind to bacterial cell walls and cell membranes, destruction happens, which changes the membrane working capacity and decreases its stiffness, which can cause transport imbalancing, respirational problems, energy disruption, and cell lysis, leading to cell death (Wang et al., 2017). ROS is involved in both in vivo and in vitro cytotoxicity of nanomaterials (Nathan and Cunningham-Bussel, 2013). The high concentrations of ROS cause stress, protein structural changes, enzyme inhibition, nucleic acid damage, and even cell lysis, but mutations occur due to low concentration of ROS (Wang et al., 2011a). Nanomaterial toxicity is photocatalytic when ROS is induced by ultraviolet (UV) or visible light, then it causes respiratory trouble and death of *E. coli*. Enzyme inhibition, induction of nitrogen reactive species (NRS), and apoptotic induction are all due to nanomaterials (Yudovin-Farber et al., 2008). The action of carbon nanoparticles comprises many physiochemical mechanisms (Ji et al., 2016). In microorganisms, damage to the cell wall and cell membrane is due to carbon nanostructures (CNSs), i.e., graphene sheets cause cell death (Prasad et al., 2017). Toxic substances, i.e.,

ROS, may generate due to chemical connection between CNSs and the micro-organism that is the cause of oxidative stress. The electron transferring process occurs due to interactions in which electrons from the outer surface of microbes are devastated, which causes cell death (Li et al., 2014).

However, the exact antibacterial mechanism of sulfur nanoparticles is poor, but they produce toxic H_2S gas that reacts with thiol present in proteins and lipids, and as a result interferes with DNA, denatures it, and ultimately causes the death of cells (Rai et al., 2016). Shankar et al. (2018) verified that the antimicrobial action of chitosan-covered stabilized sulfur nanoparticles (SNPs) and rod-shaped orthorhom-bic SNPs are more effective than spherical orthorhombic SNPs and non-stabilizing SNPs. But another study (Choudhury et al., 2013) stated that tetrapod-like b-SNPs have lower antimicrobial activity than spherical SNPs.

Nitric oxide (NO) is an endogenously produced molecule, and its nanomate-rials have antibacterial action due to low resistance (Schairer et al., 2012). The antibacterial activity of metal nanoparticles depends on their shape and size. Despite all of its advantages, the clinical importance of NO is restricted due to high reactivity capsulation, and focal transfer can oppress its antimicrobial abil-ity (Kutner and Friedman, 2013). Nanomaterials of nitric oxide differ from metal nanomaterials due to RNS instead of ROS. These nanomaterials are active in skin infections to destroy *S. aureus* (MRSA), which is methicillin resistant, and are also involved in the healing process in normal and diabetic mice (Barraud et al., 2006a).

6.4 METHODS OF PREPARATION OF NON-METALLIC NANOPARTICLES

Top-down and bottom-up are two altered means for synthesizing nanoparticles (Panigrahi et al., 2004). The top-down method involves breaking large structures into small structures. Lithography (Tapaszto et al., 2008), laser ablation (Balasooriya et al., 2017), sputtering deposition, pulsed electrochemical etching (Nissinen et al., 2016), and vapor deposition vapor (Duc Vu Quyen et al., 2019) are techniques gener-ally used in the top-down method. The bottom-up method consists of manufacturing material molecule by molecule, atom by atom, or cluster by cluster using the pro-cesses of sol-gel (Schwartz et al., 1990), laser heating transformation (Lacour et al., 2007), chemical vapor deposition (Luo et al., 2012), plasma spray synthesis (Mädler et al., 2002), and microemulsion (Darbandi et al., 2005).

Sulfur nanoparticles (SNPs) can be made through biological and physicochemi-cal methods, e.g., (i) sodium thiosulfate acidification and SNP stabilization with capping agents or surfactants (Shankar et al., 2018), (ii) various polysulfide solu-tions acidification, (iii) water-in-oil microemulsification process through chemical contact among hydrochloric acid and sodium polysulfide by reverse microemulsion using surfactants (Soleimani et al., 2013), (iv) sublimation of sulfur and nucleation of polyethylene glycol (Salem et al., 2015), (v) mechanical and ultrasonic dispersing modification of sulfur surface (Roy Choudhury et al., 2011), and (vi) taking mono-clinic SNPs through addition of acidified plant extract to sodium thiosulfate by the biological method (Khairan et al., 2019). The size and shape of SNPs depend upon the raw materials and manufacturing method (Roy Choudhury et al., 2011).

6.4.1 GREEN SYNTHESIS

Some advantages of green synthesis are cost-effectiveness, low contamination risk, safe and environment-friendly, low waste production, and biomedical and pharmaceutical usability (Nishanthi et al., 2019). Green synthesis can be done through honey usage. Honey-mediated synthesis is a fast method having an advantage over the microbial method due to microorganism culturing and separation of nanoparticles from microorganisms. Honey behaves as a reducing agent and preservative for green synthesis of nanomaterials. Wu and colleagues synthesized carbon-based nanoparticles by using honey as a precursor for real-time photoacoustic imaging (Balasooriya et al., 2017). Surface-coated polysorbate and polyethylene glycol sentinel lymph node (SLN) imaging is done with carbon nanoparticles, which are lighter in weight ($\sim$7 nm) than gold nanoparticles coated with silica (20 nm), Cu, and single-walled nanotubes. These carbon nanoparticles are the reason for fast signal enhancement ($\sim$2 min) (Luke et al., 2013). Honey is used for synthesizing high-fluorescent carbon dots of 2 nm with 19.8% quantum yield (QY), which can be used as a sensor. These dots showed high constancy, reduced toxicity, and photostability. They were used as sensors for Fe^{3+} detection, cell imaging, fluorescent staining, and biosensing (Dong et al., 2010).

6.4.2 CHEMICAL VAPOR DEPOSITION (CVD) METHOD

Chemical vapor deposition (CVD) involves deposition of gaseous reactant as a thin film on the substrate through reaction compartments. A chemical reaction occurs when gas interacts with intense substrate. This film is improved and recycled for further usage. Advantages of this vapor deposition method are purified, hardened, strengthened, and uniform nanoparticles. CVD has some disadvantages, including its requirement of a special apparatus and the release of toxic gases as by-products (Bhaviripudi et al., 2007).

Synthesis of surface-functionalized SNPs is carried out by two different methods of wet chemical, i.e., liquid phase precipitation and water-in-oil microemulsion (Deshpande et al., 2008). Liquid phase precipitation involves conversion of microsized sulfur particles into nanosized particles through precipitation of weak acids, including formic acid or acetic acid, by encapsulation of polymeric coverings. Deshpande et al. (2008) showed synthesis of SNPs by various initial substrates such as surfactants and co-surfactants, and the water and oil phase was carried out with water-in-oil microemulsion method. Synthesis, stabilizing, and concentration estimation processes of orthorhombic and monoclinic nanoallotropes of sulfur are carried out with changes in surfactants in the oil stage and environmental factors, i.e., heat and pH. The standard concentration of a-SNPs was valued as 5194·31 g/ml and b-SNPs 18,000 g/ml. Manufacturing of such sets of SNPs is done at normal temperature with minimal apparatus requirement (Roy and Goswami, 2013).

6.5 SHAPES AND CHARACTERIZATION EFFECTS ON ANTIBACTERIAL ACTIVITY OF NON-METAL NANOPARTICLES

Nanoparticles are categorized for their allotropic composition with X-ray diffraction patterning, shape with scanning electron microscopy (SEM), size with transmission electron microscopy (TEM), surface topology with atomic force microscopy, surface modification with Fourier-transform infrared (FTIR) spectroscopy, thermal stability with thermogravimetric analysis, and purity with energy dispersive X-ray spectroscopy (Rasouli et al., 2019; Azizi-Lalabadi et al., 2020; Fatemi et al., 2020; Samak et al., 2020). TEM micrographs are used for confirmation of both size and shape of SNPs. Per TEM, monoclinic SNPs behaved like nanorods and tetrapod structure formation occurred by combining nanorods. SNP antibacterial activity depends on shape and size, and capping agents that are used for stabilizers in SNP synthesis (Shankar et al., 2018). Huge SNP production needs a cost-effective scheme within economical means.

6.6 IN VITRO AND IN VIVO EVALUATION OF NON-METALLIC NANOPARTICLES

Non-metallic nanoparticles like carbon possess low cytotoxicity and have particular antiviral properties. Though they are comparatively new in nanoscience, they have been encouraged for their use to control the synthesis and function of viral surfaces (Elias et al., 2019; Salesa et al., 2019; Saleemi et al., 2020; Samak et al., 2020). Graphene oxide (GO) and reduced graphene oxide (rGO) are also non-metallic nanoparticles that have been used in different fields like photonics and electronics but limited in nanomedicine. Different experiments have been performed to check how GO inactivates the virus before its entry into the cell. Pointed ends of the GO layer damage the structure of a virus by direct interaction. The antiviral activity of GO was efficient for both RNA and DNA viruses but depends upon time of incubation and concentration of GO. It has been also observed that rGO also showed the same antiviral activity as GO. The negative charge on GO facilitates the electrostatic communication with positive charge on the virus. However, high interaction leads to the inactivation and destruction of the virus (Innocenzi and Stagi, 2020).

Silica nanoparticles are reported to eliminate microbial biofilms by releasing nitric oxide in vitro (Slomberg et al., 2013). The same observations were reported by Hetrick et al. (2009), who treated staphylococcal biofilms with potentiated silica nanoparticles and reported strong anti-biofilm action of silica nanoparticles in vitro. Barraud et al. (2006b) observed biofilm dispersion in *Pseudomonas aeruginosa* when treated with nitric oxide nanoparticles. Nanoneedles of alumina along with stabilized silver carbonate nanoparticles exhibited very strong antibacterial characteristics against bacteria (Buckley et al., 2008). Aluminum oxide nanoparticles also hinder cell membranes of *E. coli* and their biomolecules, causing growth inhibition of microbes (Ansari et al., 2014). Mobile polycationic multilayers also have the tendency to inhibit microbial growth over the tested surfaces (Lichter and Rubner, 2009). Strong antibacterial action of polylysine nanoparticles against *Bacillus* species is also evident of non-metallic nanoparticles implementation as potent anti-microbials (Hiraki, 1995; Jain et al., 2014).

Successful treatment of skin and soft tissue ailments of bacterial origin is reported by Kutner and Friedman (2013) utilizing nitric oxide nanoparticles in clinics. Abscesses caused by *Staphylococcus aureus* have been treated successfully with nitric oxide nanoparticles in model animals (Han et al., 2009a). Antimicrobial efficacy of non-metallic nanoparticles against bacterial ailments is also reported by Martinez et al. (2009) who treated common bacterial skin infections with nitric oxide nanoparticles. Nanoparticles of alumina are reported to promote horizontal plasmid transfer in multiresistant bacteria, modifying their antibiogram, an emerging and reliable way to combat them (Qiu et al., 2012) (Table 6.1).

6.7 NON-METAL NANOPARTICLES AND BACTERIAL VACCINES

Attenuated pathogens (live vaccines) or inactivated pathogens have strong immune response, but development of such vaccines is hard work because it is an expensive method having many disadvantages such as inflammation and multiple dosages are given for boosting (Jiang et al., 2017). While subunit vaccines are safe for use, they have low immunogenicity and cannot cross intestinal mucosal tissues because metabolic enzymes are degraded (Kammer et al., 2007). Adjuvants are added for improving immunogenicity because they increase dendritic cells activation and also develop strong immunity. Adjuvants are of two types, i.e., immunostimulatory effect or working as transport methods. Some adjuvants can be used in animals and humans, i.e., Alum, MF59, virus-like particles, monophosphoryl lipid A, immuno-potentiating reconstituted influenza virosomes (IRIVs), and toxins of cholera, but high toxicity limit their usage (Reed et al., 2009; Bolduc et al., 2018; Yadav et al., 2018; Bhardwaj et al., 2020).

Nanoparticle linkage with antigens produce strong immune response as compared to antigens alone (Vallet-Regi et al., 2001). Nanoparticles are good vectors such as MSNs to carry genes, medicine, proteins, and DNA because of their easy production process, biocompatibility, and stability (Xu et al., 2014). Additionally, MSNs can easily control the release of drugs due to their size modification. Non-metallic nanoparticles have been used in various vaccines to enhance their efficacy (Figure 6.3) (Table 6.2).

6.7.1 MESOPOROUS SILICA NANOPARTICLES AS ANTIGEN CARRIERS

There is incomplete data on MSNs as vectors. Protein transfer can be done by surface-modified MSNs (Sharif et al., 2012). Aminosilane-functionalized ordered silica materials (SBA-15) can be used for adsorption and bovine serum albumin release proteins. Hollow mesoporous silica nanoparticles have extraordinary drug loading capacity. They have importance for bovine serum albumin and goat IgG intracellular distribution. MSNs can behave as adjuvants in vaccine preparation for immunogenicity (Lee et al., 2008).

TABLE 6.1

Antimicrobial Activity of Some Non-metallic Nanoparticles

Non-metal Nanoparticle	Size Range	Microorganism Tested	Mode of Action	Reference
Green-synthesized silicon NP	20–50 nm	*Bacillus subtilis*	Reduce ATPase activity and cellular metabolism	(Tiwari et al., 2019)
Sulfur NP	20–86 nm	*Escherichia coli*	Cell membrane rupture and oxidative stress	(Paralikar et al., 2019)
Euporium-doped silicon NP	470–617 nm	*Bacillus anthracis*	Internalization in cell by disrupting ATP synthesis	(Na et al., 2020)
Silica (Si) NP	20–400 nm	Methicillin-resistant *Staphylococcus aureus*	Mechanical disruption of cell membrane and ROS production	(Dizaj et al., 2014)
Nitric oxide–silica NP	2–15 nm	*Escherichia coli/Staphylococcus aureus/ Staphylococcus epidermidis/Candida albicans*	Disruption of cell membrane, DNA damage, protein degradation, and ROS generation	(Hetrick et al., 2009)
C-60 fullerene	150–320 nm	*Pseudomonas aeruginosa, Escherichia coli, Candida albicans*	Alter membrane permeability	(Su et al., 2010)
SWCNT	1–3 nm	*Escherichia coli*	Membrane damage	(Grinholc et al., 2015)
Graphene	205 nm	*Escherichia coli, Candida albicans*	Attachment layers alteration	(Kang et al., 2007)
Multilayer film	1–2 nm	*Staphylococcus aureus, Pseudomonas aeruginosa*	Biofilm disruption, attachment inhibition	(Robertson et al., 2017)

TABLE 6.2

List of Antigens Delivered by Nanocarriers for the Treatment of Bacterial Infections

Antigen	Nanocarrier	Bacterial Infection	Reference
Antigenic protein	Poly(D,L-lactic-co-glycolic acid) nanospheres, nanoemulsion	Anthrax	(Na et al., 2020)
DNA encoding T cell epitopes of Esat-6 and FL	Chitosan nanoparticle	Tuberculosis	(Na et al., 2020)
Mycobacterium lipids			
Polysaccharides	Liposomes	Pneumonia	(Robertson
Bacterial toxic and parasitic protein		Cholera and malaria	et al., 2017)
Fusion protein		*Helicobacter pylori* infection	
Mycobacterium fusion protein		Tuberculosis	

6.7.2 Carbon-Based Nanodelivery Systems

Carbon-based nanomaterials and carbon nanotubes have importance as antigen carriers (Scheinberg et al., 2013). Carbon-based nanodelivery systems are intricate, non-degradable, less toxic, and intrinsically non-immunogenic. Zeinali et al. (2009) observed usage as delivery vector in vaccine distribution. The purified form of tuberculin antigen was bound onto carboxyl group SWCNT. The antigen SWCNT conjugate can cause the production of cytokines through T helper (Th1) cells, e.g., IFN-g and IL-12, through the subcutaneous process, which can be compared from conventional tuberculosis bacille Calmette–Guerin (BCG) vaccine. Carbon nanotubes can induce immunogenicity (Parra et al., 2013) as fungicide azoxystrobin was transferred on carbon nanotubes. In rabbits, maximum anti-azoxystrobin IgG antibody titers can be obtained. Carbon nanoparticles can be used in oral vaccination (Wang et al., 2011b). Carbon nanoparticles having diameter of 470 nm with 40–60 nm pore size, manufactured by Si model and encapsulated in bovine serum albumin, can prevent devastation in GIT.

6.7.3 Carbon Nanotubes Stimulate Specific Immune Responses

A good vaccine should prompt immunity and stimulate cytotoxic T cells, which eliminate damaged cells at the humoral level by incentive construction through antibody neutralization, which help in microbial destruction. Vaccines against HIV or malaria must be helpful in avoiding infection and removal of pathogens. Innate immune response is significant in antigen arrangement and recruitment of resistant cells at infected places (Douradinha and Doolan, 2011). Macrophages, monocytes, natural killer cells, dendritic cells, and T and B cells can be linked with CNTs. CNTs are harmless to the cells function. Toxicity of functionalized CNTs is lower than original CNTs. CNTs can trigger innate immune cells (Villa et al., 2011).

Microchip reporting of THP-1 displayed functionalized and non-functionalized CNTs that trigger those genes that are tangled in monocyte reaction, and non-functionalized CNTs improved expression of genes that are involved in programmed cell death. The functionalized CNTs were reacted and improved for ammonium incorporation; CNTs with ammonium and without ammonium both cannot produce chemokines in monocytes that have a diameter of, respectively, 9.5 nm and 30 nm (Pescatori et al., 2013). These nanotubes show no toxicity to human primary monocytes and THP-1 cells and help in chemokine production in IL-1β, IL-6, TNF-α, and IL-10 cells. The chemokines involved in T cells may present at infection sites and swelling. CNTs can induce a fictionalization form based on the innate immune response. THP-1 cells macrophages show the same appearance as monocytes (Chou et al., 2008).

Antigen presentation provides immunity against foreign substances and tumors. Antigen-presenting cells uptake the antigen, and destroy and present to T cells by surface antigen molecules. MHC class I prompts cytotoxic CD8+ T cell reaction and MHC class II alters immunity for helper CD4+ T cells that stimulate a body response (Douradinha and Doolan, 2011). CNTs stimulate MHC class I and class II. CNTs having immunogens of pathogens are immunogenic and defensive in experiments in which animals were used as models. CNTs having peptide for B cell epitope from FMDV prompted large antibody titer (Rodriguez and Gay, 2011).

CNTs can stimulate MHC class I receptors and Th1-based cell response for prompting cytotoxicity and produce cytokines, i.e., IFN-γ, TNF-α, and IL-12. SWNT was coated with tuberculin antigen for testing in mice (Zeinali et al., 2009). Balb/c mice were vaccinated with tuberculin pure protein derivative (PPD), which results in two types of immunity in two weeks. One group shows immunity with complete Freund's adjuvant and the second group is without adjuvant. The scientists can only compare the cytokine profile with tuberculin by SWNT with BCG. Both groups, which were immunized with BCG and tuberculin PPD, have high levels of IFN-γ. The second group had higher Th2-type cytokine (IL-5 and IL-10) levels than BCG. There is limited data on the CNT mechanism in the immune system. CNTs activate an innate immune response that is present in inflammation, infection response, vaccination, and chemokines release. Macrophages and peptides of dendritic cells combine with carbon nanotubes and display them on surface antigens, stimulating a humoral reaction for antigens (Pescatori et al., 2013).

6.8 NON-METAL NANOPARTICLES AND ANTIBIOTIC PRODUCTS

Antibiotic resistance is problematic nowadays. Improper disease diagnosis, excessive prescriptions, inadequate usage of antibiotics, incomplete antibiotic courses, antibiotics spectrum, and second- and third-generation cephalosporins are the main causes of resistance that develop MRSA (Anwar et al., 2020; Naseer et al., 2020). A combination of antibiotics with nanoparticles is of great importance because it protects the drug to the target tissue (Allahverdiyev et al., 2011). High concentrations of antibiotics reach the targeted cells while reducing the dosage quantity, resistance to antibiotics, and side effects, refining pharmacokinetics (Anwar et al., 2020).

Toxicity of nanoparticles can be observed when treatment is prolonged. It is compulsory to check the effect on tissues and organs when they are intravenously injected (Nabeshi et al., 2011). Silica nanoparticles are used in drug delivery due to thermal stability and large surface area (Echazú et al., 2016). Colloidal silica is very useful because it can be synthesized in different structures from inexpensive substrates and easily modified. Nanosized silica particles and silica-modified nanoparticles have great importance in dealing with infections.

6.8.1 Nitric Oxide-Releasing Silica Nanoparticles

NO is very important due to its antimicrobial action and it has an effective host reaction against infection (Figure 6.3). Manufacturing of NO through synthase isoform is regulated using proinflammatory cytokines and through lipopolysaccharide and lipoteichoic acid. The compounds that give off NO destroy microbes (Fang, 1997). N-Methylaminopropyltrimethoxysilane (MAP3) and N-(6-inohexyl) aminopropyltrimethoxysilane (AHAP3) were used in manufacturing of NPs. Efficacy of MAP3 silica nanoparticles is 1000 times better than AHAP3 nanoparticles against *P. aeruginosa* biofilms (Hetrick et al., 2009) because they can easily penetrate into the biofilm matrix due to small size and rapid delivery. Nanoparticles that released NO can be manufactured through combining tetraalkoxysilane with an aminoalkoxysilane modified with diazeniumdiolate. Nanoparticle linkage with *P. aeruginosa* cells explain the differential toxicity. The association is linked electrostatically or hydrophobically, which causes an increase in NO concentrations and effective NO distribution (Hetrick et al., 2008).

Carpenter et al. (2011) studied NO-releasing silica nanoparticles. Functionalized silica nanoparticles, synthesized through the microemulsion method, and N-diazeniumdiolate nitric oxide (NO) nanoparticles have efficacy for *P. aeruginosa*. Smaller particles have a quicker dispersion rate, which is helpful in rapid association and penetration into biofilm (Shin et al., 2007).

6.8.2 Metal-Modified Silica Nanoparticles

Copper and silver nanoparticles due to biocidal effects are effective (Figure 6.3). It is described that antibacterial action of silver ion concentrations was two to four times greater than Si and Cu. The metal ion concentration is reduced when Si and Cu are fused (Albers et al., 2013). There are two major metals that enhance activity of Si due to various mechanisms: silver and copper (Figure 6.2).

6.8.2.1 Silver–Silica Nanoparticles

Ionizing silver nanoparticles such as Ag–SiO_2 have great importance due to antimicrobial action, which results in dissolution of ions. There are three mechanisms that

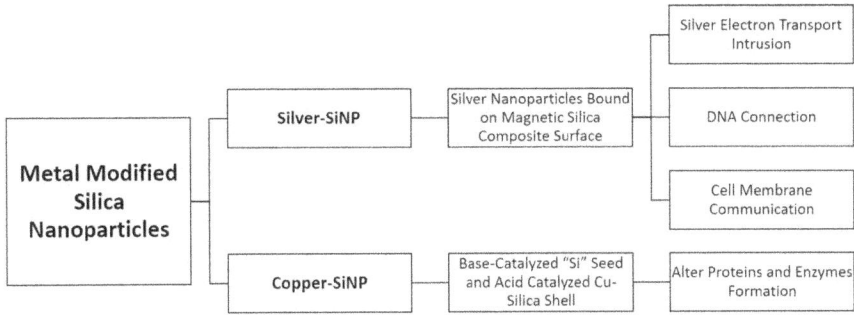

FIGURE 6.2 Mode of action of metal modified Silica Nanoparticles against bacteria.

inhibit bacteria development: silver electron transport intrusion, DNA connection, and cell membrane communication (Lok et al., 2007). Silver binds with thiol groups present in enzymes, which results in deactivation of energy. Ag ions can link with DNA to make an increase in constancy of DNA. Hollow silica nanotubes were manufactured for the silver holding. Tubular Ag hollow structures have higher antibacterial activity than spherical hollow structures due to a large concentration of silver (Wang et al., 2006). Silver and silica combinations are used to synthesize nanoparticles. Silver coating spherical nanoparticles were easily made by tetraethoxysilane and silver nitrate with ascorbic acid. They have an antibacterial mechanism against *E. coli* and *S. aureus* (Pan et al., 2014). Silver nanoparticles were bound on magnetic silica composite surface for preparation of magnetic disinfectant having stability and antibacterial activity. SiNPs with magnetic composite (Fe3O4–SiO2–Ag) were made and used in water disinfection (Malekzadeh et al., 2019). Silver nanoparticles having diameter of 10 nm were fixed then firmly bound on Si coating of Fe_3O_4–SiO_2 magnetic NPs to avoid silver aggregation. The concentrations of Fe_3O_4–SiO_2–Ag to *E. coli* was inhibited 15.625 mg/L and for *S. aureus* 31.25 mg/L. The smallest bactericidal concentrations for *E.coli* were 250 mg/L and for *S. aureus* 500 mg/L. About 150 mg in 150 ml of Fe_3O_4–SiO_2–Ag disinfectant can kill 99.9% of bacteria in 60 min. The silica coat behaves like a supporting matrix and is helpful in increasing firmness of the antiseptic. The Fe_3O_4–SiO_2–Ag composite has 75 emu/g attractive saturation showing its recovery from water by magnetic separation.

6.8.2.2 Copper–Silica Nanoparticles

Copper has importance due to antibacterial action, which was examined by the disk diffusion method (Kim et al., 2006). It can change protein and enzyme construction so its function can vary, which results in bacterial inactivation (Michels et al., 2005). Copper can be added on the surface such as SiNPs, but the release of Cu can be hindered. Maniprasad and Santra (2012) described the Cu–Si nanoparticle manufacturing process, in which Si "seeds" were made through the base-catalyzed Stöber process, followed by Cu–Si shell growth through acid-catalyzation. The seed size was 380 nm and thickness 35 nm with 98 ng loading capacity per mg of $CuSiO_2$.

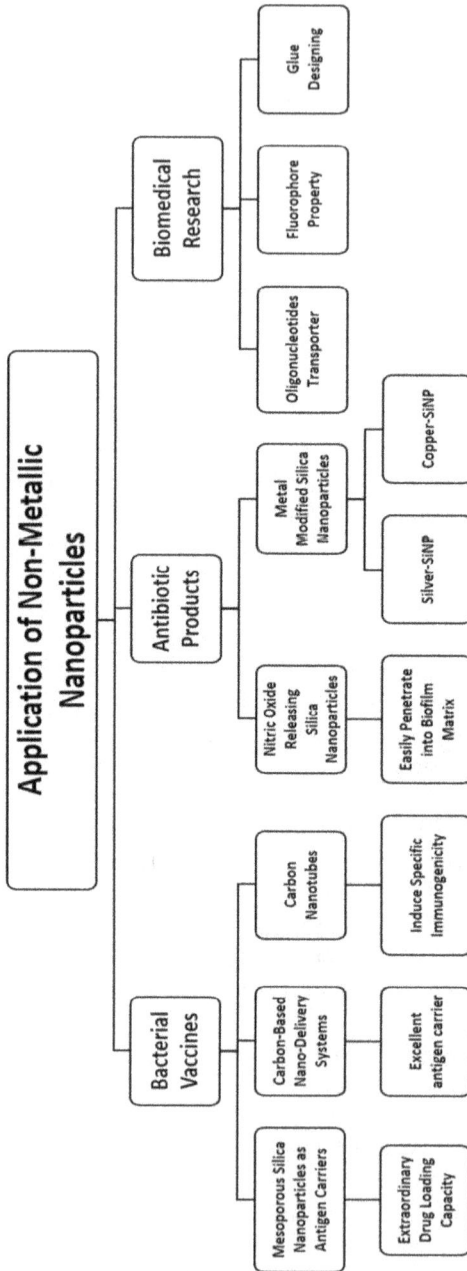

FIGURE 6.3   An overview of salient applications of non-metallic nanoparticles.

6.9 BIOMEDICAL APPLICATIONS OF NON-METALLIC NANOPARTICLES

Nanoparticles play a significant role in delivery of drugs at target sites because they act as oligonucleotides transporter in antisense treatment. Non-metallic nanoparticles have significant roles in biomedical research (Figure 6.3). Mesoporous silica nanoparticles are used in drug delivery due to mesoporous structure with large surface area. Agnihotri et al. (2015) reported the use of SiO_2 NPs as vectors for medicine distribution. Hydrophobic drugs are not active because of low absorption (Aughenbaugh et al., 2001). Silica nanoparticles increase drug absorption and reduce lethality (Yang et al., 2017; Liu et al., 2020). The therapeutic index is helpful in absorption of medications (Yao et al., 2017).

Biomarkers have specificity for biomolecule detection and have stable signal transducers. Nanoparticles are used for detection of small amounts of biomolecules (Lv et al., 2019). Currently, quantum dots have great importance in medical applications and their nanoprobes are being used in imaging. Dense metal present in quantum dots creates toxicity problems because it damages cells. Quantum dots show insolubility in water and makes connection with polymers being used. Studies show that the fluorophore property is increased by covalent linking of fluorophores in silica nanoparticle core (Naghavi et al., 2020). C dots, made through covalent connection as tetramethyl rhodamine isothiocyanate dye in the core shell of silica nanoparticles, are water soluble and non-toxic and helpful in increasing illumination and dye firmness (Yang et al., 2017). The surface of SiNPs has strong bonding power, so SiNPs are used in glue designing. Silica nanoparticles can differentiate cells. They show osteogenesis, which is inversely linked to adipogenesis, and adipogenic differentiation is decreased by silica nanoparticles so they are used in obesity treatment. Nanotechnology has significant value because it is being used in screening, microarray, disease diagnosis, drug distribution, cancer and diabetes treatment, sensitivity, and infections (Li et al., 2018). An enzymatic hydrolysis study showed quinizarin diester adsorbtion in silica nanoparticles and dispersion in neighboring areas (Zhang et al., 2008). Antibiotics are used to kill bacteria due to antibacterial agents (Zou et al., 2020). Quaternary ammonium and silica nanoparticles kill bacteria through contact by using UV antibacterial coating (Yamada et al., 2012). Encapsulation of SiNPs with peptide is active in contrast to *P. aeruginosa* in lung infection by giving protection to peptides. To overcome multidrug resistance, Valetti et al. (2017) renewed testing of clofazimine drug, which is reactive for multidrug-resistant tuberculosis. The drug encapsulation by silica nanoparticles helps in providing stability and increasing solubility (Kwon et al., 2017). An SiNP–gentamicin nanohybrid manufactured for in vitro gentamicin antibiotic release in a controlled manner can inhibit bacterial growth (Valetti et al., 2017).

6.10 CHALLENGES TO USING NON-METAL NANOPARTICLES AS ANTIBACTERIALS

Nanoantibiotics are in ascendance in many ways and their dominance has been well researched and documented, but less work has been done on their disadvantages. This is due to several hurdles including production problems, economic issues, and negative health impacts. The standard method of dosage of nanoparticles adminis-tration in different organs needs to be developed. Administration of nanoantibiotics may cause multiorgan nanotoxicity (Anwar et al., 2020). Nanoparticles and their degraded products can obstruct the coagulation pathways by implicating the hemoly-sis. Non-metal nanoparticles like fullerenes and nanotubes may cause toxicity if not administered at the desired concentration (Salesa et al., 2019; Saleemi et al., 2020). NPs when administered through the IV route may cause direct damage to the spleen, lung, liver, and bone marrow. Similarly, due to their small size and more cell vitality, NPs may cause damage to the brain, heart, lung, and liver when inhaled. More stud-ies on stability, dispersion, biodistribution, final product, and possible side effects of nanoparticles need to be carried out before putting them in the clinical trials on animal models (Manzano and Vallet, 2020).

The use of CNTs in water disinfection is a challenging task. Cytotoxicity of CNTs is based on the synthesis process, structure, solubility, and type of microbial cells. Release of CNTs plays a role in the ecosystem. They can be moved from water to humans and cause problems. Another challenge is minimizing the cost. Conventional antimicrobials are advantageous due to cost-effectiveness. SWNTs and MWNTs are more expensive than conventional antimicrobials. So, nanomaterial usage is signifi-cant because nanomaterials synthesize CNTs at low prices (Liu et al., 2018).

6.11 CONCLUSIONS

Indiscriminate use of antibiotics is causing serious problems in the medical field like multidrug resistance, and due to drug residues, new diseases are emerging, so it becomes difficult to treat many diseases. To overcome these problems, non-metal nanoparticles have been shown to exhibit antimicrobial activity. A combination of antibiotics with nanoparticles is of great importance because it protects drug to the target tissue. Nanovaccines are also effective because nanoparticle linkage with anti-gens produce strong immune response as compared to antigens alone. In the medical field, this approach has been trending. However, less information is available about their toxicity and harmful effects, nevertheless they have been getting more attention toward resolving problems in the field of medicine.

REFERENCES

Abuyeva, B., M. Burkitbayev, G. Mun, B. Uralbekov, N. Vorobyeva, D. Zharlykasimova and F. Urakaev. 2018. Preparation of ointment materials based on sulfur nanoparticles in water-soluble polymers. *Mater. Today Proc.* 5:22894–22899.
Agnihotri, S., R. Pathak, D. Jha, I. Roy, H.K. Gautam, A.K. Sharma and P. Kumar. 2015. Synthesis and antimicrobial activity of aminoglycoside-conjugated silica nanoparticles against clinical and resistant bacteria. *New J. Chem.* 39:6746–6755.
Albers, C.E., W. Hofstetter, K.A. Siebenrock, R. Landmann and F.M. Klenke. 2013. In vitro cytotoxicity of silver nanoparticles on osteoblasts and osteoclasts at antibacterial con-centrations. *Nanotoxicology.* 7:30–36.
Allahverdiyev, A.M., K.V. Kon, E.S. Abamor, M. Bagirova and M. Rafailovich. 2011. Coping with antibiotic resistance: Combining nanoparticles with antibiotics and other antimi-crobial agents. *Expert Rev. Anti. Infect. Ther.* 9:1035–1052.

Ansari, M.A., H.M. Khan, A.A. Khan, S.S. Cameotra, Q. Saquib and J. Musarrat. 2014. Interaction of Al2 O3 nanoparticles with *Escherichia coli* and their cell envelope biomolecules. *J. Appl. Microbiol.* 116:772–783.

Anwar, M.A., A.I. Aqib, K. Ashfaq, F. Deeba, M.K. Khan, S.R. Khan, I. Muzammil, M. Shoaib, M.A. Naseer, T. Riaz, Q. Tanveer, M. Sadiq, F.L. Lodhi and F. Ashraf. 2020. Antimicrobial resistance modulation of MDR *E. coli* by antibiotic coated ZnO nanoparticles. *Microb. Pathog.* 148:104450.

Astefanei, A., O. Núñez and M.T. Galceran. 2015. Characterisation and determination of fullerenes: A critical review. *Anal. Chim. Acta.* 882:1–21.

Aughenbaugh, W., S. Radin and P. Ducheyne. 2001. Silica sol-gel for the controlled release of antibiotics. II. The effect of synthesis parameters on the in vitro release kinetics of vanomycin. *J. Biomed. Mater. Res.* 57:321–326.

Azizi-Lalabadi, M., H. Hashemi, J. Feng and S.M. Jafari. 2020. Carbon nanomaterials against pathogens; the antimicrobial activity of carbon nanotubes, graphene/graphene oxide, fullerenes, and their nanocomposites. *Adv. Colloid Interface Sci.* 284:102250.

Balasooriya, E.R., C.D. Jayasinghe, U.A. Jayawardena, R.W.D. Ruwanthika, R. Mendis de Silva and P.V. Udagama. 2017. Honey mediated green synthesis of nanoparticles: New era of safe nanotechnology. *J. Nanomater.* 2017:5919836.

Barraud, N., D.J. Hassett, S.-H. Hwang, S.A. Rice, S. Kjelleberg and J.S. Webb. 2006. Involvement of Nitric Oxide in biofilm dispersal of *Pseudomonas aeruginosa*. *J. Bacteriol.* 188:7344–7353.

Bhardwaj, P., E. Bhatia, S. Sharma, N. Ahamad and R. Banerjee. 2020. Advancements in prophylactic and therapeutic nanovaccines. *Acta Biomater.* 108:1–21.

Bhaviripudi, S., E. Mile, S.A. Steiner, A.T. Zare, M.S. Dresselhaus, A.M. Belcher and J. Kong. 2007. CVD synthesis of single-walled carbon nanotubes from gold nanoparticle catalysts. *J. Am. Chem. Soc.* 129:1516–1517.

Bolduc, M., M. Baz, M.-È. Laliberté-Gagné, D. Carignan, C. Garneau, A. Russel, G. Boivin, P. Savard and D. Leclerc. 2018. The quest for a nanoparticle-based vaccine inducing broad protection to influenza viruses. *Nanomedicine Nanotechnology, Biol. Med.* 14:2563–2574.

Buckley, J.J., P.L. Gai, A.F. Lee, L. Olivi and K. Wilson. 2008. Silver carbonate nanoparticles stabilised over alumina nanoneedles exhibiting potent antibacterial properties. *Chem. Commun.* 34:4013–4015.

Carpenter, A.W., D.L. Slomberg, K.S. Rao and M.H. Schoenfisch. 2011. Influence of scaffold size on bactericidal activity of nitric oxide-releasing silica nanoparticles. *ACS Nano.* 5:7235–7244.

Chou, C.-C., H.-Y. Hsiao, Q.-S. Hong, C.-H. Chen, Y.-W. Peng, H.-W. Chen and P.-C. Yang. 2008. Single-walled carbon nanotubes can induce pulmonary injury in Mouse Model. *Nano Lett.* 8:437–445.

Choudhury, S.R., A. Mandal, M. Ghosh, S. Basu, D. Chakravorty and A. Goswami. 2013. Investigation of antimicrobial physiology of orthorhombic and monoclinic nanoallotropes of sulfur at the interface of transcriptome and metabolome. *Appl. Microbiol. Biotechnol.* 97:5965–5978.

Chung, K. and H. Kim. 2018. Effect of sulfur on the rheological properties and phase structure of polycarbonate/ poly(styrene-co-acrylonitrile)/multiwalled carbon nanotube composites. *Korea-Australia Rheol. J.* 30:309–316.

Darbandi, M., R. Thomann and T. Nann. 2005. Single quantum dots in silica spheres by microemulsion synthesis. *Chem. Mater.* 17:5720–5725.

Deshpande, A.S., R.B. Khomane, B.K. Vaidya, R.M. Joshi, A.S. Harle and B.D. Kulkarni. 2008. Sulfur nanoparticles synthesis and characterization from H(2)S gas, using novel biodegradable iron chelates in W/O microemulsion. *Nanoscale Res. Lett.* 3:221–229.

Dizaj, S.M., F. Lotfipour, M. Barzegar-Jalali, M.H. Zarrintan and K. Adibkia. 2014. Antimicrobial activity of the metals and metal oxide nanoparticles. *Mater. Sci. Eng. C. Mater. Biol. Appl.* 44:278–284.

Dong, L., A. Henderson and C. Field. 2012. Antimicrobial activity of single-walled carbon nanotubes suspended in different surfactants. *J. Nanotechnol.* 12: 1–7.

Dong, Y., N. Zhou, X. Lin, J. Lin, Y. Chi and G. Chen. 2010. Extraction of electrochemilumines-cent oxidized carbon quantum dots from activated carbon. *Chem. Mater.* 22:5895–5899.

Douradinha, B. and D.L. Doolan. 2011. Harnessing immune responses against *Plasmodium* for rational vaccine design. *Trends Parasitol.* 27:274–283.

Duc Vu Quyen, N., D. Quang Khieu, T.N. Tuyen, D. Xuan Tin and B. Thi Hoang Diem. 2019. Carbon Nanotubes: Synthesis via chemical vapour deposition without hydrogen, surface modification, and application. *J. Chem.* 2019:4260153.

Echazú, M.I.A., M.V. Tuttolomondo, M.L. Foglia, A.M. Mebert, G.S. Alvarez and M.F. Desimone. 2016. Advances in collagen, chitosan and silica biomaterials for oral tissue regeneration: from basics to clinical trials. *J. Mater. Chem. B.* 4:6913–6929.

Elias, L., R. Taengua, B. Frígols, B. Salesa and Á. Serrano-Aroca. 2019. Carbon Nanomaterials and LED irradiation as antibacterial strategies against gram-positive multidrug-resistant pathogens. *Int. J. Mol. Sci.* 20:3603.

Fang, F.C. 1997. Perspectives series: Host/pathogen interactions. Mechanisms of nitric oxide-related antimicrobial activity. *J. Clin. Invest.* 99:2818–2825.

Fatemi, H., B. Esmaiel Pour and M. Rizwan. 2020. Isolation and characterization of lead (Pb) resistant microbes and their combined use with silicon nanoparticles improved the growth, photosynthesis and antioxidant capacity of coriander (*Coriandrum sativum* L.) under Pb stress. *Environ. Pollut.* 266:114982.

Gatoo, M.A., S. Naseem, M.Y. Arfat, A. Mahmood Dar, K. Qasim and S. Zubair. 2014. Physicochemical properties of nanomaterials: Implication in associated toxic manifestations. *Biomed Res. Int.* 2014(1):498420.

Grinholc, M., J. Nakonieczna, G. Fila, A. Taraszkiewicz, A. Kawiak, G. Szewczyk, T. Sarna, L. Lilge and K.P. Bielawski. 2015. Antimicrobial photodynamic therapy with fullero-pyrrolidine: Photoinactivation mechanism of *Staphylococcus aureus*, in vitro and in vivo studies. *Appl. Microbiol. Biotechnol.* 99:4031–4043.

Hajipour, M.J., K.M. Fromm, A. Akbar Ashkarran, D. Jimenez de Aberasturi, I.R. de Larramendi, T. Rojo, V. Serpooshan, W.J. Parak and M. Mahmoudi. 2012. Antibacterial properties of nanoparticles. *Trends Biotechnol.* 30:499–511.

Han, G., L.R. Martinez, M.R. Mihu, A.J. Friedman, J.M. Friedman and J.D. Nosanchuk. 2009a. Nitric oxide releasing nanoparticles are therapeutic for *Staphylococcus aureus* abscesses in a murine model of infection. *PLoS One.* 4:e7804.

Han, L., D. Ghosh, W. Chen, S. Pradhan, X. Chang and S. Chen. 2009b. Nanosized carbon particles from natural gas soot. *Chem. Mater.* 21:2803–2809.

Hemeg, H.A. 2017. Nanomaterials for alternative antibacterial therapy. *Int. J. Nanomedicine.* 12:8211–8225.

Hetrick, E.M., J.H. Shin, H.S. Paul and M.H. Schoenfisch. 2009. Anti-biofilm efficacy of nitric oxide-releasing silica nanoparticles. *Biomaterials.* 30:2782–2789.

Hetrick, E.M., J.H. Shin, N.A. Stasko, C.B. Johnson, D.A. Wespe, E. Holmuhamedov and M.H. Schoenfisch. 2008. Bactericidal efficacy of nitric oxide-releasing silica nanoparticles. *ACS Nano.* 2:235–246.

Hiraki, J. 1995. Basic and applied studies on ε-polylysine. *J Antibact Antifung. Agents.* 23:349–354.

Huh, A.J. and Y.J. Kwon. 2011. "Nanoantibiotics": A new paradigm for treating infectious diseases using nanomaterials in the antibiotics resistant era. *J. Control. Release.* 156:128–145.

Innocenzi, P. and L. Stagi. 2020. Carbon-based antiviral nanomaterials: Graphene, C-dots, and fullerenes. A perspective. *Chem. Sci.* 11:6606–6622.

Jain, A., L.S. Duvvuri, S. Farah, N. Beyth, A.J. Domb and W. Khan. 2014. Antimicrobial polymers. *Adv. Healthc. Mater.* 3:1969–1985.

Ji, H., H. Sun and X. Qu. 2016. Antibacterial applications of graphene-based nanomaterials: Recent achievements and challenges. *Adv. Drug Deliv. Rev.* 105:176–189.

Jiang, J., G. Liu, V.A. Kickhoefer, L.H. Rome, L.-X. Li, S.J. McSorley and K.A. Kelly. 2017. A protective vaccine against chlamydia genital infection using vault nanoparticles without an added adjuvant. *Vaccines.* 5:3.

Kammer, A.R., M. Amacker, S. Rasi, N. Westerfeld, C. Gremion, D. Neuhaus and R. Zurbriggen. 2007. A new and versatile virosomal antigen delivery system to induce cellular and humoral immune responses. *Vaccine.* 25:7065–7074.

Kang, S., M. Pinault, L.D. Pfefferle and M. Elimelech. 2007. Single-walled carbon nanotubes exhibit strong antimicrobial activity. *Langmuir.* 23:8670–8673.

Khairan, K., Zahraturriaz and Z. Jalil. 2019. Green synthesis of sulphur nanoparticles using aqueous garlic extract (*Allium sativum*). *Rasayan J. Chem.* 12:50–57.

Khan, I., K. Saeed and I. Khan. 2019. Nanoparticles: Properties, applications and toxicities. *Arab. J. Chem.* 12:908–931.

Kim, Y.H., D.K. Lee, H.G. Cha, C.W. Kim, Y.C. Kang and Y.S. Kang. 2006. Preparation and characterization of the antibacterial Cu nanoparticle formed on the surface of SiO2 nanoparticles. *J. Phys. Chem. B.* 110:24923–24928.

Kutner, A.J. and A.J. Friedman. 2013. Use of nitric oxide nanoparticulate platform for the treatment of skin and soft tissue infections. *Wiley Interdiscip. Rev. Nanomedicine Nanobiotechnology.* 5:502–514.

Kwon, E.J., M. Skalak, A. Bertucci, G. Braun, F. Ricci, E. Ruoslahti, M.J. Sailor and S.N. Bhatia. 2017. Porous Silicon nanoparticle delivery of tandem peptide anti-infectives for the treatment of pseudomonas aeruginosa lung infections. *Adv. Mater.* 29(35):10.1002.

Lacour, F., O. Guillois, X. Portier, H. Perez, N. Herlin Boime and C. Reynaud. 2007. Laser pyrolysis synthesis and characterization of luminescent silicon nanocrystals. *Phys. E Low-Dimensional Syst. Nanostructures.* 38:11–15.

Lawrence, J.R., M.J. Waiser, G.D.W. Swerhone, J. Roy, V. Tumber, A. Paule, A.P. Hitchcock, J.J. Dynes and D.R. Korber. 2016. Effects of fullerene (C60), multi-wall carbon nanotubes (MWCNT), single wall carbon nanotubes (SWCNT) and hydroxyl and carboxyl modified single wall carbon nanotubes on riverine microbial communities. *Environ. Sci. Pollut. Res. Int.* 23:10090–10102.

Lee, J.W., P.T. Tra, S. Il Kim and S.H. Ron. 2008. Adsorption properties of proteins on SBA-15 nanoparticles functionalized with aminosilanes. *Journal of Nanoscience and Nanotechnology.* 10: 5152–5157.

Lee, N.-Y., W.-C. Ko and P.-R. Hsueh. 2019. Nanoparticles in the treatment of infections caused by multidrug-resistant organisms. *Front. Pharmacol.* 10:1153.

Li, C., Y. Wang, H. Jiang and X. Wang. 2020. Biosensors based on advanced sulfur-containing nanomaterials. *Sensors (Switzerland).* 20:1–27.

Li, J., G. Wang, H. Zhu, M. Zhang, X. Zheng, Z. Di, X. Liu and X. Wang. 2014. Antibacterial activity of large-area monolayer graphene film manipulated by charge transfer. *Sci. Rep.* 4:4359.

Li, T., T. Inose, T. Oikawa, M. Tokunaga, K. Hatoyama, K. Nakashima, T. Kamei, K. Gonda and Y. Kobayashi. 2018. Fabrication and dual imaging properties of quantum dot/silica core-shell particles immobilized with gold nanoparticles. *Mater. Technol.* 33:737–747.

Lichter, J.A. and M.F. Rubner. 2009. Polyelectrolyte multilayers with intrinsic antimicrobial functionality: the importance of mobile polycations. *Langmuir.* 25:7686–7694.

Liu, D., L. Liu, L. Yao, X. Peng, Y. Li, T. Jiang and H. Kuang. 2020. Synthesis of ZnO nanoparticles using radish root extract for effective wound dressing agents for diabetic foot ulcers in nursing care. *J. Drug Deliv. Sci. Technol.* 55:101364.

Liu, D., Y. Mao and L. Ding. 2018. Carbon nanotubes as antimicrobial agents for water disinfection and pathogen control. *J. Water Health.* 16:171–180.

Liu, S., L. Wei, L. Hao, N. Fang, M.W. Chang, R. Xu, Y. Yang and Y. Chen. 2009. Sharper and faster "Nano Darts" kill more bacteria: A study of antibacterial activity of individually dispersed pristine single-walled carbon nanotube. *ACS Nano.* 3:3891–3902.

Lok, C.-N., C.-M. Ho, R. Chen, Q.-Y. He, W.-Y. Yu, H. Sun, P.K.-H. Tam, J.-F. Chiu and C.-M. Che. 2007. Silver nanoparticles: Partial oxidation and antibacterial activities. *J. Biol. Inorg. Chem. JBIC a Publ. Soc. Biol. Inorg. Chem.* 12:527–534.

Luke, G.P., A. Bashyam, K.A. Homan, S. Makhija, Y.-S. Chen and S.Y. Emelianov. 2013. Silica-coated gold nanoplates as stable photoacoustic contrast agents for sentinel lymph node imaging. *Nanotechnology.* 24:455101.

Luo, Y., X. Wang, M. He, X. Li and H. Chen. 2012. Synthesis of high-quality carbon nanotube arrays without the assistance of water. *J. Nanomater.* 2012:542582.

Lv, Q., Y. Wang, C. Su, T. Lakshmipriya, S.C.B. Gopinath, K. Pandian, V. Perumal and Y. Liu. 2019. Human papilloma virus DNA-biomarker analysis for cervical cancer: Signal enhancement by gold nanoparticle-coupled tetravalent streptavidin-biotin strategy. *Int. J. Biol. Macromol.* 134:354–360.

Lyon, D.Y., L.K. Adams, J.C. Falkner and P.J.J. Alvarez. 2006. Antibacterial activity of Fullerene water suspensions: Effects of preparation method and particle size. *Environ. Sci. Technol.* 40:4360–4366.

Mädler, L., H.K. Kammler, R. Mueller and S. Pratsinis. 2002. Controlled synthesis of nanostructured particles by flame spray pyrolysis. *J. Aerosol Sci.* 33:369–389.

Malekzadeh, M., K.L. Yeung, M. Halali and Q. Chang. 2019. Preparation and antibacterial behaviour of nanostructured Ag@SiO2–penicillin with silver nanoplates. *New J. Chem.* 43:16612–16620.

Maniprasad, P. and S. Santra. 2012. Novel copper (Cu) loaded core-shell silica nanoparticles with improved Cu bioavailability: Synthesis, characterization and study of antibacterial properties. *J. Biomed. Nanotechnol.* 8:558–566.

Martinez, L.R., G. Han, M. Chacko, M.R. Mihu, M. Jacobson, P. Gialanella, A.J. Friedman, J.D. Nosanchuk and J.M. Friedman. 2009. Antimicrobial and healing efficacy of sustained release nitric oxide nanoparticles against Staphylococcus aureus skin infection. *J. Invest. Dermatol.* 129:2463–2469.

Mejías Carpio, I.E., C.M. Santos, X. Wei and D.F. Rodrigues. 2012. Toxicity of a polymer-graphene oxide composite against bacterial planktonic cells, biofilms, and mammalian cells. *Nanoscale.* 4:4746–4756.

Michels, H.T., S. Wilks, J. Noyce and C. Keevil. 2005. Copper alloys for human infectious disease control. *Mater. Sci. and Tech.* 1: 3–13.

Mocan, T., C.T. Matea, T. Pop, O. Mosteanu, A.D. Buzoianu, S. Suciu, C. Puia, C. Zdrehus, C. Iancu and L. Mocan. 2017. Carbon nanotubes as anti-bacterial agents. *Cell. Mol. Life Sci.* 74:3467–3479.

Manzano, M., and Vallet-Regí, M. 2020. Mesoporous silica nanoparticles for drug delivery. *Adv. Funct. Mater.* 30: 3–5.

Na, M., S. Zhang, J. Liu, S. Ma, Y. Han, Y. Wang, Y. He, H. Chen and X. Chen. 2020. Determination of pathogenic bacteria——Bacillus anthrax spores in environmental samples by ratiometric fluorescence and test paper based on dual-emission fluorescent silicon nanoparticles. *J. Hazard. Mater.* 386:121956.

Nabeshi, H., T. Yoshikawa, K. Matsuyama, Y. Nakazato, K. Matsuo, A. Arimori, M. Isobe, S. Tochigi, S. Kondoh, T. Hirai, T. Akase, T. Yamashita, K. Yamashita, T. Yoshida,

K. Nagano, Y. Abe, Y. Yoshioka, H. Kamada, et al. 2011. Systemic distribution, nuclear entry and cytotoxicity of amorphous nanosilica following topical application. *Biomaterials*. 32:2713–2724.

Naghavi, F., A. Morsali, M.R. Bozorgmehr and S.A. Beyramabadi. 2020. Quantum molecular study of mesoporous silica nanoparticle as a delivery system for troxacitabine anti-cancer drug. *J. Mol. Liq*. 310:113155.

Naseer, M.A., A.I. Aqib, A. Ashar, M.I. Saleem, M. Shoaib, M.F.-A. Kulyar, K. Ashfaq, Z.A. Bhutta and S. Nighat. 2020. Detection of altered pattern of antibiogram and biofilm character in Staphylococcus aureus isolated from dairy milk. *Pakistan J. Zool*.

Nath, D. and P. Banerjee. 2013. Green nanotechnology - A new hope for medical biology. *Environ. Toxicol. Pharmacol*. 36:997–1014.

Nathan, C. and A. Cunningham-Bussel. 2013. Beyond oxidative stress: An immunologist's guide to reactive oxygen species. *Nat. Rev. Immunol*. 13:349–361.

Nishanthi, R., S. Malathi, S. John-Paul and P. Palani. 2019. Green synthesis and characterization of bioinspired silver, gold and platinum nanoparticles and evaluation of their synergistic antibacterial activity after combining with different classes of antibiotics. *Mater. Sci. Eng. C*. 96:693–707.

Nissinen, T., T. Ikonen, M. Lama, J. Riikonen and V.-S. Lehto. 2016. Improved production efficiency of mesoporous silicon nanoparticles by pulsed electrochemical etching. *Powder Technol*. 288:360–365.

Pan, K.-Y., C.-H. Chien, Y.-C. Pu, C.-M. Liu, Y.-J. Hsu, J.-W. Yeh and H.C. Shih. 2014. Studies on the annealing and antibacterial properties of the silver-embedded aluminum/silica nanospheres. *Nanoscale Res. Lett*. 9:307.

Panigrahi, S., S. Kundu, S. Ghosh, S. Nath and T. Pal. 2004. General method of synthesis for metal nanoparticles. *J. Nanoparticle Res*. 6:411–414.

Paralikar, P., A.P. Ingle, V. Tiwari, P. Golinska, H. Dahm and M. Rai. 2019. Evaluation of antibacterial efficacy of sulfur nanoparticles alone and in combination with antibiotics against multidrug-resistant uropathogenic bacteria. *J. Environ. Sci. Heal. Part A*. 54:381–390.

Parra, J., A. Abad-Somovilla, J. V Mercader, T.A. Taton and A. Abad-Fuentes. 2013. Carbon nanotube-protein carriers enhance size-dependent self-adjuvant antibody response to haptens. *J. Control. Release*. 170:242–251.

Pelgrift, R.Y. and A.J. Friedman. 2013. Nanotechnology as a therapeutic tool to combat microbial resistance. *Adv. Drug Deliv. Rev*. 65:1803–1815.

Pescatori, M., D. Bedognetti, E. Venturelli, C. Ménard-Moyon, C. Bernardini, E. Muresu, A. Piana, G. Maida, R. Manetti, F. Sgarrella, A. Bianco and L.G. Delogu. 2013. Functionalized carbon nanotubes as immunomodulator systems. *Biomaterials*. 34:4395–4403.

Prasad, K., G.S. Lekshmi, K. Ostrikov, V. Lussini, J. Blinco, M. Mohandas, K. Vasilev, S. Bottle, K. Bazaka and K. Ostrikov. 2017. Synergic bactericidal effects of reduced graphene oxide and silver nanoparticles against Gram-positive and Gram-negative bacteria. *Sci. Rep*. 7:1591.

Qiu, Z., Y. Yu, Z. Chen, M. Jin, D. Yang, Z. Zhao, J. Wang, Z. Shen, X. Wang and D. Qian. 2012. Nanoalumina promotes the horizontal transfer of multiresistance genes mediated by plasmids across genera. *Proc. Natl. Acad. Sci*. 109:4944–4949.

Rai, M., A.P. Ingle and P. Paralikar. 2016. Sulfur and sulfur nanoparticles as potential antimicrobials: From traditional medicine to nanomedicine. *Expert Rev. Anti. Infect. Ther*. 14:969–978.

Rasouli, R., A. Barhoum, M. Bechelany and A. Dufresne. 2019. Nanofibers for biomedical and healthcare applications. *Macromol. Biosci*. 19:1800256.

Reed, S.G., S. Bertholet, R.N. Coler and M. Friede. 2009. New horizons in adjuvants for vaccine development. *Trends Immunol*. 30:23–32.

Robertson, S.N., D. Gibson, W.G. MacKay, S. Reid, C. Williams and R. Birney. 2017. Investigation of the antimicrobial properties of modified multilayer diamond-like carbon coatings on 316 stainless steel. *Surf. Coatings Technol.* 314:72–78.

Rodriguez, L.L. and C.G. Gay. 2011. Development of vaccines toward the global control and eradication of foot-and-mouth disease. *Expert Rev. Vaccines.* 10:377–387.

Roy Choudhury, S., M. Ghosh, A. Mandal, D. Chakravorty, M. Pal, S. Pradhan and A. Goswami. 2011. Surface-modified sulfur nanoparticles: An effective antifungal agent against *Aspergillus niger* and *Fusarium oxysporum. Appl. Microbiol. Biotechnol.* 90:733–743.

Roy, S. and A. Goswami. 2013. Supramolecular reactive sulphur nanoparticles: A novel and efficient antimicrobial agent. *J. Appl. Microbiol.* 114:1–10.

Saedi, S., M. Shokri and J.-W. Rhim. 2020. Antimicrobial activity of sulfur nanoparticles: Effect of preparation methods. *Arab. J. Chem.* 13:6580–6588.

Saha, B., J. Bhattacharya, A. Mukherjee, A. Ghosh, C. Santra, A.K. Dasgupta and P. Karmakar. 2007. In vitro structural and functional evaluation of gold nanoparticles conjugated antibiotics. *Nanoscale Res. Lett.* 2:614.

Saleemi, M.A., P.V.C. Yong and E.H. Wong. 2020. Investigation of antimicrobial activity and cytotoxicity of synthesized surfactant-modified carbon nanotubes/polyurethane electrospun nanofibers. *Nano-Structures & Nano-Objects.* 24:100612.

Salem, N., L. Al-Banna, A. Awwad, Q. Ibrahim and A. Abdeen. 2015. Green synthesis of nano-sized sulfur and its effect on plant growth. *J. Agric. Sci.* 8:188.

Salesa, B., M. Martí, B. Frígols and Á. Serrano-Aroca. 2019. Carbon nanofibers in pure form and in calcium alginate composites films: new cost-effective antibacterial biomaterials against the life-threatening multidrug-resistant *Staphylococcus epidermidis. Polymers (Basel).* 11:453.

Samak, D.H., Y.S. El-Sayed, H.M. Shaheen, A.H. El-Far, M.E. Abd El-Hack, A.E. Noreldin, K. El-Naggar, S.A. Abdelnour, E.M. Saied, H.R. El-Seedi, L. Aleya and M.M. Abdel-Daim. 2020. Developmental toxicity of carbon nanoparticles during embryogenesis in chicken. *Environ. Sci. Pollut. Res.* 27:19058–19072.

Schairer, D.O., J.S. Chouake, J.D. Nosanchuk and A.J. Friedman. 2012. The potential of nitric oxide releasing therapies as antimicrobial agents. *Virulence.* 3:271–279.

Scheinberg, D.A., M.R. McDevitt, T. Dao, J.J. Mulvey, E. Feinberg and S. Alidori. 2013. Carbon nanotubes as vaccine scaffolds. *Adv. Drug Deliv. Rev.* 65:2016–2022.

Schwartz, R., Z. Xu, D. Payne, T. Detemple and M. Bradley. 1990. Preparation and characterization of Sol-Gel derived PbTiO3 thin layers on GaAs. *MRS Proc.* 200: 167–172.

Şen Karaman, D., S. Manner and J.M. Rosenholm. 2018. Mesoporous silica nanoparticles as diagnostic and therapeutic tools: How can they combat bacterial infection? *Ther. Deliv.* 9:241–244.

Shankar, S., R. Pangeni, J.W. Park and J.-W. Rhim. 2018. Preparation of sulfur nanoparticles and their antibacterial activity and cytotoxic effect. *Mater. Sci. Eng. C. Mater. Biol. Appl.* 92:508–517.

Sharif, F., F. Porta, A.H. Meijer, A. Kros and M.K. Richardson. 2012. Mesoporous silica nanoparticles as a compound delivery system in zebrafish embryos. *Int. J. Nanomedicine.* 7:1875–1890.

Shin, J.H., S.K. Metzger and M.H. Schoenfisch. 2007. Synthesis of nitric oxide-releasing silica nanoparticles. *J. Am. Chem. Soc.* 129:4612–4619.

Slomberg, D.L., Y. Lu, A.D. Broadnax, R.A. Hunter, A.W. Carpenter and M.H. Schoenfisch. 2013. Role of size and shape on biofilm eradication for nitric oxide-releasing silica nanoparticles. *ACS Appl. Mater. Interfaces.* 5:9322–9329.

Soleimani, M., F. Aflatouni and A. Khani. 2013. A new and simple method for sulfur nanoparticles synthesis. *Colloid J.* 75:112–116.

Su, X.J., Z. Qi, S. Wang and A. Bendavid. 2010. Modification of diamond-like carbon coatings with fluorine to reduce biofouling adhesion. *Surf. Coatings Technol.* 204:2454–2458.

Tapaszto, L., G. Dobrik, P. Lambin and L. Biro. 2008. Tailoring the atomic structure of graphene nanoribbons by scanning tunnelling microscope lithography. *Nat. Nanotechnol.* 3:397–401.

Teng, Y., Q. Zhou and P. Gao. 2019. Applications and challenges of elemental sulfur, nanosulfur, polymeric sulfur, sulfur composites, and plasmonic nanostructures. *Crit. Rev. Environ. Sci. Technol.* 49:2314–2358.

Tiwari, A., Y.L. Sherpa, A.P. Pathak, L.S. Singh, A. Gupta and A. Tripathi. 2019. One-pot green synthesis of highly luminescent silicon nanoparticles using *Citrus limon* (L.) and their applications in luminescent cell imaging and antimicrobial efficacy. *Mater. Today Commun.* 19:62–67.

Valetti, S., X. Xia, J. Costa-Gouveia, P. Brodin, M.-F. Bernet-Camard, M. Andersson and A. Feiler. 2017. Clofazimine encapsulation in nanoporous silica particles for the oral treatment of antibiotic-resistant *Mycobacterium* tuberculosis infections. *Nanomedicine (Lond).* 12:831–844.

Vallet-Regi, M., A. Rámila, R.P. del Real and J. Pérez-Pariente. 2001. A new property of MCM-41: Drug delivery system. *Chem. Mater.* 13:308–311.

Villa, C.H., T. Dao, I. Ahearn, N. Fehrenbacher, E. Casey, D.A. Rey, T. Korontsvit, V. Zakhaleva, C.A. Batt, M.R. Philips and D.A. Scheinberg. 2011. Single-walled carbon nanotubes deliver peptide antigen into dendritic cells and enhance IgG responses to tumor-associated antigens. *ACS Nano.* 5:5300–5311.

Wang, L., C. Hu and L. Shao. 2017. The antimicrobial activity of nanoparticles: present situation and prospects for the future. *Int. J. Nanomedicine.* 12:1227.

Wang, S., R. Lawson, P.C. Ray and H. Yu. 2011a. Toxic effects of gold nanoparticles on *Salmonella typhimurium* bacteria. *Toxicol. Ind. Health.* 27:547–554.

Wang, T., M. Zou, H. Jiang, Z. Ji, P. Gao and G. Cheng. 2011b. Synthesis of a novel kind of carbon nanoparticle with large mesopores and macropores and its application as an oral vaccine adjuvant. *Eur. J. Pharm. Sci. Off. J. Eur. Fed. Pharm. Sci.* 44:653–659.

Wang, J.-X., Wen, L.-X., Wang, Z.-H. and Chen, J.-F. 2006. Immobilization of silver on hollow silica nanospheres and nanotubes and their antibacterial effects. *Mater. Chem. Phys.* 96:90–97.

Winkin, N., Gierth, U., Mokwa, W. and Schneider M. 2016. Nanomaterial-modified flexible micro-electrode array by electrophoretic deposition of carbon nanotubes. *J. Biochips Tissue Chips.* 6:1–6

Xu, C., Y. Niu, A. Popat, S. Jambhrunkar, S. Karmakar and C. Yu. 2014. Rod-like mesoporous silica nanoparticles with rough surfaces for enhanced cellular delivery. *J. Mater. Chem. B.* 2:253–256.

Yadav, H.K.S., M. Dibi, A. Mohammad and A.E. Srouji. 2018. Nanovaccines formulation and applications-a review. *J. Drug Deliv. Sci. Technol.* 44:380–387.

Yamada, H., C. Urata, Y. Aoyama, S. Osada, Y. Yamauchi and K. Kuroda. 2012. Preparation of colloidal mesoporous silica nanoparticles with different diameters and their unique degradation behavior in static aqueous systems. *Chem. Mater.* 24:1462–1471.

Yang, X., X. Liu, Y. Li, Q. Huang, W. He, R. Zhang, Q. Feng and D. Benayahu. 2017. The negative effect of silica nanoparticles on adipogenic differentiation of human mesenchymal stem cells. *Mater. Sci. & Eng. C, Mater. Biol. Appl.* 81:341—348.

Yao, X., X. Niu, K. Ma, P. Huang, J. Grothe, S. Kaskel and Y. Zhu. 2017. Graphene quantum dots-capped magnetic mesoporous silica nanoparticles as a multifunctional platform for controlled drug delivery, magnetic hyperthermia, and photothermal therapy. *Small.* 13(2): 1602225.

Yudovin-Farber, I., N. Beyth, A. Nyska, E.I. Weiss, J. Golenser and A.J. Domb. 2008. Surface characterization and biocompatibility of restorative resin containing nanoparticles. *Biomacromolecules.* 9:3044–3050.

Zaidi, S., L. Misba and A.U. Khan. 2017. Nano-therapeutics: A revolution in infection control in post antibiotic era. *Nanomedicine.* 13:2281–2301.

Zeinali, M., M. Jammalan, S.K. Ardestani and N. Mosaveri. 2009. Immunological and cyto-toxicological characterization of tuberculin purified protein derivative (PPD) conjugated to single-walled carbon nanotubes. *Immunol. Lett.* 126:48–53.

Zhang, L., F.X. Gu, J.M. Chan, A.Z. Wang, R.S. Langer and O.C. Farokhzad. 2008. Nanoparticles in medicine: Therapeutic applications and developments. *Clin. Pharmacol. Ther.* 83:761–769.

Zhang, L., D. Pornpattananangkul, C.-M. Hu and C.-M. Huang. 2010. Development of nanoparticles for antimicrobial drug delivery. *Curr. Med. Chem.* 17:585–594.

Zou, Y., R. Xie, E. Hu, P. Qian, B. Lu, G. Lan and F. Lu. 2020. Protein-reduced gold nanoparticles mixed with gentamicin sulfate and loaded into konjac/gelatin sponge heal wounds and kill drug-resistant bacteria. *Int. J. Biol. Macromol.* 148:921–931.

7 Biological Nanomaterials for Toxicity of Bacteria

*Diptikanta Acharya, Sagarika Satapathy,
Prasanna Kumar Dixit, Gitanjali Mishra, Padmaja
Mohanty, Jayashankar Das and Sushma Dave*

CONTENTS

7.1 General Consideration of Nanotechnology 187
7.2 Mechanisms Involved in Nanoparticle Synthesis 188
7.3 General Toxicity of NPs on Bacteria 190
7.4 Antibacterial Activity of NPs 190
7.5 Mechanisms of Nanoparticle-Induced Toxicity 191
7.6 Challenges and Toxicity 192
7.7 Various Nanomaterials and Their Toxicity on Bacteria 193
    7.7.1 Silver Nanoparticles (AgNPs) 193
    7.7.2 Gold Nanoparticles (AuNPs) 194
    7.7.3 Iron Oxide Nanoparticles 195
    7.7.4 Titanium Oxide (TiO_2) 195
    7.7.5 Copper Oxide (CuO) 195
    7.7.6 Zinc Oxide (ZnO) 196
7.8 Conclusion 197
References 197

7.1 GENERAL CONSIDERATION OF NANOTECHNOLOGY

In the last decades, nanotechnology has emerged as an important field. The vital component of nanotechnology is nanoparticles, which are considered as the key foundation of assorted nanomaterials or nanodevices [1]. Nanoparticles can be synthesized either naturally or artificially [2]. In the sectors of science and technology, metal nanoparticles (NPs) are considered as one of the most diversified materials having enormous applications in the fields of agriculture, medicine, environment, electronics, and structural engineering [3].

In nanotechnology, nanoparticles are distinct, minute entities that behaves as a complete unit by way of their transport and properties. Thus, nanoparticles have been classified based on their variable diameters [4]. In general, the size range of coarse particles are from 10,000 to 2500 nm, fine particle are from 2500 to 100 nm, and silver nanoparticles (AgNPs) vary from 1 to 100 nm. In recent years, the

DOI: 10.1201/9781003126256-7

progress in nanomaterials research and their industrial applications are due to the advancement in biological synthesis and modern characterization tools [5, 6].

Although, there may some lack of knowledge regarding the molecular interaction mechanism of silver nanoparticles in biological or nanoscience systems, the demand for silver nanoparticles in industry as marketable products has been considerably high. As human health and environmental safety are a concern, the physicochemical properties of AgNPs, like greater potent reactivity and large specific surface area, have elevated subsequent questions [7]. The eradication of associated risks also possible by using nanoparticles to make the nanotechnology field more sustainable. However, the affluence of nanoparticles in soil and water remain limited [8]. In the environment, as far as the food chain is concerned, silver nanoparticles might be bioaccumulated in an organism's body and transfer to each trophic level [9]. Plants act as producers in the food chain and are mainly responsible for the transport pathway system for bioaccumulated silver nanoparticles into the food chain of ecological systems [10, 11].

In the standard chemical and physical process, metal nanoparticles can be synthesized, but they are considered to be more toxic to biological organisms and to the environment. On the contrary, an alternative safe approach using available natural resources has been associated with the biological union of nanoparticles, where plant parts and microorganisms are used [12]. For the blend of metal nanoparticles, plants are observed as the best natural precursor because they have fewer biohazards in comparison to microbes, reduce the risk of isolation, maintenance of the microorganism, and their culture media cost [13]. Therefore, the cost capacity for the use of microbes is growing. The designed concord has provided valuable information about the size and dispersal of metal/metal oxide nanoparticles and their management [14]. Without any adverse effects on the environment, researchers have given more attention toward the green approach, i.e., nanoparticle synthesis using plant extracts in aqueous conditions [15]. Nanoparticles are significant in various processes, especially in photocatalysis and biomedical approaches. Nanoparticles have vast applications in many domains such as antimicrobial activity, antioxidants, bioimaging, fuel cells, catalysts, sensors, and cosmetics, because of their free radical scavenging properties and small size [14].

The possible role as reducing agents from different phytochemicals, such as flavonoids, alkaloids, terpenoids, amides, and aldehydes, and other biological constituents have been utilized to characterize the biophysical properties along with the rate-limiting steps during green synthesis. Furthermore, the utilization of various organic entities like microorganisms, parasites, and plant extricates have benefits for the synthesis of metal and metal oxide nanoparticles. There are several advantages of green synthesis approaches of nanoparticles, which will be helpful to investigators and researchers associated with this budding research area [16, 17].

7.2   MECHANISMS INVOLVED IN NANOPARTICLE SYNTHESIS

Diverse biomolecules like proteins, amino acids, enzymes, polysaccharides, and vitamins present in the biological extract are used for the reduction of metal ions, which is an environment-friendly but chemically intricate process [18]. A well-known enzyme

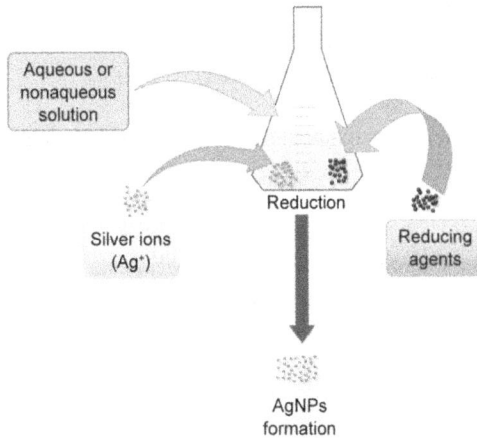

FIGURE 7.1 A hypothetical diagram for formation of silver nanoparticles.

for synthesis of AgNPs is nitrate reductase. This enzyme is responsible for conversion of nitrate to nitrite in the nitrogen cycle. This reduction process is performed by nitrate reductase. An enzyme called NADPH-dependent nitrate reductase is used for in vitro AgNP synthesis, as it helps in eliminating the downstream processing needed for the application of AgNPs in homogenous catalysis and nonlinear optics. In catalysis process conversion of nitrate into nitrite results in the shuttling of an electron and forms the silver ion. The microorganism *Bacillus licheniformis* secretes the co-factor NADH and enzyme NADH-dependent nitrate reductase, which biologically reduce Ag^+ to Ag^0, subsequently leading to AgNP production [19].

Among these undertakings, concrete evidence for the formation of nanoparticles was specified. The microorganism *Fusarium oxysporum* produces a purified enzyme nitrate reductase that helps in silver nanoparticles production. A test tube containing a reaction mixture of silver nitrate and NADPH with an enzyme nitrate reductase, slowly exhibited a brown color produced by the AgNPs. This came as primary confirmation for AgNPs synthesis by enzyme nitrate reductase [20].

Although synthesis of AgNPs is an "ability" of microorganisms, more precisely it is a defense mechanism of microorganisms for extremely reactive silver ions. Silver ions attach with different biomolecules/components of an microorganism to induce its death, called silver ions antimicrobial activity. The real mechanism is to alter the very reactive silver ions to inert AgNPs [21] (Figure 7.1).

Remarkably, "apoptosis" also occurs in microorganisms, like cells of higher organisms. Some of the different pathways of silver ions antimicrobial activity include the following:

1. They can bind to the negatively charged DNA due to an absence of histone proteins in prokaryotic organisms, causing a loss of the native structure of DNA and stopping the replication process.

2. They can also bind to the protein have a thiol-group and stop their activity.
3. They induce synthesis of reactive oxygen species (ROS), which produce highly sensitive radicals that devastate cells [22].

Silver ions also halt function of the NADH dehydrogenase II enzyme of the respiratory system, a site of reactive oxygen species production in vivo. The free radical hydroxyl (OH–) produced from H_2O_2 is a by-product of Fenton reaction [23]. Catalases are known to be scavengers of reactive oxygen species in microbes, which can help in an indirect analysis of the aforementioned mechanism. Experiments involving *B. licheniformis* showed that the silver ion concentration rises when catalase synthesis increases, helping cells to survive until minimal inhibitory concentration is reached. This is evident in *E. coli* UM1 (katEkat G) mutated strain, which because it lacks catalase, becomes more sensitive to silver nitrate and silver zeolite from its parent, so silver ions do apoptosis in *E. coli* UM1. For protection, *E. coli* UM1 converts reactive silver ions to an inactive Ag^0 form. A supplementary entry of silver ions with electrons produce Ag crystals [24].

7.3   GENERAL TOXICITY OF NPS ON BACTERIA

From the literature data, it has been seen that NPs discharged into the environment may cause harmful reactions to the natural microbial alliance and other organisms.

NPs can change the character of bacterial cell film, impact the penetrability of the bacterial cells, can cause damage to DNA, and deliver the harmful particles [25]. Graphene-based silver NPs were tried comparable to *Pseudomonas aeruginosa* and tracked down the smallest bactericidal fixation (MBC) at 20 mg/ml, and after one hour deterioration of the cell wall [26]. Silver and copper NPs also pass through the wall and cell membrane of bacteria *Listeria monocytogenes*, which led to the partition of the cytoplasmic membrane of the cell wall [27]. Many NPs have antibacterial characteristics against bacteria like *Aeoromonas hydrophila* (nano-Al_2O_3, nano-ZrO_2); *Salmonella typhimurium* (Cu NPs/chitosan); *Klebsiella pneumoniae* (nano-ZnO); *P. aeruginosa* (nano-Ag); *Enterobacter cloacae* (Cu-SiO_2 nanocomposite); *L. monocytogenes* (Cu NPs); *Enterococcus jaecalis* (nano-TiO_2); and *Staphylococcus aureus*, *Escherichia coli*, and *Bacillus subtilis* (nano-Cu, nano-ZnO, and nano-Ag) [28, 29].

Also, NPs demonstrate an apathetic effect on microorganisms associated with wastewater treatment. It has been seen that zinc oxide NPs had troublesome effects on nitrogen evacuation by natural cycles, just as on denitrifying microbes and microorganisms liable for phosphorus removal called PAO (polyphosphate aggregating living beings) [30]. It has likewise been accounted for that during wastewater treatment, 1 rng/l of nanosilver repressed the development of autotrophic nitrifying microscopic organisms by about 80%. There is also poisonousness impacts of aluminum oxide NPs (nano-Al_2O_3, <50 nm) against *P. putida* microbes [31].

7.4   ANTIBACTERIAL ACTIVITY OF NPS

Scientists have done extensive research on the bactericidal effects of NPs against anaerobic and aerobic bacteria. Although AgNPs with very little concentration do not produce any harm on human cells, they can prove fatal for the majority of viruses and bacteria. AgNPs show activity to reduce cell toxicity without having an effect on their antibacterial efficiency [32]. Nanoparticles possess a finely smooth surface and can penetrate through the cell membrane, adversely affecting intracellular processes. The formation of free radicals on AgNPs surface renders them a higher antibacterial effect [33]. Various mechanisms explaining the activity of inhibition of bacteria cells by AgNPs have been recognized. AgNPs have a high affinity for phosphorus and sulfur, which is a prime reason for their antibacterial properties. Many proteins present on the cell membrane contain sulfur. The AgNPs and the sulfur-containing amino acids show reaction both inside and outside of the cells destroying the viability of the microorganism [34]. It has also been reported that the reaction of silver ions and phosphorous result in inhibition and replication of DNA. The reaction of proteins having sulfur inhibit the function of several enzymes. AgNPs smaller than ≤ 10 nm have been reported to generally attack sulfur-containing bacterial proteins, increasing the penetrability of plasma membrane of the cell and eventually dying of cells takes place [35]. It has been shown that these AgNPs can create pores on the walls of bacterial cells, which releases cytoplasmic contents of the cell resulting in cell death, and these particles do not interact with proteins and nucleic acids inside the cell. The addition of antibiotics to AgNPs has been observed to have synergic effects against various microbes [36].

The modification of Ag sulfadiazine with the addition of dendrimers has been seen to improve its antibacterial efficacy. The AgNPs also have anti-inflammation activity. The polymers containing a lower quantity of AgNPs enhance their bactericidal function. An example of such a polymer is cationic chitosan. Composites of chitosan with AgNPs have a higher antibacterial effect than chitosan and AgNPs individually. The capture of a matrix with positive charge chitosan and negative charge bacterial surface results is formation of pores, thereby leading bacteria cells disintegrate. The bactericidal function of AgNPs depends on the shape and size. As the bactericidal function of nanoparticles are heavily dependent on the surface-to-volume ratio and the penetration ability, these smaller particles have been shown to have higher antibacterial activity than larger particles [37, 38, 39].

7.5 MECHANISMS OF NANOPARTICLE-INDUCED TOXICITY

AgNP-induced toxicity in cells is revealed by the interaction among silver nanoparticles and cells followed by cellular uptake and toxic reaction [40]. AgNP uptake into cells is mostly by the endocytosis mechanism and it completely depends upon the occurrence of main organelles, i.e., endosomes and lysosomes [41]. In an acidic environment, lysosomes form Ag^+ ions due to exposure of silver nanoparticles inside and induce ROS production [42]. ROS are highly reactive molecules, comprised of superoxide anions, hydroxyl radicals, and hydrogen peroxide. The major factor behind Ag^+ discharge in vivo is due to reactions between hydrogen peroxide and AgNPs. As ROS are highly reactive, they enhance oxidative damage to proteins and DNA, and stimulate mitochondrial dysfunction within the cells. In affected cells, the cell membrane is ruptured, cytoplasmic contents leak out, and eventually necrosis

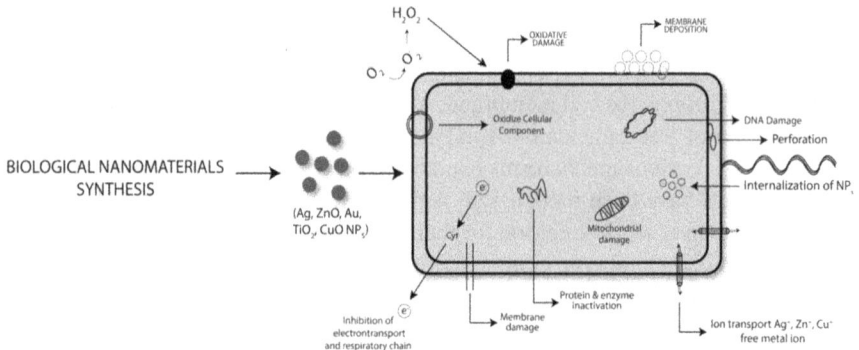

FIGURE 7.2 Toxicity mechanism of biogenic nanoparticles on bacteria.

occur. In addition, the mitochondrial dysfunction impairs electron transfer, therefore mitochondrion-dependent apoptosis is also identified in cells [43]. It has been also reported that through nuclear pores, AgNPs could willingly diffuse and translocate to the nucleus, and thereby inducing the formation of ROS, which directly activate chromosomal abnormality and damage DNA [41] (Figure 7.2).

7.6 CHALLENGES AND TOXICITY

The transition of AgNPs from the laboratory to commercial applications is one of the major challenges of the upcoming decade. A vast majority of research scholars and scientists all over the world are dedicated to the production and analysis of nanoparticles. Although AgNPs are reported to display very little toxicity to the body, they could become life threatening [44]. More research should be aimed for synthesis of AgNPs to overcome these problems. Then they could be applied for making active drug delivery agents for the treatment of lethal diseases with higher safety and efficacy. But, the aggregation and toxicity of AgNPs limit their usage for certain applications. Medical care for a burn patient using Ag$^+$ ions causes a hypersensitivity response. Research has also found that AgNPs cause harm for particular cell lines. It has been reported that the toxicity of AgNPs is inversely related to the size of the AgNPs. Aggregation of silver nanoparticles is caused by their high surface energy. AgNPs show easy contamination or oxidization with the air due to their upper surface energy. However, this difficulty has been completely controlled by AgNPs integrated with biological polymer matrices like chitosan, alginate, and gelatin [45].

As the particles are of nano size, they can easily permeate through the cell membrane and enter into tiny capillaries present in the body. The size, surface area, and morphology of nanoparticles have been recognized to be important and determine their toxicity. However, information on the toxic implications of AgNPs is limited. The toxic nature of AgNPs is identified in vitro with particles 1–100 nm in size [46].

AgNPs also possess the ability to attach with various tissues, including proteins and enzymes, in cells of mammals and create toxicity. The adhesive interaction with cell plasma membrane resulting in the creation of highly reactive toxic radicals, i.e.,

ROS, creates skin inflammation and mitochondria toxicity. The most specific organs of nanosilver toxic effect include the liver and the immune system [21].

7.7 VARIOUS NANOMATERIALS AND THEIR TOXICITY ON BACTERIA

Silver (Ag), gold (Au), titanium oxide (TiO_2), copper oxide (CuO), zinc oxide (ZnO), and iron oxide (Fe_3O_4) are among those metal and metal oxide nanoparticles that are acknowledged for their extreme toxic effects. Details of the toxic effects are discussed in the following sections.

7.7.1 SILVER NANOPARTICLES (AGNPS)

AgNPs are well known for their amazing antimicrobial properties upon various microbes, like microorganisms, growths, and infections. Notwithstanding, the systems responsible for the bactericidal impact of AgNPs remain uncertain. There is continuing debate about whether an AgNP or silver particle shows a cytotoxic impact on microorganisms [46]. The toxic effects of AgNPs have been expected upon immediate contact of nanoparticles to the bacterial cell wall, followed by infiltration into the cytoplasm. AgNPs with large surfaces reach straightforwardly on a bacterial cell wall and incite harm to the film, resulting in the removal of cell material and is fatal to cell [33]. It has been seen that AgNPs with sizes under 10 nm are more poisonous for microorganisms. AgNPs pass into the cytoplasm and connect with biomolecules like proteins, lipids, and DNA. AgNPs additionally connect with the respiratory chemicals and cause the decaying of ROS, for example, hydrogen peroxide (H_2O_2), hydroxyl (OH–), and superoxide (O_2–) revolutionaries, which actuate oxidative strain and damage to proteins and nucleic acids [21]. It has been accounted for that the buildup of AgNPs of 12 nm size on the cell mass of *E. coli* cause the formation of pits. These pits cause a deficiency of dependability in the external film, leading to the arrival of lipopolysaccharide particles and layer proteins, and thus possible cell death happens [33]. It has been seen that AgNPs (1–10 nm) collaborate on the cell mass of *E. coli* and upset its typical activities, like respiration and penetrability. The nanoparticles likewise go into cytoplasm and along with protein and DNA cause cell death. Furthermore, AgNPs can likewise release silver particles, prompting further harm in cells [36]. Recently, it is additionally seen that AgNPs (10 nm) interfere in the penetrability and metabolic pathways of cholera, by joining to its cell wall and causing cell death [47].

The cytotoxic activity of AgNPs against microscopic organisms is on the grounds that the oxidative disintegration of Ag particles from AgNPs. As we probably are aware, metals are synthetically oxidized in fluid for creating metallic particles. Subsequently, AgNPs consistently oxidize in liquid for assembly of Ag particles [35]. It has been shown that silver particles created from AgNPs in high-impact conditions are totally responsible for antibacterial action. In any case, in an anaerobic condition, very little Ag particles are formed, so for this situation AgNPs are nonharmful to microbes [36]. It has been seen that Ag^+ particles are firmly related to their communication with thiol (sulfhydryl), which causes expected antimicrobial activity. Consequently, Ag particles can without much effort associate with the –SH gatherings of catalysts and proteins present in the cell divider, prompting interference with the respiratory chain of microscopic organisms, cause problems of the bacterial cell divider. Also, the particles yielded from AgNPs can enter the cell divider into the cytoplasm, in this manner debasing chromosomal DNA, or partner with thiol gatherings of the proteins present in the cytoplasm. Thus, the duplication of DNA will be halted and the proteins important to the ATP proliferation get inactivated. As a rule, nanosilver particles can enter the cytoplasm habitually than bulk particles. Direct contact among cells and nanoparticles propel the release of silver particles from AgNPs, hence raising the amount of cell take-up of molecule-related Ag particles [48]. It is additionally known that positively charged particles released from AgNPs tend to impede with the ability of the bacterial electron transport chain of *E. coli*, creating ROS. ROS cause bacterial death since they work with lipid peroxidation, as well as hinder DNA replication and ATP fabrication. In this way, increased ROS levels in bacterial cells can cause oxidative strain [49].

7.7.2 Gold Nanoparticles (AuNPs)

AuNPs have wide potential for drug delivery frameworks. The release of numerous proteins, vaccines, and nucleotides to their objective are conceivable by AuNPs [50]. It has been seen that AuNP–drug forms show better antibacterial action than individual nanoparticles and medications. Kanamycin and AuNP–kanamycin forms have likely antibacterial activity against gram-positive *Staphylococcus* epidermidis and gram-negative *Enterobacter aerogenes* [51]. Likewise the antibacterial nature of gallic acid and AuNP–gallic acid against the food-borne pathogenic bacterial species *Plesiomonas shigelloides* and *Shigella flexneri* B has been observed. AuNP–gallic modifies lipids, proteins, and nucleic acids of the bacterial cell film. The AuNP conjugates generally have improved antimicrobial efficacy due to the adhesion and delivery ability with their conjugates, probably due to their smaller diameter nanoparticles that exhibit surface effects and large surface area-to-mass ratios that can adsorb large quantities of molecules [52].

So, the mutation of AuNPs through their conjugates stabilizes the particles and thus enhances delivery efficacy. It has been reported that AuNPs interact with bacterial components, such as lysosomes, ribosomes, and enzymes, and mutate the permeability of the cell membrane, causing the interruption of enzyme inhibition, electrolyte balance, and protein deactivation [33]. Also it has been reported that AuNP conjugates pass the cell wall of bacteria, resulting in changes of the cellular atmosphere and causing cell lysis and the leakage of cellular components [33]. It is reported that 45 nm AuNPs enter cells via clathrin-mediated endocytosis, while 13 nm AuNPs penetrated mostly through phagocytosis. The exact mechanisms of antimicrobial activities of AuNPs are not clarified, but the size, shape, and AuNP surface modifications activate the antimicrobial activity, and further research is required to resolve the accurate processes involved in AuNP–bacteria interactions [53].

7.7.3 Iron Oxide Nanoparticles

Iron oxide nanoparticles are a special type of metal oxide nanomaterials, having uncommon high adsorbing characteristics, magnetic properties, and superior bio-activities. Iron oxide nanoparticles have an widespread history of application in the environmental and biomedical engineering fields, magnetic fluid hyperthermia, drug delivery, magnetic separation of immune cells, tissue engineering applications, orthopedic insert infection prevention, arsenic(V) removal, environmental redress, and wastewater treatment [54]. It has been seen that the biological preparation of iron oxide nanoparticles executed significant antibacterial activity upon gram-positive *Staphylococcus aureus* and gram-negative *Pseudomonas aeruginosa* [55]. It has been described that there is significant antibacterial efficacy of plant-extract-mediated iron oxide nanoparticles against *Proteus mirabilis* MTCC 425, *Escherichia coli* MTCC 443, and *Bacillus subtilis* MTCC 441 [56]. Antibacterial efficiency has also seen with iron oxide nanoparticles against the biofilms of *Escherichia coli*, *Staphylococcus aureus, and Pseudomonas aeruginosa* [57].

7.7.4 Titanium Oxide (TiO₂)

From numerous observations it has been seen that metal oxide has exceptional anti-fungal and antibacterial activities against a wide range of gram-positive and gram-negative microbes. Among these, biosynthesis of titanium dioxide nanoparticles (TiO_2 NPs) has critical harmfulness against microscopic organisms [58]. Titanium dioxide when exposed to ultraviolet light breaks down to produce free oxygen radicals that attack anything attached to their surface. Basically it attacks the organic material, like bacterial cell membrane [59]. Titanium dioxide nanomaterials through biological synthesis have been successful against a wide range of microorganisms, like protozoa, algae, parasites, microbes, and viruses. Titanium nanoparticles have antibacterial properties against gram-positive *Staphylococcus aureus* and gram-negative *Escherichia coli* [58]. Especially, titanium dioxide has been estimated as a brilliant antimicrobial compound because of its photocatalytic nature, and it is an artificially steady, non-poisonous, modest, and generally recognized as safe (GRAS) substance. It is reported that infections and microbes can be reduced utilizing titanium dioxide by means of photocatalysis (openness to bright light) [60, 61].

7.7.5 Copper Oxide (CuO)

Copper oxide nanoparticles (CuO NPs) perform splendid antimicrobial action against a few bacterial strains. CuO NPs have potential as antibacterial specialists in surface coatings on various film to prevent microorganisms from connecting, colonizing, spreading, and framing biofilms. Nano-copper oxide has among the best antimicrobial outcome upon dental caries bacteria [62]. Due to huge surface-to-volume ratio of CuO NPs, they can be used in diverse ways, including catalysts, sensors, electronics, medicines, and bioremediation of pollutants. CuO NPs possess in vitro and in vivo toxicity against bacteria, algae, fish, rats, and human cell lines. The major details that control the toxicity of CuO NPs are particle shape, size, and surface functionalization [63]. The research data shows that CuO NP contact to the living systems is in the formation of ROS resulting in oxidative stress in bacteria. But, the physiochemical properties of CuO NPs, concentration, mode of disclosure, and assessment are the main outlooks that describe the toxicology of CuO NPs [64]. It has been seen that the CuO NPs have potential antimicrobial activity against *Streptococcus mutans*, *Lactobacillus acidophilus*, and *Lacticaseibacillus casei* [65]. CuO NPs are also harmful for native soil bacteria. CuO NPs interact with cell wall components and can change the cell morphology, disrupting the function of membrane proteins. Thus, the cytotoxicity of CuO NPs are owing to the generation of reactive species through oxide-reduction reactions upon the cell membrane [63].

7.7.6 Zinc Oxide (ZnO)

The iniquity of zinc oxide has been seen in numerous organisms and plants. The harmful impact of zinc oxide nanoparticles depends on the fixation and dissolvability. The harmfulness of zinc oxide nanoparticles is because of the onset of Zn^{2+} particles in the fluid medium. The antimicrobial impacts of metal oxide nanoparticles are exceptionally reasonable to bacterial groups, which is reflected by the inhibition zones [66].

Zinc oxide nanoparticles can get into food material and kill microorganisms. In this manner, zinc oxide is for the most part utilized as an additive and incorporated in polymeric wrapping material to prevent food material from being harmed by organisms [66]. It has been seen that *Salmonella typhi* and *S. aureus* can be restrained by zinc oxide nanoparticles in vitro [67].

It is for the most part realized that zinc oxide nanoparticles are poisonous to microbes and hinder the intensification and advancement of microorganisms by getting into the cell film. Along these lines it produces oxidative pressure in microscopic organisms resulting in harming lipids, sugars, proteins, and DNA. Lipid peroxidation brings about change in the cell layer, which ultimately upsets the fundamental cell capacities [68]. It has been advised that oxidative strain happens in *Escherichia coli* through zinc oxide nanoparticles. In any case, in mass zinc oxide suspension, concentration of H_2O_2 has been reported to clarify the counter bacterial properties. Strangely, the zinc oxide is amphoteric in nature, that it can responds with the two acids and antacids and can create Zn^{2+} particles [33].

Although the mutagenic capability of zinc oxide has not been efficiently concentrated in microscopic organisms, its DNA-harming potential has been accounted for. Zinc oxide nanoparticles are generally set off by retention of ultraviolet (UV) light and can turn out to be more receptive. The antibacterial movement of zinc oxide nanoparticles might be catalyzed by daylight, yet hopefully, they can impede the development of ROS. Zinc oxide nanoparticles and zinc nanoparticles covered with solvent polymeric material are helpful for treating wounds, ulcers, and numerous microbial contaminations. They have tremendous potential as a secured antibacterial medication, which may substitute for anti-infection agents in the future [69, 66] (Table 7.1).

TABLE 7.1
Toxicity of Some Biogenic Nanomaterials on Bacteria

NP	Name of the Bacteria	Effect	Reference
AgNP	*Azotobacter vinelandii*	Reduces cell number, induces apoptosis, causes structural damage, inhibits biological nitrogen fixation (BNF), ROS generation	[70]
AgNP	*Escherichia coli*	33–45% *E. coli* cells are destroyed	[71]
AgNP	Soil microbial activity	Reduction in microbial metabolic activity and its nitrifying ability	[72]
AuNP	*Escherichia coli* and *Staphylococcus aureus*	Toxicity	[73]
AuNP	*P. aeruginosa*	Antibiofilm activity and toxicity	[74]
CuO NP	Soil nitrifying bacteria	Suppressed the nitrification kinetics	[75]
CuO	*Bacillus subtilis*	Toxicity	[76]
FeO	*Staphylococcus aureus* and *Pseudomonas aeruginosa*	Bactericidal activity	[77]
FeO	*Bacillus subtilis Escherichia coli*	ROS production, bactericidal activity	[78]
TiO_2	*Vibrio fischeri*	Toxicity	[79]
TiO_2	*Pseudomonas aeruginosa*	Toxicity	[80]
ZnO	*Escherichia coli* O157:H7	960 mg/l causes 100 % inhibition	[81]
ZnO	*Escherichia coli*	20 mg/l ZnO causes 100% mortality	[82]
ZnO	*Escherichia coli*	Causes 92% inhibition	[83]

7.8 CONCLUSION

Despite the fact that there are some risks to the climate and their outcome on living life forms, research on the impact of nanomaterials on microorganisms has to be clarified. Expanding utilization of nanomaterials in customary applications has elevated the concern for of their risks because of their ominous and toxicological consequences for the microbial community.

REFERENCES

1. Jeevanandam, J., Barhoum, A., Chan, Y. S., Dufresne, A. and Danquah, M. K. (2018). Review on nanoparticles and nanostructured materials: History, sources, toxicity and regulations. *Beilstein Journal of Nanotechnology*, 9, 1050–1074.
2. Griffin, S., Masood, M., Nasim, M., Sarfraz, M., Ebokaiwe, A., Schäfer, K. H., Keck, C. and Jacob, C. (2017). Natural nanoparticles: A particular matter inspired by nature. *Antioxidants*, 7(1), 3.
3. Shang, Y., Hasan, M. K., Ahammed, G. J., Li, M., Yin, H. and Zhou, J. (2019). Applications of nanotechnology in plant growth and crop protection: A review. *Molecules*, 24(14), 2558.

4. Das, J. and Velusamy, P. (2013). Antibacterial effects of biosynthesized silver nanoparticles using aqueous leaf extract of *Rosmarinus officinalis* L. *Materials Research Bulletin*, 48, 4531–4537.

5. David, L., Moldovan, B., Vulcu, A., Olenic, L., Schrepler, M. P., Fodor, E. F, Florea, A., Crisan, M., Chiorean, I., Clichici, S. and Filip, G. A. (2014). Green synthesis, characterization and anti-inflammatory activity of silver nanoparticles using European black elderberry fruits extract. *Colloids and Surfaces B: Biointerfaces*, 122, 767–777.

6. Dhas, T. S., Kumar, V. G., Abraham, L. S., Karthick, V. and Govindaraju, K. (2012). *Sargassum myriocystum* mediated biosynthesis of gold nanoparticles. *Spectrochimica Acta Part A*, 99, 97–101.

7. Zhang, X. F., Liu, Z. G., Shen, W. and Gurunathan, S. (2016). Silver nanoparticles: Synthesis, characterization, properties, applications, and therapeutic approaches. *International Journal of Molecular Sciences*, 17(9), 1534.

8. Elumalai, E., Prasad, T., Hemachandran, J., Therasa, S. V., Thirumalai, T. and David, E. (2010). Extracellular synthesis of silver nanoparticles using leaves of *Euphorbia hirta* and their antibacterial activities. *Journal of Pharmaceutical Sciences and Research*, 2, 549–554.

9. Fayaz, A. M., Balaji, K., Kalaichelvan, P. T. and Venkatesan, R. (2009). Fungal based synthesis of silver nanoparticles: An effect of temperature on the size of particles. *Colloids and Surfaces B: Biointerfaces*, 74, 123–126.

10. Fayaz, A. M., Tiwary, C. S., Kalaichelvan, P. T. and Venkatesan, R. (2010). Blue orange light emission from biogenic synthesized silver nanoparticles using *Trichoderma viride*. *Colloids and Surfaces B: Biointerfaces*, 75, 175–178.

11. Dave, S. (2018). Electrochemical and spectral characterization of silver nanoparticles synthesized employing root extract of *Curculigo orchioides*. *Indian Journal of Chemical Technology (IJCT)*, 25(2), 201–207.

12. Singh, J., Dutta, T., Kim, K. H., Rawat, M., Samddar, P. and Kumar, P. (2018). Green synthesis of metals and their oxide nanoparticles: Applications for environmental remediation. *Journal of Nanobiotechnology*, **16,** 84 (1).

13. Dave, S., Dave, S., Mathur, A. and Das, J. (2021). Biological synthesis of magnetic nanoparticles. In *Nanobiotechnology* Edited by Ghosh et al. (pp. 225–234). Elsevier.

14. Annu, A. A. and Ahmed, S. (2018). Green synthesis of metal, metal oxide nanoparticles, and their various applications. In *Handbook of Ecomaterials* In: Martínez L., Kharissova O., Kharisov B. (eds) Handbook of Ecomaterials. Springer, Cham. https://doi.org/10.1007/978-3-319-48281-1_115-1 (pp. 1–45). Springer.

15. Dave, S., Dave, S. and Das, J. (2021). Biological synthesis of platinum, palladium, copper, and zinc nanostructures. In *Nanobiotechnology* (pp. 211–223). Elsevier. https://doi.org/10.1016/B978-0-12-822878-4.00013-4.

16. Acharya, D., Satapathy, S., Somu, P., et al. (2021). Apoptotic effect and anticancer activity of biosynthesized silver nanoparticles from marine algae *Chaetomorpha linum* extract against human colon cancer cell HCT-116. *Biological Trace Element Research*, 199, 1812–1822. https://doi.org/10.1007/s12011-020-02304-7.

17. Acharya, D., Satapathy, S., Thathapudi, J. J., Somu, P. and Mishra, G. (2020). Biogenic synthesis of silver nanoparticles using marine algae *Cladophora glomerata* and evaluation of apoptotic effects in human colon cancer cells. *Materials Technology*, 2020, 1–12. https://doi.org/10.1080/10667857.2020.1863597.

18. Sharma, D., Kanchi, S. and Bisetty, K. (2015). Biogenic synthesis of nanoparticles: A review. *Arabian Journal of Chemistry*, 12(8), 3576–3600.

19. Lee, S. and Jun, B. H. (2019). Silver nanoparticles: Synthesis and application for nanomedicine. *International Journal of Molecular Sciences*, 20(4), 865.

20. Birla, S. S., Gaikwad, S. C., Gade, A. K. and Rai, M. K. (2013). Rapid synthesis of silver nanoparticles from *Fusarium oxysporum* by optimizing physicocultural conditions. *The Scientific World Journal*, 2013, 1–12.

21. Dakal, T. C., Kumar, A., Majumdar, R. S. and Yadav, V. (2016). Mechanistic basis of antimicrobial actions of silver nanoparticles. *Frontiers in Microbiology*, 7, 1831.

22. Dave, S. and Tarafdar, J. C. (2016). Phytofabrication of copper nanoparticles using rhizomes of *Curculigo orchioides* (Kali Musli) and its antimicrobial activity, *International Journal of Advanced Research* (IJAR)4, 152–158.

23. Zorov, D. B., Juhaszova, M. and Sollott, S. J. (2014). Mitochondrial reactive oxygen species (ROS) and ROS-induced ROS release. *Physiological Reviews*, 94(3), 909–950.

24. Ravindra, K. S. and Imlay, J. A. (2013). How *Escherichia coli* tolerates profuse hydrogen peroxide formation by a catabolic pathway. *Journal of Bacteriology*, 195(20), 4569–4579.

25. Tripathi, D. K., Tripathi, A., Shweta, S, S., Singh, Y., Vishwakarma, K., Yadav, G., Sharma, S., Singh, V. K., Mishra, R. K., Upadhyay, R. G., Dubey, N. K., Lee, Y. and Chauhan, D. K. (2017). Uptake, accumulation and toxicity of silver nanoparticle in autotrophic plants, and heterotrophic microbes: A concentric review. *Frontiers in Microbiology*, 8, 07. https://doi.org/10.3389/fmicb.2017.00007.

26. Chauhan, N., Tyagi, A. K., Kumar, P. and Malik, A. (2016). Antibacterial potential of *Jatropha curcas* synthesized silver nanoparticles against food borne pathogens. *Frontiers in Microbiology*, 7, 1748. https://doi.org/10.3389/fmicb.2016.01748.

27. Dave, S., Khan, A. M., Purohit, S. D. and Suthar, D. L. (2021). Application of green synthesized metal nanoparticles in the photocatalytic degradation of dyes and its mathematical modelling using the Caputo–Fabrizio fractional derivative without the singular kernel. *Journal of Mathematics*, 9948422, 8 pages, https://doi.org/10.1155/2021/9948422.

28. Baptista, P. V., McCusker, M. P., Carvalho, A., Ferreira, D. A., Mohan, N. M., Martins, M. and Fernandes, A. R. (2018). Nano-strategies to fight multidrug resistant bacteria-A battle of the titans. *Frontiers in Microbiology*, 9, 1441. https://doi.org/10.3389/fmicb.2018.01441.

29. Sirelkhatim, A., Mahmud, S., Seeni, A., Kaus, N., Ann, L. C., Bakhori, S., Hasan, H. and Mohamad, D. (2015). Review on zinc oxide nanoparticles: Antibacterial activity and toxicity mechanism. *Nano-Micro Letters*, 7(3), 219–242. https://doi.org/10.1007/s40820-015-0040-x.

30. Dave, S., Dave, S. and Das, J. (2021). Photocatalytic degradation of dyes in textile effluent: A green approach to eradicate environmental pollution. In *The Future of Effluent Treatment Plants* (pp. 199–214).Edited by maulin P shah. Elsevier.

31. Doskocz, N., Affek, K., Załęska-Radziwiłł, M., Kaźmierczak, B., Kutyłowska, M., Piekarska, K., and Trusz-Zdybek, A. (2017). Effects of aluminium oxide nanoparticles on bacterial growth. *E3S Web of Conferences*, 17, 00019. https://doi.org/10.1051/e3sconf/20171700019.

32. Chen, Z., Yang, P., Yuan, Z. and Guo, J. (2017). Aerobic condition enhances bacteriostatic effects of silver nanoparticles in aquatic environment: An antimicrobial study on *Pseudomonas aeruginosa*. *Scientific Reports*, 7(1), 7398.

33. Wang, L., Hu, C. and Shao, L. (2017). The antimicrobial activity of nanoparticles: Present situation and prospects for the future. *International Journal of Nanomedicine*, 12, 1227–1249.

34. Rudakiya, D. M. and Pawar, K. (2017). Bactericidal potential of silver nanoparticles synthesized using cell-free extract of *Comamonas acidovorans*: In vitro and in silico approaches. *Biotech*, 7, 92.

35. Qing, Y., Cheng, L., Li, R., Liu, G., Zhang, Y., Tang, X., Wang, J., Liu, H. and Qin, Y. (2018). Potential antibacterial mechanism of silver nanoparticles and the optimization of orthopedic implants by advanced modification technologies. *International Journal of Nanomedicine*, 13, 3311–3327.

36. Jagtap, P., Nath, H., Kumari, P. B., Dave, S., Mohanty, P., Das, J. and Dave, S. (2021). Mycogenic fabrication of nanoparticles and their applications in modern agricultural practices & food industries. In *Fungi Bio-Prospects in Sustainable Agriculture, Environment and Nano-Technology* edited by Dr Vineet Sharma,Published by Elsevier (pp. 475–488).

37. Mihai, M. M., Dima, M. B., Dima, B. and Holban, A. M. (2019). Nanomaterials for wound healing and infection control. *Materials*, 12(13), 2176.

38. Cinteza, L., Scomoroscenco, C., Voicu, S., Nistor, C., Nitu, S., Trica, B., Jecu, M. L. and Petcu, C. (2018). Chitosan-stabilized Ag nanoparticles with superior biocompatibility and their synergistic antibacterial effect in mixtures with essential oils. *Nanomaterials*, 8(10), 826.

39. Akmaz, S., Dilaver A. E., Yasar, M. and Erguven, O. (2013). The effect of Ag content of the chitosan-silver nanoparticle composite material on the structure and antibacterial activity. *Advances in Materials Science and Engineering*, 2013, 1–6.

40. Zhang, X. F., Shen, W. and Gurunathan, S. (2016). Silver nanoparticle-mediated cellular responses in various cell lines: An in vitro model. *International Journal of Molecular Sciences*, 17(10), 1603.

41. Panzarini, E., Mariano, S., Carata, E., Mura, F., Rossi, M. and Dini, L. (2018). Intracellular transport of silver and gold nanoparticles and biological responses: An update. *International Journal of Molecular Sciences*, 19(5), 1305.

42. Abdal, D. A., Hossain, M., Lee, S., Kim, K., Saha, S., Yang, G. M., Choi, H. and Cho, S. G. (2017). The role of reactive oxygen species (ROS) in the biological activities of metallic nanoparticles. *International Journal of Molecular Sciences*, 18(1), 120.

43. Phaniendra, A., Jestadi, D. B. and Periyasamy, L. (2015). Free radicals: Properties, sources, targets, and their implication in various diseases. *Indian Journal of Clinical Biochemistry*, 30(1), 11–26.

44. Mao, B. H., Chen, Z. Y., Wang, Y. J. and Yan, S. J. (2018). Silver nanoparticles have lethal and sublethal adverse effects on development and longevity by inducing ROS-mediated stress responses. *Scientific Reports*, 8(1), 2445.

45. Ray, P. C., Yu, H. and Fu, P. P. (2009). Toxicity and environmental risks of nanomaterials: Challenges and future needs. *Journal of Environmental Science and Health, Part C*, 27(1), 1–35.

46. Liao, C., Li, Y. and Tjong, S. C. (2019). Bactericidal and cytotoxic properties of silver nanoparticles. *International Journal of Molecular Sciences*, 20(2), 449. https://doi.org/10.3390/ijms20020449.

47. Dakal, T. C., Kumar, A., Majumdar, R. S. and Yadav, V. (2016). Mechanistic basis of antimicrobial actions of silver nanoparticles. *Frontiers in Microbiology*, 7, 1831. https://doi.org/10.3389/fmicb.2016.01831.

48. Vassallo, A., Silletti, M. F., Faraone, I. and Milella, L. (2020). Nanoparticulate antibiotic systems as antibacterial agents and antibiotic delivery platforms to fight infections. *Journal of Nanomaterials*, 2020, 6905631. https://doi.org/10.1155/2020/6905631.

49. Nita, M. and Grzybowski, A. (2016). The role of the reactive oxygen species and oxidative stress in the pathomechanism of the age-related ocular diseases and other pathologies of the anterior and posterior eye segments in adults. *Oxidative Medicine and Cellular Longevity*, 2016, 3164734. https://doi.org/10.1155/2016/3164734.

50. Kong, F. Y., Zhang, J. W., Li, R. F., Wang, Z. X., Wang, W. J. and Wang, W. (2017). Unique roles of gold nanoparticles in drug delivery, targeting and imaging applications. *Molecules (Basel, Switzerland)*, 22(9), 1445. https://doi.org/10.3390/molecules22091445.

51. Payne, J. N., Waghwani, H. K., Connor, M. G., Hamilton, W., Tockstein, S., Moolani, H., Chavda, F., Badwaik, V., Lawrenz, M. B. and Dakshinamurthy, R. (2016). Novel synthesis of kanamycin conjugated gold nanoparticles with potent antibacterial activity. *Frontiers in Microbiology*, 7, 607. https://doi.org/10.3389/fmicb.2016.00607.

52. Rattanata, N., Klaynongsruang, S., Leelayuwat, C., Limpaiboon, T., Lulitanond, A., Boonsiri, P., Chio-Srichan, S., Soontaranon, S., Rugmai, S. and Daduang, J. (2016). Gallic acid conjugated with gold nanoparticles: Antibacterial activity and mechanism of action on foodborne pathogens. *International Journal of Nanomedicine*, 11, 3347–3356. https://doi.org/10.2147/IJN.S109795.

53. Tao, C. (2018). Antimicrobial activity and toxicity of gold nanoparticles: Research progress, challenges, and prospects. *Letters in Applied Microbiology*, 67(6), 537–543. https://doi.org/10.1111/lam.13082.

54. Mondal, S., Manivasagan, P., Bharathiraja, S., Santha Moorthy, M., Kim, H. H., Seo, H., Lee, K. D. and Oh, J. (2017). Magnetic hydroxyapatite: A promising multifunctional platform for nanomedicine application. *International Journal of Nanomedicine*, 12, 8389–8410. https://doi.org/10.2147/IJN.S147355.

55. Armijo, L. M., Wawrzyniec, S. J., Kopciuch, M. et al. (2020). Antibacterial activity of iron oxide, iron nitride, and tobramycin conjugated nanoparticles against *Pseudomonas aeruginosa* biofilms. *Journal of Nanobiotechnology*, 18, 35. https://doi.org/10.1186/s12951-020-0588-6.

56. Swarnavalli, G., Jemima, C., Dinakaran, S., Raman, N., Jegadeesh, R. and Pereira, C. (2015). Bio inspired synthesis of monodispersed silver nano particles using *Sapindus emarginatus* pericarp extract – Study of antibacterial efficacy. *Journal of Saudi Chemical Society*, 21(2), 172–179. https://doi.org/10.1016/j.jscs.2015.03.004.

57. Thukkaram, M., Sitaram, S., Kannaiyan, S. K. and Subbiahdoss, G. (2014). Antibacterial efficacy of iron-oxide nanoparticles against biofilms on different biomaterial surfaces. *International Journal of Biomaterials*, 2014, 716080. https://doi.org/10.1155/2014/716080.

58. Lopez de Dicastillo, C., Correa, M., Martínez, F., Streitt, C. and Galotto, M. (2020). Antimicrobial effect of titanium dioxide nanoparticles. https://doi.org/10.5772/intechopen.90891.

59. Gerba, C. P. (2013). Titanium dioxide as disinfectant. In Kretsinger, R. H., Uversky, V. N., Permyakov, E. A. (eds), *Encyclopedia of Metalloproteins*. Springer. https://doi.org/10.1007/978-1-4614-1533-6_530.

60. Kumaravel, V., Nair, K. M., Mathew, S., Bartlett, J., Kennedy, J. E., Manning, H. G., Whelan, B. J., Leyland, N. S. and Pillai, S. C. (2021). Antimicrobial TiO2 nanocomposite coatings for surfaces, dental and orthopaedic implants. *Chemical Engineering Journal*, 416, 129071. https://doi.org/10.1016/j.cej.2021.129071.

61. Foster, H. A., Ditta, I. B., Varghese, S. and Steele, A. (2011). Photocatalytic disinfection using titanium dioxide: Spectrum and mechanism of antimicrobial activity. *Applied Microbiology and Biotechnology*, 90(6), 1847–1868. https://doi.org/10.1007/s00253-011-3213-7.

62. Ahamed, M., Alhadlaq, H A., Majeed Khan, M. A., Karuppiah, P. and Al-Dhabi, N. A. (2014). Synthesis, characterization, and antimicrobial activity of copper oxide nanoparticles. *Journal of Nanomaterials*, 2014, 637858. https://doi.org/10.1155/2014/637858.

63. Naz, S., Gul, A. and Zia, M. (2020). Toxicity of copper oxide nanoparticles: A review study. *IET Nanobiotechnology*, 14(1), 1–13. https://doi.org/10.1049/iet-nbt.2019.0176.

64. Chang, Y. N., Zhang, M., Xia, L., Zhang, J. and Xing, G. (2012). The toxic effects and mechanisms of CuO and ZnO nanoparticles. *Materials*, 5(12), 2850–2871. https://doi.org/10.3390/ma5122850.

65. Amiri, M., Etemadifar, Z., Daneshkazemi, A. and Nateghi, M. (2017). Antimicrobial effect of copper oxide nanoparticles on some oral bacteria and *Candida* species. *Journal of Dental Biomaterials*, 4(1), 347–352.

66. Siddiqi, K. S., Ur Rahman, A., Tajuddin and Husen, A. (2018). Properties of zinc oxide nanoparticles and their activity against microbes. *Nanoscale Research Letters*, 13(1), 141. https://doi.org/10.1186/s11671-018-2532-3.

67. Ifeanyichukwu, U.L., Fayemi, O. E. and Ateba, C. N. (2020). Green synthesis of zinc oxide nanoparticles from pomegranate (*Punica granatum*) extracts and characterization of their antibacterial activity. *Molecules*, 25(19), 4521. https://doi.org/10.3390/molecules25194521.

68. Sirelkhatim, A., Mahmud, S., Seeni, A., et al. (2015). Review on zinc oxide nanoparticles: Antibacterial activity and toxicity mechanism. *Nano-Micro Letters*, 7, 219–242. https://doi.org/10.1007/s40820-015-0040-x.

69. Singh, R., Cheng, S. and Singh, S. (2020). Oxidative stress-mediated genotoxic effect of zinc oxide nanoparticles on *Deinococcus radiodurans*. *3 Biotech* 10, 66. https://doi.org/10.1007/s13205-020-2054-4.

70. Zhang, L., Wu, L., Si, Y. and Shu, K. (2018). Size-dependent cytotoxicity of silver nanoparticles to *Azotobacter vinelandii*: Growth inhibition, cell injury, oxidative stress and internalization. *PLoS One*, 13. https://doi.org/10.1371/journal.pone.0209020.

71. Beddow, J., Stolpe, B., Cole, P., Lead, J., Sapp, M., Lyons, B., Colbeck, I. and Whitby, C. (2014). Effects of engineered silver nanoparticles on the growth and activity of ecologically important microbes. *Environmental Microbiology Reports*, 6, 448–458.

72. He, S., Feng, Y., Ni, J., Sun, Y., Xue, L., Feng, Y., Yu, Y., Lin, X., and Yang, L. (2016). Different responses of soil microbial metabolic activity to silver and iron oxide nanoparticles. *Chemosphere*, 147, 195–202. https://doi.org/10.1016/j.chemosphere.2015.12.055.

73. Murphin Kumar, P. S., MubarakAli, D., Saratale, R. G., Saratale, G. D., Pugazhendhi, A., Gopalakrishnan, K. and Thajuddin, N. (2017). Synthesis of nano-cuboidal gold particles for effective antimicrobial property against clinical human pathogens. *Microbial Pathogenesis*, 113, 68–73. https://doi.org/10.1016/j.micpath.2017.10.032.

74. Ali, S. G., Ansari, M. A., Alzohairy, M. A., Alomary, M. N., AlYahya, S., Jalal, M., Khan, H. M., Asiri, S., Ahmad, W., Mahdi, A. A., El-Sherbeeny, A. M. and El-Meligy, M. A. (2020). Biogenic gold nanoparticles as potent antibacterial and antibiofilm nano-antibiotics against *Pseudomonas aeruginosa*. *Antibiotics (Basel, Switzerland)*, 9(3), 100. https://doi.org/10.3390/antibiotics9030100.

75. Vandevoort, A. R. and Arai, Y. (2018). Macroscopic observation of soil nitrification kinetics impacted by copper nanoparticles: Implications for micronutrient nanofertilizer. *Nanomaterials*, 8, 927. https://doi.org/10.3390/nano8110927.

76. Baek, Y. W. and An, Y. J. (2011). Microbial toxicity of metal oxide nanoparticles (CuO, NiO, ZnO, and Sb2O3) to *Escherichia coli*, *Bacillus subtilis*, and *Streptococcus aureus*. *Science of the Total Environment*, 409, 1603–1608.

77. Mohan, P. and Mala, R. (2019). Comparative antibacterial activity of magnetic iron oxide nanoparticles synthesized by biological and chemical methods against poultry feed pathogens. *Materials Research Express*, 6(11), 115077. https://doi.org/10.1088/2053-1591/ab4964.

78. Arakha, M., Pal, S., Samantarrai, D., et al. (2015). Antimicrobial activity of iron oxide nanoparticle upon modulation of nanoparticle-bacteria interface. *Scientific Reports*, 5, 14813. https://doi.org/10.1038/srep14813.

79. Heinlaan, M., Ivask, A., Blinova, I., Dubourguier, H. C. and Kahru, A. (2008). Toxicity of nanosized and bulk ZnO, CuO and TiO2 to bacteria *Vibrio fischeri* and crustaceans *Daphnia magna* and *Thamnocephalus platyurus*. *Chemosphere*, 71, 1308–1316.

80. Hessler, C. M., Wu, M. Y., Xue, Z., Choi, H. and Seo, Y. (2012). The influence of capsular extracellular polymeric substances on the interaction between TiO2 nanoparticles and planktonic bacteria. *Water Research*, 46, 4687–4696.
81. Applerot, G., Lipovsky, A., Dror, R., Perkas, N., Nitzan, Y., Lubart, R. and Gedanken, A. (2009). Enhanced antibacterial activity of nanocrystalline ZnO due to increased ROS-mediated cell injury. *Advanced Functional Materials*, 19, 842–852.
82. Jiang, W., Mashayekhi, H. and Xing, B. (2009). Bacterial toxicity comparison between nano- and micro-scaled oxide particles. *Environmental Pollution*, 157, 1619–1625.
83. Liu, Y., He, L., Mustapha, A., Li, H., Hu, Z. Q. and Lin, M. (2009). Antibacterial activities of zinc oxide nanoparticles against *Escherichia coli* O157:H7. *Journal of Applied Microbiology*, 107, 1193–1201.

8 Metallic and Non-metallic Nanoparticles against Viruses

*Aqsa Ahmad, Iqra Muzammil, Tariq Munir,
Muhammad Aamir Naseer, Amjad Islam
Aqib, Muhammad Muddassir Ali, Imran
Khan Sohrani, Arslan Rasool, Ibrahim Hakki
Cigerci and Muhammad Imran Arshad*

CONTENTS

Abbreviations/Acronyms 205
8.1   Introduction about Viruses 206
8.2   Nanotechnology/Nanoparticles (NPs) 208
8.3   Nanosensors for Virus Detection..209
      8.3.1   Working of Biosensors..209
      8.3.2   Applications of Biosensors for Viral Diseases 209
8.4   Metallic and Non-metallic Nanoparticles 210
8.5   Antiviral Mechanism of Nanoparticles 211
8.6   Nanovaccines Against Viral Diseases 213
8.7   Approved Nanomedicines...214
      8.7.1   *In vitro* Studies of Metallic and Non-metallic Nanoparticles as
              Antivirals 215
8.8   Challenges Using Nanoparticles as Antivirals 218
8.9   Conclusions and Future Outlooks...219
References..220

ABBREVIATIONS/ACRONYMS

NPs: nanoparticles
AgNPs   silver nanoparticles
DNA   deoxyribonucleic acid
RNA   ribonucleic acid
HIV   human immunodeficiency virus
ELISA   enzyme-linked immunosorbent assay
PCR   Polymerase Chain Reaction

DOI: 10.1201/9781003126256-8

RT-PCR	Reverse transcription Polymerase Chain Reaction
HSV-1	herpes simplex virus I
HSV-2	herpes simplex virus II
Zn	zinc
Ti	titanium
Cu	copper
Ag	silver
Au	gold
gp120	glycoprotein 120
TiO$_2$	titanium dioxide
GO	graphene oxide
rGO	reduced graphene oxide
RSV	respiratory syncytial virus
MPV	monkey pox virus
HBV	hepatitis B virus
Au-MES NPs	gold capping mercaptoethane sulfate nanoparticles
Bcl-2	B-cell lymphoma 2
Bax	Bcl2-associated X protein
TGEV	transmissible gastroenteritis virus
COVID-19	coronavirus disease
FDA	Food and Drug Administration
HCV	hepatitis C virus
PEG-ADA	pegylated adenosine deaminase
PEG-ASP	pegylated asparaginase
AIDS	acquired immunodeficiency syndrome
SCID	severe combined immunodeficiency
EDIII	domain III of envelope glycoprotein
PVP	N-vinyl-2-pyrrolidone
NS1	non-structural protein 1
NPL	neoproteoliposomes
TCRV	Tacaribe complex viruses

8.1   INTRODUCTION ABOUT VIRUSES

Infectious diseases caused by bacteria, fungi, viruses and parasites are responsible for 15 million deaths annually all over the world (Qasim *et al.* 2014). Viruses are small microscopic particles that depend on a host for their reproduction. Viruses contain either single-stranded or double-stranded deoxyribonucleic acid (DNA) or ribonucleic acid (RNA) in their genome. The complete virus particle is known as a "virion", which consists of an outer protein coat and genome. The major function of a virion is to incorporate its genome into the host cell for regulation, i.e., transcription and translation. Viruses are classified in different families on the basis of their size, genome structure, physiochemical properties, molecular processes and morphology (Kane and Golovkina 2010). There are 26 families of viruses involved in causing human diseases in which humans are a clear reservoir for some viruses and in others humans serve as an incidental host (Siegel 2018). Viral infections pose significant global health challenges due to resistant strains and side effects caused by the long-term use of antiviral drugs. Viral diseases exert substantial health issues worldwide by infecting a large number of people globally and also impart severe negative impacts on socioeconomic development (Mehendale, Joshi and Patravale 2013).

A pandemic is known as a "global disease outbreak" and is frequently caused by a new virus or its strain that has not previously spread between people for a long time. People have no immunity against pandemics and the virus spreads rapidly from person to person all over the world, like the current COVID-19 pandemic. Some of the deadliest pandemics of history are shown in Table 8.1. The current COVID-19 pandemic was caused by a new coronavirus SARS-CoV-2 (Di Gennaro *et al.* 2020) and

TABLE 8.1
History of Pandemics

Disease	Time Frame of the Disease	Death Toll	Reference
Antonine plague	165–180	5 million	Scheidel and W. (2002)
Japanese smallpox epidemic	735–737	1 million	Collcutt and M. (1987)
Small pox	1520	56 million	Henderson and D. A. (2011)
Yellow fever	Late 1800s	100,000–150,000	Robertson *et al.* (1996)
Russian flu	1889–1890	1 million	Gregg *et al.* (1978)
Spanish flu	1918–1919	40–50 million	Tognotti and E. (2009)
Asian flu	1957–1958	1.1 million	Tognotti and E. (2009)
Hong Kong flu	1968–1970	1 million	Cockburn *et al.* (1969)
HIV/AIDS	1981–present	25–35 million	CDC (1997)
SARS	2002–2003	770	Chan-Yeung *et al.* (2003)
Swine flu	2009–2010	200,000	Dhama *et al.* (2012)
Ebola	2014–2016	11,300	Gostin *et al.* (2014)
MERS	2015–present	850	Donnelly *et al.* (2019)
Novel coronavirus (COVID-19)	2019–present	1.5 million (it is hard to forecast the death rate as data is still coming in)	CDC (2020)

as of this writing there had been 4.8 million deaths globally. The virus spreads faster than its ancestors SARS-CoV-1 and MERS-CoV (Singh *et al.* 2020).

Most viruses develop resistance against antiviral drugs restricting effective treatment, specifically to infections linked with influenza and human immunodeficiency virus (HIV) (Maseko *et al.* 2016; Hayden 2009). Subsequently, there is an urgent need for alternative methods and advanced techniques to overcome the resistance and to treat viral damages (Bamrungsap *et al.* 2012). So, scientists have considered nanoparticles.

8.2 NANOTECHNOLOGY/NANOPARTICLES (NPS)

Nanotechnology is an emergent scientific field that built its working principles at the atomic and molecular level. It is the study of very small particles of size 0.1 to 100 nm in which magnetic, optical, structural and electronic features are generated. In 1959, the concept of nanotechnological strategy was first presented by physicist Richard P. Feynman (Kargozar 2018). Later in 1986, E. Drexler developed and published Feynman's ideas.

Nanoparticles can be categorized on the basis of their composition and dimensions. There are three categories of dimension for these particles in which the one-dimensional group includes thin layers, surfaces and films; the two-dimensional group include graphene sheets; and the three-dimensional group includes quantum dots, graphite sheets, dendrimers, fullerenes and nanoparticles. Graphene sheets can be rolled into nanotubes and nanowires (Peng *et al.* 2014). Depending on composition, nanoparticles are divided into single-phase solids, multiphase solids and multiphase systems. All types of nanosized materials are characterized as composites, polymers, ceramics and metallic (Hou, Westerhoff and Posner 2013).

During the past few years, nanotechnology has been used in medicine as nanotherapeutics and nanodelivery systems to increase drug efficiency, and has been used in many medical and research fields (Zhang *et al.* 2018). Other implementations of nanotechnology can be seen in disease diagnosis and screening, environment and energy, electronics, space exploration, transportation, agri-food production and cosmetics. In spite of the progress in conventional therapeutics and vaccines, developments are needed because of weak immunogenicity, toxicity, intrinsic instability *in vivo* and the several-administrations regime for desired results. To combat these problems, nanotechnology has been used in vaccine development and therapeutic modifications. Main aspects of using nanoparticles include identification and treatment of many viral diseases like hepatitis, herpes simplex virus I (HSV-1) and herpes simplex virus II (HSV-2), influenza virus and HIV. For the development and optimization of vaccines, some nanoparticle (NP) systems are used to increase the delivery of antigens and effective induction of immunity (Cojocaru *et al.* 2020). The benefits of using NPs in vaccine are that they bring proteins or peptides to a specific tissue for a stronger immune response; protect the applied antigen from enzymatic degradation; and they are easy to produce, biodegradable and biocompatible. These characteristics of NPs enhance the *in vivo* bioavailability of antigens and stimulate both an innate and adaptive immune response, involving the generation of memory cells, which play an important role in the success of vaccines (Sulczewski *et al*. 2018).

8.3 NANOSENSORS FOR VIRUS DETECTION

Detection of pathogens is necessary for the treatment of diseases. Different methods can detect viruses, including enzyme-linked immunosorbent assay (ELISA), polymerase chain reaction (PCR), reverse transcription polymerase chain reaction (RT-PCR), cell culturing and gene sequencing. These methods are labor-intensive, time-consuming and quite expensive. Hence, there is the need to develop simple, sensitive, accurate and rapid methods for detection of pathogens.

A biosensor is an analytical device used to measure very low analyte concentration in a sample. It is made up of biological elements like antibodies, nucleic acid, microorganisms and cell receptor enzymes. It is comprised of an analyte, bioreceptor, transducer and measurable signals (S. Singh *et al*. 2020). On the basis of transduction principles, biosensors are further categorized into five classes: piezoelectric, optic, electrochemical, magnetic and thermal.

8.3.1 WORKING OF BIOSENSORS

A nanosensor generally consists of a biosensitive layer that can either contain biological recognition elements or be made of biological recognition elements covalently attached to the transducer. The contact between the sample (analyte) and bioreceptor produces physiochemical changes, which consequently lead to the generation of measurable signals. Bioreceptors allow specific binding of component from the sample to reduce the interaction of unwanted components (Odobašić, Šestan and Begić 2019) (Figure 8.1).

8.3.2 APPLICATIONS OF BIOSENSORS FOR VIRAL DISEASES

Biosensors have many different clinical and medical applications for viral diseases that include

*   Early detection of viral diseases
*   Measuring analyte in a biological sample

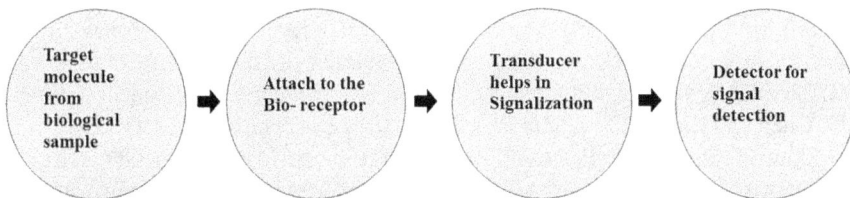

FIGURE 8.1   Schematic diagram of biosensor.

- Assessment of viral drug activity
- Measurement of a viral drug and its metabolites
- Early diagnosis of viral diseases
- Doping analysis

8.4 METALLIC AND NON-METALLIC NANOPARTICLES

Nanotechnology gives a platform to develop and enhance the characteristics of metals by driving them into metal nanoparticles (Deschênes and Ells, 2020). Different metallic nanoparticles like zinc (Zn), titanium (Ti), copper (Cu), silver (Ag) and gold (Au) have antimicrobial, antiviral and antifungal properties (Malarkodi *et al.* 2014). Silver nanoparticles possess strong antiviral properties against some viral families, including poxviridae, herpesviridae, arenaviridae, retroviridae, paramyxoviridae, hepadnaviridae and orthomyxoviridae (Galdiero *et al.* 2011; Akbarzadeh *et al.* 2018; Youssef *et al.* 2019). The silver nanoparticles bind with the viral genome by interacting with glycoprotein 120 (gp120), inhibiting its entry into the host cell (Khandelwal *et al.* 2014). Similarly, other metallic nanoparticles like gold and copper nanoparticles possess their antiviral characteristics by damaging the viral capsule (Broglie *et al.* 2015). Moreover, their antiviral efficacy can be increased by combining with other nanoparticles (Zhong *et al.* 2019). Many studies reported that metal oxide nanosubstances, like titanium dioxide (TiO_2) polylysine, efficiently inhibit influenza virus (Levina *et al.* 2016). At 365 nm, nanoformulations of TiO_2 inhibiting H9N2 influenza virus were reported (Cui *et al.* 2010). Similarly, zinc oxide nanocomposites inhibit the entry of HSV-1 virus in corneal fibroblasts (Mishra *et al.* 2011). Moreover, zinc oxide nanoparticles also possess antiviral characteristics against chikungunya virus (R. Kumar *et al.* 2018). Many other negatively charged nanosubstances are used for antiviral treatment against HSV-1 virus by preventing its attachment with the host cell inhibiting viral replication (Tavakoli *et al.* 2018).

Non-metallic nanoparticles, like carbon, possess low cytotoxicity and have particular antiviral properties (Shimabuku *et al.* 2018; Youssef *et al.* 2019). Though they are comparatively new in nanoscience, their use has been encouraged in research to control the synthesis and function of viral surfaces. Graphene oxide (GO) and reduced graphene oxide (rGO) are also non-metallic nano particles that have uses in different fields, like photonics and electronics, but are limited in nanomedicine (Tapaszto *et al.* 2008; Ramazanpour Esfahani *et al.* 2020). Different experiments have been performed to check how GO inactivates the virus before it enters into the cell. The sharp edges of the GO layer disrupt the structure of the virus by direct interaction. The antiviral activity of GO was effective for both RNA and DNA viruses, but dependent on time of incubation and concentration of GO (Chen and Liang, 2020; Samak *et al.* 2020). It has been also observed that rGO also showed same antiviral activity like GO. The negative charge on GO facilitates the electrostatic communication with a positive charge on the virus. However, high interaction leads to the inactivation and destruction of the virus (Innocenzi and Stagi 2020).

8.5 ANTIVIRAL MECHANISM OF NANOPARTICLES

Nanoparticles are utilized nowadays for their antiviral activities, which hinder the viral number using numerous mechanisms. Many findings have verified the activity of metal nanoparticles against a lot of well-known viruses, including influenza virus, respiratory syncytial virus (RSV), HIV-1, monkey pox virus (MPV), HBV and HSV-1 (Galdiero et al. 2011). Several nanoparticles are imitators of heparin sulfate, a receptor present on the cell surface, so mercaptoethane-sulfate capping gold nanoparticles (Au-MES NPs) play the role of operative HSV-1 disease inhibitors. Au-MES NPs function in a multimodal way, like hindering the attachment of viruses to the cell surface, interfering with their entry or restricting the cellular spreading (Figure 8.2). Nanoparticle carriers achieve greater strength and higher efficiency due to the fast multiplication of ligands, which enhances their interfaces with the virus polyvalently and also increases the specificity. By observing the consequences, it was emphasized that this multivalent nanoparticle-induced inhibition has high potential as a substitute method for antiviral therapy (Baram-Pinto et al. 2010).

As a carrier of antiviral therapy, silver nanoparticles have been broadly considered. Numerous *in vitro* studies illuminated the efficiency of silver nanoparticles against a large number of HIV-1 strains including T and M tropic strains, laboratory strains and resistant strains (Lara et al. 2010). However, the details of the findings about AgNPs as antivirals are still unknown. Limited information related to the mechanism of nanoparticles against viral infection treatment could be due to the complex structure of viruses. AgNPs deal with viruses in two ways: (1) these

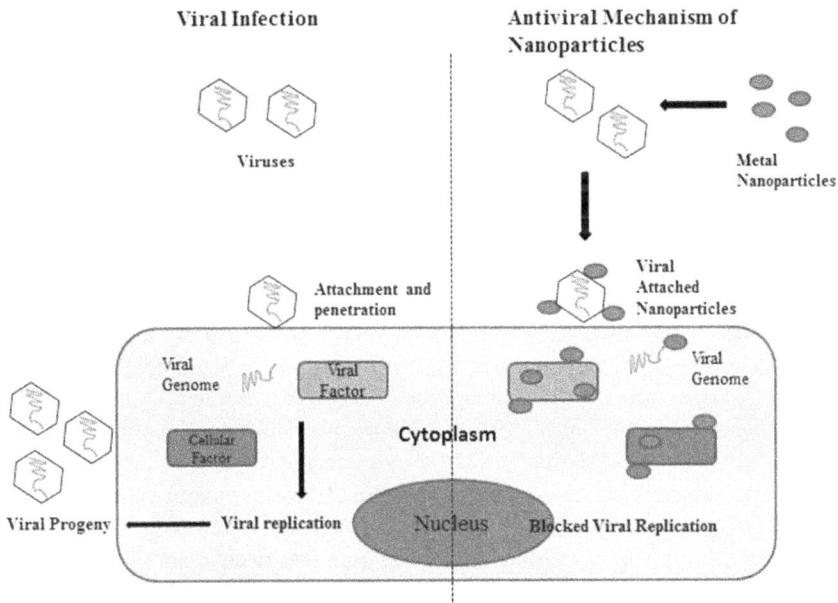

FIGURE 8.2 Antiviral mechanism of metal nanoparticles.

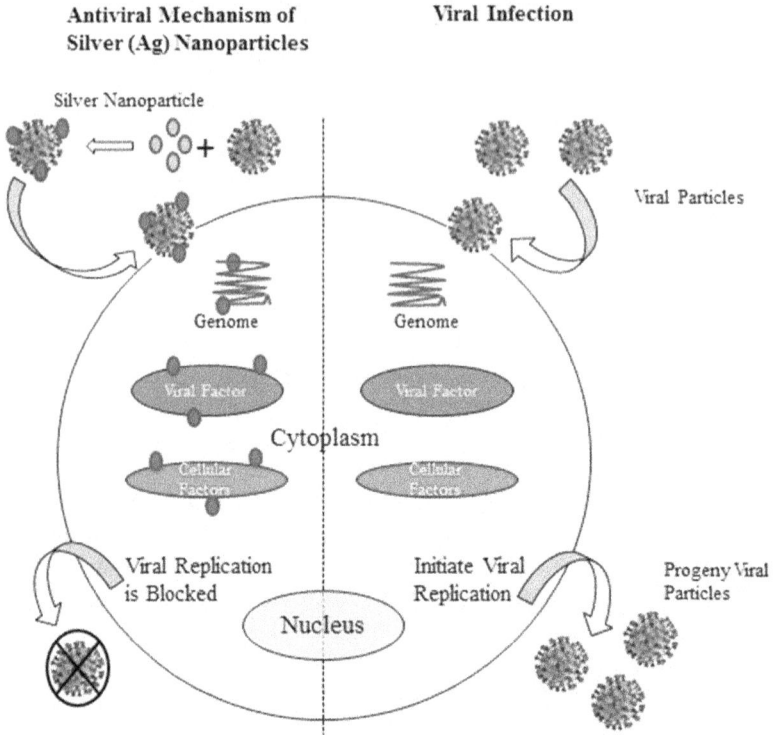

FIGURE 8.3 Antiviral mechanism of silver nanoparticles.

nanoparticles fix the outer covering of the virus and prevent its binding with receptors of the cells, and (2) AgNPs also restrict the viral propagation or replication within host cells by interacting with the viral RNA or DNA (Figure 8.3). By knowing the antiviral activity of this nanoparticle against various kinds of viruses one could improve the existing antiviral therapies with nanotechnologies (Salleh *et al.* 2020).

During the viral replication process, viruses initiate the apoptosis pathway in the host cells. Due to this pathway, the host showed severe symptoms and can lead to death if untreated (Kane and Golovkina, 2010; Lara *et al.* 2010; Park *et al.* 2018). A signaling pathway that helps in mediating the cell apoptosis involved in depolarization of mitochondrial membrane potential is known as the mitochondrial-induced apoptosis pathway. This apoptotic pathway consists of two proteins: B-cell lymphoma 2 (Bcl-2) and Bax. Permeabilization of the mitochondrial membrane is performed by the pro-apoptotic protein Bax, which induces cell apoptosis by facilitating the movement of mitochondrial intermembrane proteins to cytosol, whereas the anti-apoptotic protein Bcl-2 opposes the mechanism of Bax (Ren *et al.* 2012).

Many silver nanomaterials, including of AgNPs, silver colloids and two different kinds of silver nanowires against transmissible gastroenteritis virus (TGEV), have been used in research (Lv *et al.* 2014; Akbarzadeh *et al.* 2018; Youssef *et al.* 2019). This research involved pre-clinical tests by infecting TGEV to swine testicle cells and the results showed that TGEV is able to induce the mitochondrial-mediated apoptosis pathway by increasing the Bax level in the cells. In this experiment, silver nanomaterials were used to inhibit the initiation of TGEV infection by binding to the surface protein, S glycoprotein (Lv *et al.* 2014). This study also proposed that silver nanomaterials were capable of changing the structure of surface proteins, thus impeding their recognition and attachment to the receptor of the host. The antiviral activities of these nanomaterials could lower the risk of infection as well as prevent an epidemic viral disease such as coronavirus disease (COVID-19) (Lv *et al.* 2014; Park *et al.* 2018; Bayoumy *et al.* 2019).

Dynamic expansion in the development of metal nanoparticles for biomedical applications is ongoing. Along with this advancement, the potential toxicities of these novel carriers and the properties resulting in their toxic responses must be carefully tracked. For the safe and effective development of such emerging nano-technologies, a detailed assessment of the factors influencing their biocompatibility and/or toxicity is crucial.

8.6 NANOVACCINES AGAINST VIRAL DISEASES

Vaccine development with efficient immune boosting capacity and minimum limi-tations, including subunit- and nanocarrier-based vaccinal approaches, are desired tasks of scientists for diseases with potential outbreak capabilities (Du *et al.* 2016; Zhang *et al.* 2015). Challenges in this regard are the complete exploitation of viral pathogenesis, transmission and local immune response, lack of funding and insuf-ficient diagnostic tools and animal models (Cho *et al.* 2018). Vaccines are the most promising solution to mitigate new viral strains. Nanotechnology benefits modern vaccine design since nanomaterials are ideal for antigen delivery, as adjuvants and as mimics of viral structures. In fact, the first vaccine candidate launched into clini-cal trials was an mRNA vaccine delivered via lipid nanoparticles (Shin *et al.* 2020). "Nanovaccines" have been explored to elicit a strong immune response with the advantages of nanosized range, enhanced immunogenicity, high antigen loading, more retention in lymph nodes, controlled antigen presentation and promotion of patient compliance by a lower frequency of dosing. Nanovaccines have the poten-tial to induce both cell-mediated and antibody-mediated immunity and can render long-lasting immunogenic memory (Yadav *et al.* 2018; Bhardwaj *et al.* 2020). The applications of nanovaccines in infectious and non-infectious diseases, like malaria, tuberculosis, AIDS, influenza and cancers, have been well studied (Bhardwaj *et al.* 2020). Due to the similar scale between nanoparticles and pathogens, the immune system can be well stimulated, resulting in triggered cellular and humoral immu-nity responses along with better stability, longer shelf life, enhanced immune system stimulation, no booster doses, no need of a cold chain and the ability to create active targeting. In addition, nanovaccines have raised the hope to treat diseases such as rheumatoid arthritis, AIDS, malaria and chronic autoimmune (Gheibi Hayat and Darroudi 2019).

In spite of continuous work, no internationally approved MERS-CoV vaccine or therapeutic protocol was available until now. However, researchers have exploited different vaccinal candidates including nanoparticle-based approaches, which were in focus under the umbrella of the One Health concept (Xu *et al.* 2019; Alharbi 2017; Chi *et al.* 2017). Vaccinal trials focused on the spike protein showed promising results in lab animals, especially plasmid vaccine GLS-5300 and m-336 monoclonal antibodies, which will be tested as a prophylactic vaccine in human beings (Rabaan *et al.* 2017). Modified Ankara MERS-CoV vaccine has shown excellent results in camels (Haagmans *et al.* 2016).

Nanovaccines with novel delivery mechanisms could make obsolete the use of needles for administering any vaccine. But the beneficiary population and target disease selection are major questions alongside the wide range deployment of resources (Stammers 2013). Nanovaccines induce both cell-mediated as well as humoral immunity in the host, proving to be more effective and successful than conventional vaccines. These are reported to be efficient immune boosters against infections, immunity disorders (e.g., HIV, malaria, arthritis, sclerosis) and pathogen spread (Sekhon, Sekhon and Saluja 2011).

The most recent COVID-19 (SARS-CoV-2) pandemic has tested the limits of health care institutes by imposing tremendous workloads and managerial burdens. In such a scenario, nanotechnology provides the platform for formulation of novel preventive, diagnostic and therapeutic strategies against human coronavirus and other deadly outbreaks (Campos *et al.* 2020). Nanotechnology has shown highly effective nanocarriers that combat COVID-19 and are capable of countering the limitations of conventional biological and antiviral therapeutics (Editorials 2020). Nanotechnology provides potent carriers for risk-free, target-oriented and efficient immunity pathways against COVID-19 and MERS-CoV challenges (Chauhan *et al.* 2020).

8.7   APPROVED NANOMEDICINES

Most of the approved nanotherapeutics are the nanoformulations of the drugs approved by the Food and Drug Administration (FDA), and some nanotherapeutics have received national approvals. Many successful nanoformulations either belong to the extremely toxic drugs, like amphotericin B, or with paclitaxel, which has less solubility. There are many nanotherapeutics that have high efficacy and low toxicity with good pharmacokinetic properties. However, most of the drugs that didn't show increased efficacy or less toxicity were declared failures during development process. Nanoformulations that have more affinity with each other can be used as alternatives to reduce toxicity. Moreover, a large number of nanoformulations that have significant results are still under the clinical developmental process (Caster *et al.* 2017).

Nanotherapeutics that are approved or under clinical trial for treating viral infections are listed in Table 8.2 and antiviral nanopreparations in Table 8.3 (Singh *et al.* 2017).

TABLE 8.2
Approved Nanotherapeutics in Clinical and Research Trials

Name	Indication	Approval Year/Stage of Development	Reference
Inflexal[1]	Influenza	1997	Herzog *et al.* (2009)
Epaxal[1]	HAV	2003	Patravale, Dandekar and Jain (2012)
PegIntron[2]	HCV	2001	Alconcel, Baas and
Pegasys[2]	HBV, HCV	2002	Maynard (2011)
Feraheme[2]	Anemia of chronic kidney disease	2009	Lu *et al.* (2010)
VivaGel[2] (SPL 7013)	HIV, HSV	2006	Gupta and Perumal (2014)
Dermavir	HIV	In clinical trial	Rodriguez *et al.* (2013)
Doravirine (MK-1439)	HIV	In clinical trial	L. Singh *et al.* (2017)
ARB-001467 TKM-HBV	HBV	In clinical trial	Seto and Yuen (2016)

[1] Approved by WHO and the European Medicines Agency (EMA).
[2] Approved by United States FDA.

8.7.1 *In vitro* Studies of Metallic and Non-metallic Nanoparticles as Antivirals

Recent research has shown promising results and functionality of many nanoparticles as potent antiviral entities, prominent of which are metallic nanoparticles (silver, gold, zinc, copper, titanium and metal oxides), quantum dots, nanoclusters, graphene oxide, carbon dots, silicon materials, dendrimers and polymers (Lembo and Cavalli 2010; Prasad *et al.* 2017). Implementation of metallic nanoparticles or nanopreparations of metallic oxides have shown great progress for improvement of diagnostics and curative capabilities of antiviral agents alongside boosting the targeted drug delivery efficiency for infected cells. The big hurdle in the application of such potent preparations is the lack of *in vivo* experimental validation and efficacy range knowledge in various viral ailments (Yadavalli and Shukla 2016).

In spite of varying inhibitory potential and different antiviral action pathways, the functionality spectrum of nanoparticle-based antiviral entities has shown a full potential of candidates for potent antiviral candidatures. The revolutionary field of nanomedicine has provided a new therapeutic window for combating viral challenges with better success in a more effective manner. Drug delivery and targeting systems based on nanotechnology are supposed to alter the release kinetics of antiviral agents, improving their effective spectra, boosting their biologically available amount and minimizing the toxic windows and therapeutic expenditures (Lembo and Cavalli 2010). Nevertheless, drug delivery with more precision and targeting

TABLE 8.3
Antiviral Nanotherapeutics

Nanomedicine	Constituent	Antiviral Effect	Reference
Experal™	Pegylated adenosine deaminase (PEG-ADA)	Severe combined immunodeficiency disease	Anselmo and Mitragotri (2019)
Oncospar™	Pegylated asparaginase (PEG-ASP)	Lymphoblastic leukemia	Gradishar et al. (2005)
Doxil™	Liposomal doxorubicin	Ovarian and breast cancer	Mehendale et al. (2013)
PegIntron™	Pegylated interferon α-2b	Hepatitis B	Alconcel, Baas and Maynard (2011)
Pegasys™	Pegylated interferon α-2a	Hepatitis B, C	Weissig, Pettinger and Murdock (2014)
Abraxane™	Albumin bound paclitaxel	Cancer of pancreas, lungs and breast	Weissig and Guzman-Villanueva (2015)
Depocyte™	Liposomal cytarabine	Lymphomatous meningitis	Pita, Ehmann and Papaluca (2016)
Marquibo™	Liposomal vincristine	Adult acute myelogenous leukemia	Caster et al. (2017)
VivaGel™	Astodrimer dendrimers	HIV and HSV	Mehendale et al. (2013)
RSV-F™	RSV fusion protein	respiratory syncytial virus	Weissig and Guzman-Villanueva (2015)
NANOefavirenz™	Polymeric efavirenz	HIV	Weissig, Pettinger and Murdock (2014)
NANOlopinavir™	Polymeric lopinavir	HIV	Ventola (2017)
DaunoXome®	Daunorubicin citrate encapsulated in liposomes	HIV	Pita, Ehmann and Papaluca (2016)
Doxil®	Doxorubicin hydrochloride encapsulated in liposomes	AIDS	Caster et al. (2017)
Inflexal® V	Influenza virus antigens on surface of liposomes	Influenza	Ventola (2017)
PegIntron®	Pegylated interferon alfa-2b	Hepatitis C	Prasad et al. (2017)
Vyxeos®	Liposomal daunorubicin and cytarabine	Myeloid leukemia	Anselmo and Mitragotri (2019)
Adagen®	Pegademase bovine	Severe combined immunodeficiency (SCID)	Farjadian et al. (2019)
DaunoXome®	Liposomal daunorubicin	HIV	Chakravarty and Vora (2020)

the pathogenic reservoirs in the body is now possible with nanolevel manipulation of the drug carriers. These characteristics are of prime importance when dealing with higher doses of antivirals, expensive therapeutic regimes and questionable succession of therapeutic protocols for patients' endurance and tolerance (Prasad *et al.* 2017).

Nanoparticles showed important assemblies that can be successfully contrived for numerous purposes and have significant prospects by intensifying the functions of metal oxides through cross-linking, like removal of microorganisms from various environments (Vodnar *et al.* 2020). Nanoparticles provide limited physical properties that have related advantages for drug delivery. All these properties are predominantly because of a large surface area-to-volume ratio (improved solubility compared to larger particles), particle size (which effects circulation time and bioavailability), the tunable surface charge of particles with opportunities for encapsulation and huge drug payloads that can be lodged. The assets that are different from bulk resources of the same configurations make nanoparticulate drug transfer systems perfect entrants to discover better medicinal effects (L. Singh *et al.* 2017). Researchers have utilized the domain III of envelope glycoprotein derived from serotype 2 of dengue virus (EDIII)-activated gold nanoparticles in mice models for the induction of neutralization antibodies against dengue virus. The study noted higher level of antibody induction potential of gold nanoparticles, mediating sero-specified neutralization, which was a dose-dependent and nanoparticle size-related phenomenon (Quach *et al.* 2018).

Research has unveiled the attachment and entrance mechanism of herpes simplex virus-I into the host cell, which is an interaction of cell surface heparan sulfate and glycoproteins of the viral envelope. Mercaptoethane-capped silver nanoparticles are supposed to inhibit this mechanism by competing for the attachment site at cellular heparan sulfate sticking ends, resulting in inhibition of viral entry in the host cell and subsiding the infection without any side effects compared to antiviral agents. The study provided an efficient way to inhibit HSV-1 infection *in vitro* suggesting *in vivo* studies for further validation (Baram-Pinto *et al.* 2009). The silver nanoparticles significantly reduced African swine fever virus load in a pig house following a 25 ppm nanoparticle solution spray. A solution concentration of 0.78 ppm showed zero toxicity to respiratory cells of pigs and complete viral inhibition at titer 103 HAD50 (Thi Ngoc Dung *et al.* 2020).

Silver nanoparticles were activated by tannic acid in a mucoadhesive gelling system implemented to treat herpes virus infections (HSV-I, HSV-II) *in vitro* using immortal human keratinocyte cell lines and *in vivo* in a murine model. Viral attachment, penetration, transmission and evoked activity were notably altered by the nano-based mucoadhesive gelling systems (Szymańska *et al.* 2018). Photocatalytic silver-doped titanium dioxide nanoparticles were used to test their ability to inactivate bacteriophage MS2 *in vitro*. A more than fivefold boost in the phage inactivation rate was noted depending on the TiO_2 base and directly related to silver content used. Viral inactivation was caused by an enhanced amount of hydroxyl free radicals (Liga *et al.* 2011).

Early steps in the field of nanotechnology have created a wondrous base for the modification manipulation of materials at the nanolevel to develop potent and more effective active entities possessing unmatched physicochemical characteristics, higher available surface area, lower volume of active material and a better action spectrum. As per the literature, metallic nanoparticles, particularly silver, possess promising antiviral, antibacterial and antifungal properties alongside host immune boosting to combat pathogenic challenge. However, the action spectrum of these nanoparticles is still an unexploited aspect of medical research work (Rai *et al.* 2016). Silver nanoparticles showed good activity against HIV, herpes simplex virus, monkey pox virus, hepatitis B virus and respiratory syncytial virus. In spite of antiviral resistance, it is foreseen that the viral genome cannot adopt resistance against these metallic invaders because of the multitargeting mechanism shown by nanoparticles (Galdiero *et al.* 2011). There is the need to enhance ongoing prevention strategies and modify antiviral therapies keeping in view the interactions between metallic nanoparticles and viral biomolecules. Silver nanoparticles tend to bind with the external coat of lipid envelope of viruses, resulting in attachment inhibition and preventing infection development (Lara *et al.* 2011). Specified organ targeting and active molecular delivery (genes, drugs, proteins) in efficacious amounts are needed to maintain desired clinical outcomes and narrowing the adverse spectrum for other organ systems. Selective targeting and delivery system implementation not only boost the therapeutic line but also improve the diagnostic efficiency (Toita *et al.* 2015). Vero cell culture analysis for 10 nm silver nanoparticles at one-hour incubation prior infection showed that Tacaribe virus (TCRV) titer was reduced up to 50% in progeny virus titer at the concentration 10 µg/ml and zero detectable titer of progeny virus at the concentration 25 µg/ml or higher. The PS-Ag (both 10 and 25 nm) showed a similar action spectrum at 10 and 25 µg/ml, with no progeny virus detection at 50 µg/ml or more (Speshock *et al.* 2010).

8.8 CHALLENGES USING NANOPARTICLES AS ANTIVIRALS

Nanoparticles show a combination of chemical, biological and physical characteristics that determine their *in vivo* behavior. There are range of techniques available for evaluating nanoparticle biodistribution, including electron microscopy, histology, indirectly measuring drug concentrations, liquid scintillation counting, computed tomography, *in vivo* optical imaging, nuclear medicine imaging and magnetic resonance imaging (Arms *et al.* 2018). Lack of understanding about *in vivo* targeting drug delivery by nanoparticles and alterations in biological behavior in complex living bodies is a huge setback of nanomedicines (Chenthamara *et al.* 2019). Complex preparation and structural confirmatory processes and questionable stability alongside higher manufacturing costs are major drawbacks of nanomedicine (Sotropa 2018). Environmental safety and airborne hazards are also negative features of nanoparticles (Shubhika 2012; Yao *et al.* 2013). Agglomeration and safety limitations are also questionable in the case of metallic nanoparticles (Kwon, Lee and Yoon 2010; Arms *et al.* 2018).

Nanoparticles are so small they can enter the skin and bloodstream causing toxicity. A large surface can make them too reactive and explosive in some situations. They easily become airborne, and breathing them in can potentially damage the lungs (Laurentius *et al.* 2016; Akbarzadeh *et al.* 2018; Sulczewski *et al.* 2018; Youssef *et al.* 2019). Nanolevel preparations change the properties of the original material. They might be toxic to some types of cells, such as skin, bone, brain and liver cells (Laurentius *et al.* 2016; Kwon, Lee and Yoon 2010). Suboptimal delivery is achieved with most nanoparticles due to heterogeneities of vascular permeability, which reduce nanoparticle penetration. Further, a slow drug release limits bioavailability (Manzoor *et al.* 2012). Organ-based toxicity and local damage is a risk of nanomedicines, especially with the potential for crossing natural barriers of the human body (De Jong and Borm 2008).

Higher toxicity of liposomes due to the increased uptake by reticuloendothelial systems is reported. Polymeric nanomaterials have questionable biodistribution and bone targeting. Mesoporous silica has restricted practical usage because of less availability and higher cost of the template lead. Hydroxy apatite are highly brittle, very hard to process, as well as a higher risk of intravenous aggregation and embolism (Naveed, Khalid and Sajid 2018). Toxicity of metal oxides can be attributed to several routes, like size, dissolution and exposure. Zinc oxide, silicon oxide and aluminum oxide nanoparticles cause deformities in embryos, depolarization-induced neuronal injuries, and cellular toxicity due to severe oxidative stress and higher surface energy. TiO_2 nanoparticles have genotoxicity and cytotoxicity with least clearance from brain cells. Titanium nanoparticles can cause chromosomal instability, cellular transformation and apoptotic and necrotic modifications (V. Kumar *et al.* 2012). Oxidative injuries have been reported from chromium nanoparticles. $TiCl_4$ nanoparticles show higher toxicity in marine organisms. Iron oxide nanoparticles can lead to DNA damage in cell lines, liver, breast cells damage in mice models and synovial inflammation in rats. Copper oxide nanoparticles lower cell viability of human bronchial cell lines. Nanoceria can cause oxidative killing of selective cells in animal models (Girigoswami 2018).

8.9   CONCLUSIONS AND FUTURE OUTLOOKS

In conclusion, diverse nanoparticles, either metallic or non-metallic, possess many properties based on their production means, and many studies are unable to reveal their mechanism of action, which hinders the use of these nanoparticles against some viral diseases in the clinical setting. Drugs that are currently used against viral infections usually cause drug toxicity and are restricted from further usage. To overcome these limitations, researchers are moving toward the use of metal- and non-metal-based nanoparticles.

These nanoparticles have also been stated to have the possibility to conquer viral disorders, augment the effectiveness of antiviral treatments, expedient for prophylactic and tonic presentations. They possess antiviral activity by attaching to the virion by inhibiting the receptor binding site of the virus and then by inhibiting the initial stages of viral replication. However, there are only limited reports, indicating that more research is required in developing nanoparticles for treating viral diseases.

Despite the aforementioned prospective therapeutic effectiveness of these nanomaterials, some of the nanoparticles exhibited reduced biological activity, which was attributed their design, nature of the metal and poor selectivity toward the target cells. Due to the integration of these metal nanoparticles with the drug delivery mechanisms, these limitations can be overcome suggesting urgent requirement for the development of metal-associated nanoparticles with high therapeutic effects.

The efficacy, toxicological characteristics and pharmacokinetics of these nanoparticles-based molecules are yet to be studied. Like several organic molecules, a metal-based nanoparticle possesses properties to overwhelm drug resistance issues. There is no doubt that these nanoparticles are promising future therapeutics for the treatment of infectious viral diseases.

REFERENCES

Akbarzadeh, A., L. Kafshdooz, Z. Razban, A. Dastranj Tbrizi, S. Rasoulpour, R. Khalilov, T. Kavetskyy, S. Saghfi, A.N. Nasibova, S. Kaamyabi and T. Kafshdooz. 2018. "An Overview Application of Silver Nanoparticles in Inhibition of Herpes Simplex Virus." *Artificial Cells, Nanomedicine, and Biotechnology* 46:263–267.

Alconcel, Steevens N.S., Arnold S. Baas, and Heather D. Maynard. 2011. "FDA-Approved Poly(Ethylene Glycol)-Protein Conjugate Drugs." *Polymer Chemistry* 2 (7): 1442–48. https://doi.org/10.1039/c1py00034a.

Alharbi, Naif Khalaf. 2017. "Vaccines against Middle East Respiratory Syndrome Coronavirus for Humans and Camels." *Reviews in Medical Virology* 27 (2): e1917. https://doi.org/10.1002/rmv.1917.

Arms, Lauren, Doug W Smith, Jamie Flynn, William Palmer, Antony Martin, Ameha Woldu, and Susan Hua. 2018. "Advantages and Limitations of Current Techniques for Analyzing the Biodistribution of Nanoparticles." *Frontiers in Pharmacology, Aug 14;9:802.*.

Anselmo, Aaron C., and Samir Mitragotri. 2019. "Nanoparticles in the Clinic: An Update." *Bioengineering & Translational Medicine* 4 (3): e10143.

Bamrungsap, Suwussa, Zilong Zhao, Tao Chen, Lin Wang, Chunmei Li, Ting Fu, and Weihong Tan. 2012. "Nanotechnology in Therapeutics: A Focus on Nanoparticles as a Drug Delivery System." *Nanomedicine (London, England)* 7 (8): 1253–71. https://doi.org/10.2217/nnm.12.87.

Baram-Pinto, Dana, Sourabh Shukla, Aharon Gedanken, and Ronit Sarid. 2010. "Inhibition of HSV-1 Attachment, Entry, and Cell-to-Cell Spread by Functionalized Multivalent Gold Nanoparticles." *Small* 6: 1044–50. https://doi.org/10.1002/smll.200902384.

Baram-Pinto, Dana, Sourabh Shukla, Nina Perkas, Aharon Gedanken, and Ronit Sarid. 2009. "Inhibition of Herpes Simplex Virus Type 1 Infection by Silver Nanoparticles Capped with Mercaptoethane Sulfonate." *Bioconjugate Chemistry* 20 (8): 1497–1502. https://doi.org/10.1021/bc900215b.

Bhardwaj, Prateek, Eshant Bhatia, Shivam Sharma, Nadim Ahamad, and Rinti Banerjee. 2020. "Advancements in Prophylactic and Therapeutic Nanovaccines." *Acta Biomaterialia* 108 (May): 1–21. https://doi.org/10.1016/j.actbio.2020.03.020.

Broglie, Jessica Jenkins, Brittny Alston, Chang Yang, Lun Ma, Audrey F. Adcock, Wei Chen, and Liju Yang. 2015. "Antiviral Activity of Gold/Copper Sulfide Core/Shell Nanoparticles against Human Norovirus Virus-like Particles." *PLoS ONE* 10 (10): 1–14. https://doi.org/10.1371/journal.pone.0141050.

Bayoumy, S., E. Juntunen, K. Pettersson and S.M. Talha. 2019. "Detection of Antibodies to the Hepatitis C Virus Using Up-Converting Nanoparticles-Based Lateral Flow Immunoassay." *Clinica Chimica Acta* 493: S653–S654.

Campos, Estefânia V.R., Anderson E. S. Pereira, Jhones Luiz de Oliveira, Lucas Bragança Carvalho, Mariana Guilger-Casagrande, Renata de Lima, and Leonardo Fernandes Fraceto. 2020. "How Can Nanotechnology Help to Combat COVID-19? Opportunities and Urgent Need." *Journal of Nanobiotechnology* 18 (1): 125. https://doi.org/10.1186/s12951-020-00685-4.

Caster, Joseph M., Artish N. Patel, Tian Zhang, and Andrew Wang. 2017. "Investigational Nanomedicines in 2016: A Review of Nanotherapeutics Currently Undergoing Clinical Trials." *Wiley Interdisciplinary Reviews: Nanomedicine and Nanobiotechnology, 9:e1416*: 1–18. https://doi.org/10.1002/wnan.1416.

Chauhan, Gaurav, Marc J Madou, Sourav Kalra, Vianni Chopra, Deepa Ghosh, and Sergio O Martinez-Chapa. 2020. "Nanotechnology for COVID-19: Therapeutics and Vaccine Research." *ACS Nano* 14 (7): 7760–82. https://doi.org/10.1021/acsnano.0c04006.

Chenthamara, Dhrisya, Sadhasivam Subramaniam, Sankar Ganesh Ramakrishnan, Swaminathan Krishnaswamy, Musthafa Mohamed Essa, Feng-Huei Lin, and M Walid Qoronfleh. 2019. "Therapeutic Efficacy of Nanoparticles and Routes of Administration." *Biomaterials Research* 23 (1): 20. https://doi.org/10.1186/s40824-019-0166-x.

Chi, Hang, Xuexing Zheng, Xiwen Wang, Chong Wang, Hualei Wang, Weiwei Gai, Stanley Perlman, Songtao Yang, Jincun Zhao, and Xianzhu Xia. 2017. "DNA Vaccine Encoding Middle East Respiratory Syndrome Coronavirus S1 Protein Induces Protective Immune Responses in Mice." *Vaccine* 35 (16): 2069–75. https://doi.org/10.1016/j.vaccine.2017.02.063.

Cho, Heeyoun, Jean-Louis Excler, Jerome H Kim, and In-Kyu Yoon. 2018. "Development of Middle East Respiratory Syndrome Coronavirus Vaccines – Advances and Challenges." *Human Vaccines & Immunotherapeutics* 14 (2): 304–13. https://doi.org/10.1080/21645515.2017.1389362.

Chen, L. and J. Liang. 2020. "An Overview of Functional Nanoparticles as Novel Emerging Antiviral Therapeutic Agents." *Materials Science and Engineering C: Materials for Biological Applications*. 112: 110924.

Chakravarty, Malobika, and Amisha Vora. 2021. "Nanotechnology-Based Antiviral Therapeutics." *Drug Delivery and Translational Research*: 11(3):748–787.

Cojocaru, Florina-Daniela, Doru Botezat, Ioannis Gardikiotis, Cristina-Mariana Uritu, Gianina Dodi, Laura Trandafir, Ciprian Rezus, Elena Rezus, Bogdan-Ionel Tamba, and Cosmin-Teodor Mihai. 2020. "Nanomaterials Designed for Antiviral Drug Delivery Transport across Biological Barriers." *Pharmaceutics* 12 (2): 171.

Cui, Haixin, Jianfang Jiang, Wei Gu, Changjiao Sun, Donglai Wu, Tao Yang, and Guochen Yang. 2010. "Photocatalytic Inactivation Efficiency of Anatase Nano-TiO2 Sol on the H9N2 Avian Influenza Virus." *Photochemistry and Photobiology* 86 (5): 1135–39. https://doi.org/10.1111/j.1751-1097.2010.00763.x.

Du, Lanying, Wanbo Tai, Yusen Zhou, and Shibo Jiang. 2016. "Vaccines for the Prevention against the Threat of MERS-CoV." *Expert Review of Vaccines* 15 (9): 1123–34. https://doi.org/10.1586/14760584.2016.1167603.

Deschênes, L. and T. Ells. 2020. "Bacteria-nanoparticle Interactions in the Context of Nanofouling." *Advances in Colloid and Interface Science* 277: 102106.

Editorials, Nature. 2020. "Nanotechnology versus Coronavirus." *Nature Nanotechnology* 15 (8): 617. https://doi.org/10.1038/s41565-020-0757-7.

Farjadian, Fatemeh, Amir Ghasemi, Omid Gohari, Amir Roointan, Mahdi Karimi, and Michael R Hamblin. 2019. "Nanopharmaceuticals and Nanomedicines Currently on the Market: Challenges and Opportunities." *Nanomedicine* 14 (1): 93–126.

Galdiero, Stefania, Annarita Falanga, Mariateresa Vitiello, Marco Cantisani, Veronica Marra, and Massimiliano Galdiero. 2011a. "Silver Nanoparticles as Potential Antiviral Agents." *Molecules* 16: 8894–8918. https://doi.org/10.3390/molecules16108894.

Gennaro, Francesco Di, Damiano Pizzol, Claudia Marotta, Mario Antunes, Vincenzo Racalbuto, Nicola Veronese, and Lee Smith. 2020. "Coronavirus Diseases (COVID-19) Current Status and Future Perspectives: A Narrative Review." *International Journal of Environmental Research and Public Health* 17 (8): 2690.

Gheibi Hayat, Seyed Mohammad, and Majid Darroudi. 2019. "Nanovaccine: A Novel Approach in Immunization." *Journal of Cellular Physiology* 234 (8): 12530–36. https://doi.org/10.1002/jcp.28120.

Girigoswami, Koyeli. 2018. "Toxicity of Metal Oxide Nanoparticles." *Advances in Experimental Medicine and Biology* 1048: 99–122. https://doi.org/10.1007/978-3-319-72041-8_7.

Gradishar, William J., Sergei Tjulandin, Neville Davidson, Heather Shaw, Neil Desai, Paul Bhar, Michael Hawkins, and Joyce O'Shaughnessy. 2005. "Phase III Trial of Nanoparticle Albumin-Bound Paclitaxel Compared with Polyethylated Castor Oil–Based Paclitaxel in Women with Breast Cancer." *Journal of Clinical Oncology* 23 (31): 7794–7803.

Gupta, Umesh, and Omathanu Perumal. 2014. "Dendrimers and Its Biomedical Applications." In *Natural and Synthetic Biomedical Polymers*, 1st ed., 243–57. Editors: Sangamesh Kumbar Cato Laurencin Meng Deng, Elsevier Science, USA. https://doi.org/10.1016/B978-0-12-396983-5.00016-8.

Haagmans, Bart L., Judith M. A. van den Brand, V. Stalin Raj, Asisa Volz, Peter Wohlsein, Saskia L. Smits, Debby Schipper, et al. 2016. "An Orthopoxvirus-Based Vaccine Reduces Virus Excretion after MERS-CoV Infection in Dromedary Camels." *Science (New York, N.Y.)* 351 (6268): 77–81. https://doi.org/10.1126/science.aad1283.

Hayden, Frederick. 2009. "Developing New Antiviral Agents for Influenza Treatment: What Does the Future Hold ?" *Clinical Infectious Diseases* 22908 (Supplement 1): 3–13. https://doi.org/10.1086/591851.

Herzog, Christian, Katharina Hartmann, Valérie Künzi, Oliver Kürsteiner, Robert Mischler, Hedvika Lazar, and Reinhard Glück. 2009. "Eleven Years of Inflexal® V—A Virosomal Adjuvanted Influenza Vaccine." *Vaccine* 27 (33): 4381–87. https://doi.org/10.1016/j.vaccine.2009.05.029.

Hou, Wen-Che, Paul Westerhoff, and Jonathan D. Posner. 2013. "Biological Accumulation of Engineered Nanomaterials: A Review of Current Knowledge." Environmental Science: Processes & Impacts 15 (1): 103–22. https://doi.org/10.1039/C2EM30686G.

Innocenzi, Plinio, and Luigi Stagi. 2020. "Carbon-Based Antiviral Nanomaterials: Graphene, C-Dots, and Fullerenes. A Perspective." *Chemical Science* 11 (26): 6606–22. https://doi.org/10.1039/D0SC02658A.

Jong, Wim H. De, and Paul J. A. Borm. 2008. "Drug Delivery and Nanoparticles: Applications and Hazards." *International Journal of Nanomedicine* 3 (2): 133–49. https://doi.org/10.2147/ijn.s596.

Kane, Melissa, and Tatyana Golovkina. 2010. "Common Threads in Persistent Viral Infections." *Journal of Virology* 84 (9): 4116–23. https://doi.org/10.1128/JVI.01905-09.

Kargozar, Saeid. 2018. "Nanotechnology and Nanomedicine: Start Small, Think Big." *Materials Today: Proceedings* 5 (January): 15492–500. https://doi.org/10.1016/j.matpr.2018.04.155.

Khandelwal, N., G. Kaur, N. Kumar, and A. Tiwari. 2014. "Application of Silver Nanoparticles in Viral Inhibition: A New Hope for Antivirals." *Digest Journal of Nanomaterials and Biostructures* 9 (1): 175–86.

Kumar, R., G. Sahoo, K. Pandey, M.K. Nayak, R. Topno, V. Rabidas, and P. Das. 2018. "Virostatic Potential of Zinc Oxide (ZnO) Nanoparticles on Capsid Protein of Cytoplasmic Side of Chikungunya Virus." *International Journal of Infectious Diseases* 73: 368. https://doi.org/10.1016/j.ijid.2018.04.4247.

Kumar, Vineet, Avnesh Kumari, Praveen Guleria, and Sudesh Kumar Yadav. 2012. "Evaluating the Toxicity of Selected Types of Nanochemicals." *Reviews of Environmental Contamination and Toxicology* 215: 39–121. https://doi.org/10.1007/978-1-4614-1463-6_2.

Kwon, Dongwook, Song Hee Lee, and Tae Hyun Yoon. 2010. "Current Limitations and Challenges of Nanoparticle Toxicity Assessments." In 10th IEEE International Conference on Nanotechnology, 1187–88. https://doi.org/10.1109/NANO.2010.5697882.

Lara, Humberto H., Nilda V Ayala-Nuñez, Liliana Ixtepan-Turrent, and Cristina Rodriguez-Padilla. 2010. "Mode of Antiviral Action of Silver Nanoparticles against HIV-1." *Journal of Nanobiotechnology* 8: 1–10.

Lara, Humberto H., Elsa N Garza-Treviño, Liliana Ixtepan-Turrent, and Dinesh K Singh. 2011. "Silver Nanoparticles Are Broad-Spectrum Bactericidal and Virucidal Compounds." *Journal of Nanobiotechnology* 9 (August): 30. https://doi.org/10.1186/1477-3155-9-30.

Laurentius, Lars B., Nicholas A. Owens, Jooneon Park, Alexis C. Crawford, and Marc D. Porter. 2016. "Advantages and Limitations of Nanoparticle Labeling for Early Diagnosis of Infection." *Expert Review of Molecular Diagnostics* 16 (8): 883–95. https://doi.org/10.1080/14737159.2016.1205489.

Lembo, David, and Roberta Cavalli. 2010. "Nanoparticulate Delivery Systems for Antiviral Drugs." *Antiviral Chemistry & Chemotherapy* 21 (2): 53–70. https://doi.org/10.3851/IMP1684.

Levina, Asya S., Marina N. Repkova, Elena V. Bessudnova, Ekaterina I. Filippova, Natalia A. Mazurkova, and Valentina F. Zarytova. 2016. "High Antiviral Effect of TiO2·PL-DNA Nanocomposites Targeted to Conservative Regions of (-)RNA and (+)RNA of Influenza A Virus in Cell Culture." *Beilstein Journal of Nanotechnology* 7 (1): 1166–73. https://doi.org/10.3762/bjnano.7.108.

Liga, Michael V., Erika L. Bryant, Vicki L. Colvin, and Qilin Li. 2011. "Virus Inactivation by Silver Doped Titanium Dioxide Nanoparticles for Drinking Water Treatment." *Water Research* 45 (2): 535–44. https://doi.org/10.1016/j.watres.2010.09.012.

Lu, Min, Martin H. Cohen, Dwaine Rieves, and Richard Pazdur. 2010. "FDA Report: Ferumoxytol for Intravenous Iron Therapy in Adult Patients with Chronic Kidney Disease." *American Journal of Hematology* 85: 315–19. https://doi.org/10.1002/ajh.21656.

Lv, Xiaonan, Peng Wang, Ru Bai, Yingying Cong, Siqingaowa Suo, Xiaofeng Ren, and Chunying Chen. 2014. "Biomaterials Inhibitory Effect of Silver Nanomaterials on Transmissible Virus-Induced Host Cell Infections." *Biomaterials* 35: 4195–4203. https://doi.org/10.1016/j.biomaterials.2014.01.054.

Malarkodi, C., S. Rajeshkumar, K. Paulkumar, M. Vanaja, G. Gnanajobitha, and G. Annadurai. 2014. "Biosynthesis and Antimicrobial Activity of Semiconductor Nanoparticles against Oral Pathogens." In *Bioinorganic Chemistry and Applications*, edited by Imre Sovago, Hindawi Publishers, London, UK, 2014: 347167. https://doi.org/10.1155/2014/347167.

Manzoor, Ashley A., Lars H. Lindner, Chelsea D. Landon, Ji-Young Park, Andrew J. Simnick, Matthew R. Dreher, Shiva Das, et al. 2012. "Overcoming Limitations in Nanoparticle Drug Delivery: Triggered, Intravascular Release to Improve Drug Penetration into Tumors." *Cancer Research* 72 (21): 5566–75. https://doi.org/10.1158/0008-5472.CAN-12-1683.

Maseko, Sibusiso B., Satheesh Natarajan, Vikas Sharma, Neelakshi Bhattacharyya, Thavendran Govender, Yasien Sayed, Glenn E. M. Maguire, Johnson Lin, and Hendrik G. Kruger. 2016. "Protein Purification and Characterization of Naturally Occurring HIV-1 (South African Subtype C) Protease Mutants from Inclusion Bodies." *Protein Expression and Purification* 122: 90–96. https://doi.org/10.1016/j.pep.2016.02.013.

Mehendale, Rujuta, Medha Joshi, and Vandana Patravale. 2013. "Nanomedicines for Treatment of Viral Diseases." *Critical ReviewsTM in Therapeutic Drug Carrier Systems* 30: 1–49.

Mishra, Yogendra Kumar, Rainer Adelung, Claudia Röhl, Deepak Shukla, Frank Spors, and Vaibhav Tiwari. 2011. "Virostatic Potential of Micro-Nano Filopodia-like ZnO Structures against Herpes Simplex Virus-1." *Antiviral Research* 92 (2): 305–12. https://doi.org/10.1016/j.antiviral.2011.08.017.

Naveed, M., S. Khalid, and U. Sajid. 2018. "NanoparticlesA Picture Who Worth's a Thousand Words in Biotechnology." *Drug Des* 7 (160). doi: 10.4172/2169-0138.1000160.

Odobašić, Amra, Indira Šestan, and Sabina Begić. 2019. "Biosensors for Determination of Heavy Metals in Waters." In *Biosensors for Environmental Monitoring*. IntechOpen.

Park, S., Y.-S. Ko, H. Jung, C. Lee, K. Woo and G. Ko. 2018. "Disinfection of Waterborne Viruses Using Silver Nanoparticle-decorated Silica Hybrid Composites in Water Environments." *Science of the Total Environment* 625: 477–485.

Patravale, Vandana, Prajakta Dandekar, and Retnesh Jain. 2012. "Case Studies: Nano-Systems in the Market." In *Nanoparticulate Drug Delivery*, 209–20. Woodhead Publishing, UK. https://doi.org/10.1533/9781908818195.209.

Peng, Xu, Lele Peng, Changzheng Wu, and Yi Xie. 2014. "Two Dimensional Nanomaterials for Flexible Supercapacitors." *Chemical Society Reviews* 43 (10): 3303–23. https://doi.org/10.1039/C3CS60407A.

Pita, Ruben, Falk Ehmann, and Marisa Papaluca. 2016. "Nanomedicines in the EU—Regulatory Overview." *The AAPS Journal* 18 (6): 1576–82.

Prasad, Minakshi, Koushlesh Ranjan, Basanti Brar, Ikbal Shah, Upendra Lalmbe, J Manimegalai, Bhavya Vashisht, et al. 2017. "Virus-Host Interactions: New Insights and Advances in Drug Development Against Viral Pathogens." *Current Drug Metabolism* 18 (10): 942–70. https://doi.org/10.2174/1389200218666170925115132.

Qasim, Muhammad, Dong-Jin Lim, Hansoo Park, and Dokyun Na. 2014. "Nanotechnology for Diagnosis and Treatment of Infectious Diseases." *Journal of Nanoscience and Nanotechnology* 14 (10): 7374–87. https://doi.org/10.1166/jnn.2014.9578.

Quach, Quang Huy, Swee Kim Ang, Jang-Hann Justin Chu, and James Chen Yong Kah. 2018. "Size-Dependent Neutralizing Activity of Gold Nanoparticle-Based Subunit Vaccine against Dengue Virus." *Acta Biomaterialia* 78: 224–35.//doi.org/10.1016/j.actbio.2018.08.011.

Rabaan, Ali A., Ali M. Bazzi, Shamsah H. Al-Ahmed, and Jaffar A. Al-Tawfiq. 2017. "Molecular Aspects of MERS-CoV." *Frontiers of Medicine* 11 (3): 365–77. https://doi.org/10.1007/s11684-017-0521-z.

Rai, Mahendra, Shivaji D Deshmukh, Avinash P Ingle, Indarchand R Gupta, Massimiliano Galdiero, and Stefania Galdiero. 2016. "Metal Nanoparticles: The Protective Nanoshield against Virus Infection." *Critical Reviews in Microbiology* 42 (1): 46–56. https://doi.org/10.3109/1040841X.2013.879849.

Ramazanpour Esfahani, A., O. Batelaan, J.L. Hutson and H.J. Fallowfield. 2020. "Effect of Bacteria and Virus on Transport and Retention of Graphene Oxide Nanoparticles in Natural Limestone Sediments." *Chemosphere*. 248: 125929.

Ren, Haigang, Kai Fu, Chenchen Mu, Xuechu Zhen, and Guanghui Wang. 2012. "L166P Mutant DJ-1 Promotes Cell Death by Dissociating Bax from Mitochondrial Bcl-X$_L$." *Molecular Neurodegenration* 1: 1–11.

Rodriguez, Benigno, David M. Asmuth, Roy M. Matining, John Spritzler, Jeffrey M. Jacobson, Robbie B. Mailliard, Xiao Dong Li, et al. 2013. "Safety, Tolerability, and Immunogenicity of Repeated Doses of Dermavir, a Candidate Therapeutic HIV Vaccine, in HIV-Infected Patients Receiving Combination Antiretroviral Therapy: Results of the ACTG 5176 Trial." *Journal of Acquired Immune Deficiency Syndromes (1999)* 64 (4): 351–59. https://doi.org/10.1097/QAI.0b013e3182a99590.

Samak, D.H., Y.S. El-Sayed, H.M. Shaheen, A.H. El-Far, M.E. Abd El-Hack, A.E. Noreldin, K. El-Naggar, S.A. Abdelnour, E.M. Saied, H.R. El-Seedi, L. Aleya and

M.M. Abdel-Daim. 2020. "Developmental Toxicity of Carbon Nanoparticles During Embryogenesis in Chicken." *Environmental Science and Pollution Research* 27: 19058–19072.

Salleh, Atiqah, Ruth Naomi, Nike Dewi Utami, Abdul Wahab Mohammad, Ebrahim Mahmoudi, Norlaila Mustafa, and Mh Busra Fauzi. 2020. "The Potential of Silver Nanoparticles for Antiviral and Antibacterial Applications: A Mechanism of Action." *Nanomaterials* 10: 1–20.

Sekhon, Bhupinder, Bhupinder Sekhon, and Vikrant Saluja. 2011. "Nanovaccines–An Overview." *International Journal of Pharmaceutical Frontier Research* 1 (January): 101–9.

Seto, Wai Kay, and Man Fung Yuen. 2016. "New Pharmacological Approaches to a Functional Cure of Hepatitis B." *Clinical Liver Disease* 8 (4): 83–88. https://doi.org/10.1002/cld.577.

Shin, Matthew D., Sourabh Shukla, Young Hun Chung, Veronique Beiss, Soo Khim Chan, Oscar A. Ortega-Rivera, David M. Wirth, et al. 2020. "COVID-19 Vaccine Development and a Potential Nanomaterial Path Forward." *Nature Nanotechnology* 15 (8): 646–55. https://doi.org/10.1038/s41565-020-0737-y.

Shubhika, Kwatra. 2012. "Nanotechnology and Medicine–The Upside and the Downside." *International Journal of Drug Development and Research* 5: 1–10.

Siegel, Robert David. 2018. "Classification of Human Viruses." In *Principles and Practice of Pediatric Infectious Diseases*, edited by Sarah S. Long, Charles G. Prober, and Marc Fischer, 1044–1048, Elsevier, London, UK. https://doi.org/10.1016/B978-0-323-40181-4.00201-2.

Singh, Lavanya, Hendrik G. Kruger, Glenn E. M. Maguire, Thavendran Govender, and Raveen Parboosing. 2017. "The Role of Nanotechnology in the Treatment of Viral Infections." *Therapeutic Advances in Infectious Disease* 4 (4): 105–31. https://doi.org/10.1177/2049936117713593.

Singh, Simranjeet, Vijay Kumar, Daljeet Singh Dhanjal, Shivika Datta, Ram Prasad, and Joginder Singh. 2020. "Biological Biosensors for Monitoring and Diagnosis." In *Microbial Biotechnology: Basic Research and Applications*, 317–35. Springer, NY, USA. Guest Editor (s): Joginder Singh, Ashish Vyas, Shanquan Wang, and Ram Prasad.

Sotropa, Roxana-Mihaela Barbu. 2018. "The Advantages and Disadvantages of Nanotechnology." *Romanian Journal of Oral Rehabilitation* 10 (2).

Speshock, Janice L., Richard C. Murdock, Laura K. Braydich-Stolle, Amanda M. Schrand, and Saber M. Hussain. 2010. "Interaction of Silver Nanoparticles with Tacaribe Virus." *Journal of Nanobiotechnology* 8 (1): 19. https://doi.org/10.1186/1477-3155-8-19.

Stammers, Trevor. 2013. "The Promise and Challenges of Nanovaccines and the Question of Global Equity." *Nanotechnology Perceptions* 9 (March): 16–27. https://doi.org/10.4024/N02ST13A.ntp.09.01.

Sulczewski, Fernando B., Raquel B. Liszbinski, Pedro R. T. Romão, and Luiz Carlos Rodrigues Junior. 2018. "Nanoparticle Vaccines against Viral Infections." *Archives of Virology* 163 (9): 2313–25. https://doi.org/10.1007/s00705-018-3856-0.

Shimabuku, Q.L., T. Ueda-Nakamura, R. Bergamasco and M.R. Fagundes-Klen. 2018. "Chick-Watson Kinetics of Virus Inactivation with Granular Activated Carbon Modified with Silver Nanoparticles and/or Copper Oxide." *Process Safety and Environmental Protection* 117: 33–42.

Szymańska, Emilia, Piotr Orłowski, Katarzyna Winnicka, Emilia Tomaszewska, Piotr Bąska, Grzegorz Celichowski, Jarosław Grobelny, Anna Basa, and Małgorzata Krzyżowska. 2018. "Multifunctional Tannic Acid/Silver Nanoparticle-Based Mucoadhesive Hydrogel for Improved Local Treatment of HSV Infection: In Vitro and In Vivo Studies." *International Journal of Molecular Sciences* 19 (2). https://doi.org/10.3390/ijms19020387.

Tavakoli, Ahmad, Angila Ataei-Pirkooh, Gity Mm Sadeghi, Farah Bokharaei-Salim, Peyman Sahrapour, Seyed J. Kiani, Mohsen Moghoofei, Mohammad Farahmand, Davod Javanmard, and Seyed H. Monavari. 2018. "Polyethylene Glycol-Coated Zinc Oxide Nanoparticle: An Efficient Nanoweapon to Fight against Herpes Simplex Virus Type 1." *Nanomedicine* 13 (21): 2675–90. https://doi.org/10.2217/nnm-2018-0089.

Tapaszto, L., G. Dobrik, P. Lambin and L. Biro. 2008. "Tailoring the Atomic Structure of Graphene Nanoribbons by Scanning Tunnelling Microscope Lithography." *Nature Nanotechnology* 3: 397–401.

Thi Ngoc Dung, Tran, Vu Nang Nam, Tran Thi Nhan, Trinh Thi Bich Ngoc, Luu Quang Minh, Bui Thi To Nga, Van Phan Le, and Dang Viet Quang. 2020. "Silver Nanoparticles as Potential Antiviral Agents against African Swine Fever Virus." *Materials Research Express* 6 (12): 1250g9. https://doi.org/10.1088/2053-1591/ab6ad8.

Toita, Riki, Takahito Kawano, Jeong-Hun Kang, and Masaharu Murata. 2015. "Applications of Human Hepatitis B Virus PreS Domain in Bio- and Nanotechnology." *World Journal of Gastroenterology* 21 (24): 7400–7411. https://doi.org/10.3748/wjg.v21.i24.7400.

Ventola, C. Lee. 2017. "Progress in Nanomedicine: Approved and Investigational Nanodrugs." *Pharmacy and Therapeutics* 42 (12): 742.

Vodnar, Dan Cristian, Laura Mitrea, Lavinia Florina Călinoiu, Katalin Szabo, and Bianca Eugenia Ştefănescu. 2020. "Removal of Bacteria, Viruses, and Other Microbial Entities by Means of Nanoparticles." In *Micro and Nano Technologies*, edited by Lucian Baia, Zsolt Pap, Klara Hernadi, and Monica B. T., Advanced Nanostructures for Environmental Health Baia, 465–491. Elsevier. //doi.org/10.1016/B978-0-12-8158 82-1.00011-2.

Weissig, Volkmar, Tracy K. Pettinger, and Nicole Murdock. 2014. "Nanopharmaceuticals (Part 1): Products on the Market." *International Journal of Nanomedicine* 9: 4357.

Weissig, Volkmar, and Diana Guzman-Villanueva. 2015. "Nanopharmaceuticals (Part 2): Products in the Pipeline." *International Journal of Nanomedicine* 10: 1245.

Xu, Jiuyang, Wenxu Jia, Pengfei Wang, Senyan Zhang, Xuanling Shi, Xinquan Wang, and Linqi Zhang. 2019. "Antibodies and Vaccines against Middle East Respiratory Syndrome Coronavirus." *Emerging Microbes & Infections* 8 (1): 841–56. https://doi.org /10.1080/22221751.2019.1624482.

Yadavalli, Tejabhiram, and Deepak Shukla. 2016. "Could Zinc Oxide Tetrapod Nanoparticles Be Used as an Effective Immunotherapy against HSV-2?" *Nanomedicine (London, England)*. England. https://doi.org/10.2217/nnm-2016-0249.

Yadav, H.K.S., M. Dibi, A. Mohammad and A.E. Srouji. 2018. "Nanovaccines Formulation and Applications–A Review." *Journal of Drug Delivery Science and Technology* 44: 380–387.

Yao, Duoxi, Zheng Chen, Kui Zhao, Qing Yang, and Wenying Zhang. 2013. "Limitation and Challenge Faced to the Researches on Environmental Risk of Nanotechnology." *Procedia Environmental Sciences* 18: 149–56. //doi.org/10.1016/j.proenv.2013.04.020.

Youssef, F.S., H.A. El-Banna, H.Y. Elzorba and A.M. Galal. 2019. "Application of Some Nanoparticles in the Field of Veterinary Medicine." *International Journal of Veterinary Science and Medicine* 7: 78–93.

Zhang, Naru, Jian Tang, Lu Lu, Shibo Jiang, and Lanying Du. 2015. "Receptor-Binding Domain-Based Subunit Vaccines against MERS-CoV." *Virus Research* 202 (April): 151–59. https://doi.org/10.1016/j.virusres.2014.11.013.

Zhang, Rui Xue, Jason Li, Tian Zhang, Mohammad A Amini, Chunsheng He, Brian Lu, Taksim Ahmed, HoYin Lip, Andrew M Rauth, and Xiao Yu Wu. 2018. "Importance of Integrating Nanotechnology with Pharmacology and Physiology for Innovative Drug Delivery and Therapy–An Illustration with Firsthand Examples." *Acta Pharmacologica Sinica* 39 (5): 825–44.

Zhong, Jiayu, Yu Xia, Liang Hua, Xiaomin Liu, Misi Xiao, Tiantian Xu, Bing Zhu, and Hong Cao. 2019. "Functionalized Selenium Nanoparticles Enhance the Anti-EV71 Activity of Oseltamivir in Human Astrocytoma Cell Model." *Artificial Cells, Nanomedicine and Biotechnology* 47 (1): 3485–91. https://doi.org/10.1080/21691401 .2019.1640716.

9 Nanomaterials against Parasites: The Developments and the Way Forward

Tean Zaheer, Sadia Muneer, Rao Zahid Abbas,
Muhammad Kasib Khan, Muhammad Imran,
Amna Ahmed, Iqra Zaheer and Nighat Perveen

CONTENTS

9.1 Metallic Nanoparticles against Parasites 229
9.2 Metallic Nanoparticles against Endoparasites 232
    9.2.1 Use of Gold Nanoparticles against Endoparasites 232
        9.2.1.1 Use of Silver Nanoparticles against Endoparasites 233
        9.2.1.2 Use of Nickel Nanoparticles against Endoparasites 234
        9.2.1.3 Use of Palladium Nanoparticles against Endoparasites 235
    9.2.2 Metallic Nanoparticles against Ectoparasites 235
    9.2.3 Nanoparticles of Green Origin against Parasites 237
9.3 Examples of Non-metallic Nanomaterials against Parasites 239
9.4 Conclusions and Future Outlook 242
References 242

9.1 METALLIC NANOPARTICLES AGAINST PARASITES

Parasites are dangerous microorganisms to human health, either living on the host surface (ectoparasites) or within the host cell (endoparasites). They can cause a number of infectious diseases such as leishmaniasis, trypanosomiasis and malaria, resulting in high morbidity and mortality in developing countries. Among them, malaria (caused by the *Plasmodium* parasite) [1] is an acute public health concern that needs to be eradicated. Parasites include arthropods and helminths or protozoa, including ectoparasites that are arthropods, that either act as vectors transmitting alternative parasites or causing diseases, and endoparasites, which reside in body tissues causing serious health issues [2]. Parasitic diseases are some of the world's most prevalent infections, leading to high morbidity and mortality rates annually. Most of the parasitic infections and their modes of transmission are well known.

DOI: 10.1201/9781003126256-9

Human behavior, population movement and climatic changes have prominent roles in the transmission, incidence, prevalence and distribution of parasitic infections [3]. Parasitic infections have profound impact on human health due to their unique characteristics such as ability to divide rapidly in number, evolutionary variations and unpredictability [1]. For example, in 2019, more than 229 million people were infected by malaria throughout the world [4].

In ancient times, many infectious diseases were generally associated with tropical and subtropical regions. However, nowadays, vector ecological changes, the migration of humans and animals, and a significant increase in international travel and armed conflicts have particularly influenced the transmission of parasitic diseases in developed countries. It has also been reported that a number of patients who never travelled to endemic regions also suffer from blood-borne protozoal infections [5]. Among the highest rates of morbidities and mortalities throughout the world are vector-borne infections such as yellow fever, malaria, filariasis, dengue, Japanese encephalitis and many others, which are regarded as dangerous to human health. Due to serious environmental issues generated by misuse/overuse of insecticides and increasing resistance among arthropods to insecticidal substances, it is hard to effectively stop the reproduction of arthropods and diseases caused by them. Hence, a novel class of agents with minimal impact on the environment and high anti-arthropod activity is necessary to improve the control of arthropods and transmission of their infections [6].

Therefore, there is an urgent need to develop and implement new techniques to diagnose infectious disease to control the spread, ensure public health safety and promote treatment. Moreover, the ideal techniques used for infectious diseases should be rapid, sensitive, specific, low-cost, user-friendly, robust and accurate [7, 8]. So, our aim is to shed light on the use of nanotechnology against infectious diseases caused by parasites (Figure 9.1 and 9.2).

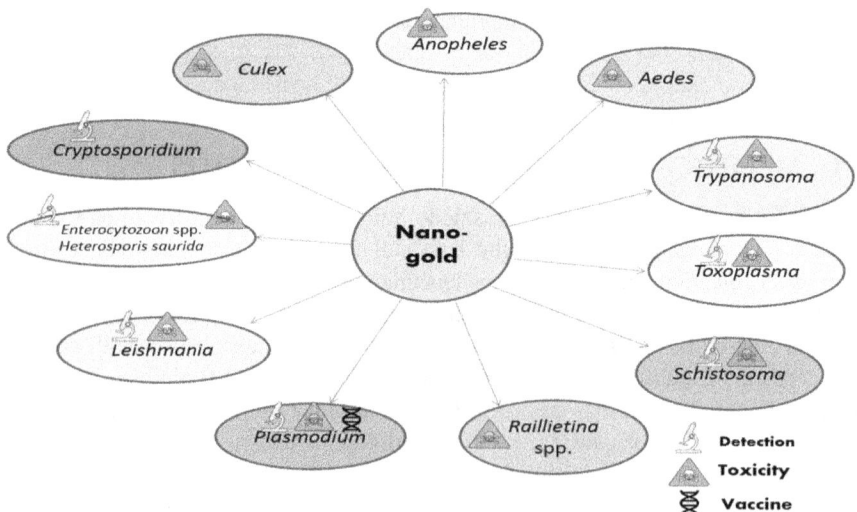

FIGURE 9.1 Summary of gold nanoparticles applications in parasitology.

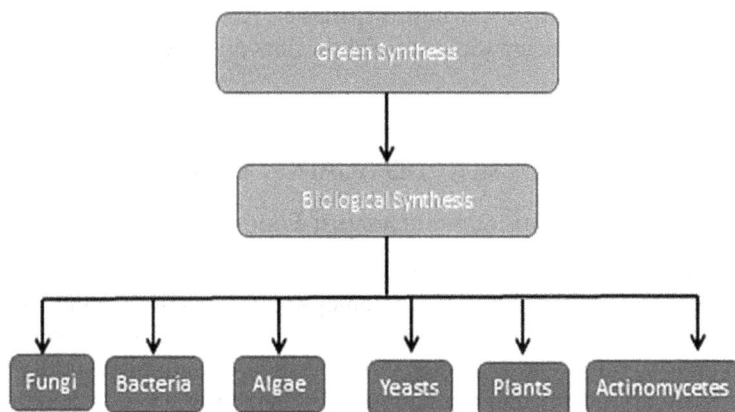

FIGURE 9.2 Brief layout of biological routes for nanoparticle synthesis.

Nanoparticles (NPs) are currently known for many environmental, biomedical and biological processes, due to their excellent chemical and physical properties, which are currently subject of much scientific research [9–12]. Nanomaterials having diameter of ≤100 nm, have wide ranges of medical applications and have attracted great attention as an alternative technique to control various infectious agents [13]. They are characterized by their large surface area as well as high particle number per unit mass as compared to bulk materials [9–12]. In aquaculture, metal and metal oxide nanomaterials showed many effective anti-microbial characteristics against pathogens of fish [13] and have been applied as an anti-microbial agent and as a water decontaminant [14–17]. Particularly, the application of green synthesis processes have allowed many advantages over classical physical and chemical nanosynthesis techniques, including being environment-friendly and not requiring the use of hazardous substances or high energy inputs [18–23].

NPs cannot only circulate widely all over the body but also have the ability to penetrate in the cells, control the drug distribution and carry it specifically to the targeted place. Therefore, these particles are regarded as a very effective solution for treatment of various infectious diseases [24]. Drug carrier nanoparticles vary in shapes and sizes as well as material. Each poses a specific drug carrying capacity, stability and cellular targeting. NPs-based methods used for the treatment and diagnosis of various diseases are expected to have remarkable status in the near future. However, in terms of human health effects, NPs also face some important challenges [25]. Due to adverse side effects and uncertainty regarding the therapeutic effectiveness of anti-parasitic drugs, the use of NPs as medication, specifically for parasitic infections, have earned special attention in the last few years. NPs have a deep history of diagnosis and treatment of various diseases [26]. Various reports indicate that NPs of gold, silver, chitosan and metal oxide have inhibitory effects on the growth of parasites like malaria, *Leishmania* and insects larvae [27]. These metal NPs can be applied singly or in combination with other metals and their oxides to eliminate various parasites. [28]. As they have inhibitory and destructive effects, the use of NPs is recommended to destroy parasites, making less harmful and more effective drugs and producing various useful vaccines for the control and prevention of parasites [29].

9.2 METALLIC NANOPARTICLES AGAINST ENDOPARASITES

9.2.1 USE OF GOLD NANOPARTICLES AGAINST ENDOPARASITES

Currently, the use of gold nanoparticles is an area of interest for researchers in parasitology and entomology [30–31]. Being non-cytotoxic in nature, having polyvalent effects and having high ability for functionalization, gold nanoparticles (Au-NPs) have a promising role in the development of various novel anti-microbial agents [16, 32–35]. Au-NPs have an additional advantage of large surface area, making their surfaces accessible for modification with targeting molecules and enhanced anti-microbial properties [36]. The bioactive potential of gold nanoparticles have been studied on some species of insects of economic importance, including some major mosquito vectors of dengue and zika virus, malarial parasites and lymphatic filariasis, such as *Aedes aegypti* L., *Aedes stephensi* Liston and *Culex quinquefasciatus* Say (Diptera: Culicidae), respectively [19, 35–40]. Moreover, the toxic effects of gold nanoparticles have been studied on *Schistosoma trematodes* [41–43] and helminthes including *Raillietina cestodes*, which is an intestinal parasite [44]. This class of newly discovered nanoparticles has a wide range of applications such as drug development and delivery against various parasitic species of public health interest. These include parasites such as *Leishmania* spp., *Plasmodium* spp., *Trypanosoma* spp., microsporidians, *Toxoplasma gondii* and *Cryptosporidium parvum*, allowing scientists to use gold nanoparticles as novel candidates for drug discovery [45–51], biomarkers and diagnostic tools [52–57] as well as adjuvants for inducing transmission blocking immunity in research of malarial vaccine [58].

Malaria is caused by protozoan parasites belonging to the genus *Plasmodium* and transmitted by female *Anopheles* mosquitoes. Malaria causes high rates of morbidity and mortality throughout the world. One of the main factors that have contributed in the prevalence of malaria disease in the world is the increase in the number of drug-resistant parasites. To overcome drug-resistance, NPs have been studied for their ability to effectively deliver overall low doses of anti-malarial drugs to kill parasites and avoid drug-resistance evolution [59]. Keeping in view the current scenario, there is the urgent and utmost need to develop reliable and effective methods for the diagnosis and treatment of malarial diseases to ensure human health safety.

The development of highly effective and reliable malaria management programs has been a long-time challenge for research in the field of parasitology [60, 61]. Although some progress in the fight against malaria parasites belonging to genus *Plasmodium* (Plasmodiidae) has been mostly achieved through the control of vectors of *Anopheles* mosquitoes, mainly using long-lasting insecticidal nets (LLIN) and indoor residual spraying (IRS) [62–65] along with the use of effective and novel anti-plasmodial drugs [66, 67]. Moreover, use of gold-based drugs have a deep history and promising effects against a wide range of parasites, particularly *Plasmodium*

[68]. For example, a colorimetric aptasensor was designed by Jeon et al. [55] for the specific and sensitive diagnosis of *Plasmodium* dehydrogenase in malaria. Therefore, Au-NPs have become a useful tool due to ease of synthesis, strong distance-dependent optical properties and high extinction coefficients. Colorimetric-based gold nanoparticles aptasensors have obtained great attention for the detection of various targets, as molecular recognition events can be easily viewed by the naked eyes without the use of any sophisticated equipment. Due to great commercial diagnostic potential of colorimetric-based gold nanoparticles aptasensors, *Plasmodium falciparum* and *Plasmodium vivax* were determined [55]. Another good example is a gold (I)-containing drug, namely auranofin, used for antiarthritic purposes and currently used as an anti-parasitic against *Leishmania infantum* Nicolle, *Leishmania major* (Schokhor and Yakimoff), *Schistosoma* and *Plasmodium* [69–72].

Varela-Aramburu et al. [73] tested binding of glucose-based ultra-small gold nanoparticles with cysteine-rich domains of *Plasmodium falciparum* surface proteins. Microscopy shows specific binding of ultra-small gold NPs to all intracellular and extracellular erythrocytic stages of *Plasmodium falciparum*. These NPs can be used as drug delivery agents for ciprofloxacin, which is weakly soluble antibiotic, and possesses low anti-malarial activity. Ciprofloxacin becomes more water-soluble than free drug when conjugated with ultra-small gold NPs. Glyco-gold NPs may be helpful for the treatment and control of malaria, as they target cysteine-rich domains on parasites [74].

In tropical areas, Schistosomiasis is a major parasitic infection that affects human health. The main anti-schistosomal drug is Praziquantel. New drugs are required due to the emergence of drug-resistant strains of *Schistosoma* and poor effects of the drug on the immature worms and parasite juveniles. Nanotechnology is one of the most important emerging methods to cope with challenges of parasitic infections. Dkhil et al. [75] tested the use of both selenium and gold nanoparticles against *S. mansoni*. The results of this study showed the protective use of gold and selenium NPs, as they significantly reduced the levels of body weight changes, histological impairment in the jejunal tissue and oxidative stress.

9.2.1.1   Use of Silver Nanoparticles against Endoparasites

Silver nanoparticles (Ag-NPs) present unique properties of anitmicrobial agents [76] improving their biocompatibility due to their nontoxicity to humans, lower costs and high ratio of surface area to volume [77]. Some studies show that Ag-NPs bind with the cell wall or membrane, proteins and DNA of microorganisms to show anti-microbial properties by different mechanisms. They bind to cell membrane proteins and disturb the cell membrane structure by inducing production of reactive oxygen species leading to cell death. They also bind to cytochrome intracellularly and interfere with nucleic acids of pathogens, subsequently inhibiting replication and cell division [78, 79]. They have also been shown to affect the pathogenicity and morphology of the protozoan parasite *Leishmania tropica* in vitro [17].

Gastrointestinal protozoan parasites, especially, *Cryptosporidium parvum* and *Entamoeba histolytica*, are among the most common etiological agents of diarrheal disease throughout the world, particularly among children. The prevalence rate of these parasites is higher in children suffering from diarrhea as compared to those without them. Therefore, NPs have gained much importance due to their anti-parasitic properties. In Egypt, Saad et al. [80] tested the anti-parasitic effects of silver and copper NPs against two of the most environmentally spread parasites (*Cryptosporidium parvum* and *Entamoeba histolytica*). A significant reduction for cysts viability ($p >$ 0.05) was noticed for Ag-NPs against *Cryptosporidium parvum* oocytes and CuO-NPs against *Entamoeba histolytica* cysts. Furthermore, LC_{50}-3 h of Ag-NPs for *Cryptosporidium parvum* and *Entamoeba histolytica* recorded 0.54 and 0.34 mg/l, while CuO-NPs were 0.72 and 0.13 mg/l, respectively. Therefore, these NPs could be considered as safe and effective new treatments due to their capability to inactivate cysts of *Cryptosporidium parvum* and *Entamoeba histolytica* parasites. These particles may also be regarded as water treatment agents and alternative drugs for *Entamoeba histolytica* and *Cryptosporidium parvum* cysts, consequently playing an important part in overcoming cryptosporidiosis and amoebiasis.

Chagas disease and leishmaniasis are the most important protozoan infections caused by intracellular parasites. Treatments of these diseases face some limitations like variable efficacy, toxicity, lengthy treatment regimens and requirements for parenteral administration. The high cost is main limitations in the treatments of visceral leishmaniasis. There is no availability of therapeutic and preventive vaccines for these parasitic diseases. An attractive alternative to fulfill the urgent demand of developing new medication is offered by the intervention of nanotechnology in the immunological and pharmaceutical field. Nanomedicines are more effective, less toxic and do not induce drug resistance in parasites and help to cope with the challenges posed by the treatment and prophylaxis of parasitic infections such Chagas disease and leishmaniasis [81, 82].

In another study, Baranwal et al. [83] reported that nanotherapeutic agents play an important role in treatment of *Leishmania donovani*. In an in vitro experimental setting, Ag-NPs were used to study the anti-leishmanial potential on clinical isolates of *Leishmania donovani* promastigote cells. The dose-dependent killing activity of the Ag-NPs was examined with an IC50 value of 51.88 ± 3.51 µg/ml and also compared with a commercially available drug, miltefosine. Moreover, the clinical suitability of Ag-NPs as a therapeutic agent against *Leishmania* was proven by testing them against a normal mammalian monocyte cell line (U937) and no significant toxicity of Ag-NPs was observed on the normal mammalian cell line.

9.2.1.2   Use of Nickel Nanoparticles against Endoparasites

Nickel nanoparticles (Ni-NPs) possess many unique characteristics, including high magnetism and surface energy, low burning point, high surface area and low melting point. They can be widely used in modern industries as catalysts, sensors and electronic applications, along with many biomedical applications such as cytotoxicity and inflammation studies [84, 85]. Nickel possesses good anticorrosion properties [86]. Some authors reported the synthesis of Ni-NPs with an average size of 150 nm. Nickel was applied against blood-sucking parasites, resulting in excellent larvicidal activity. However, they synthesized Ni-NPs using hydrazine as a precursor, which acts as a reducing agent [86].

In a study, Rajakumar et al. [87] reported the anti-parasitic potential of nickel nanoparasites against the larvae of cattle ticks *Hyalomma analoticum (a) analoticum* (Acari: Ixodidae) and *Rhipicephalus (Boophilus) micropilus*, fourth instar larvae of *Anopheles subpictus, Culex quinquefasciatus* and *Culex gelidus* (Diptera: Culicidae). The nickel nanoparticles showed maximum activity against the larvae of *C. gelidus, C. quinquefasciatus, A. subpictus, H. a. anatolicum* and *R. (B.) micropilus* with r^2 values of 0.988, 0.950, 0.992, 0.993, and 0.990, and LC_{50} values of 4.94, 5.56, 4.93, 10.81 and 10.17 mg/L, respectively. The efficacy of nickel hydrazine complexes showed the values of r^2 0.978, 0.944, 0.989, 0.986, 0.988 and LC_{50} values of 7.83, 9.69, 8.29, 22.72 20.35 mg/L, respectively. These findings showed that nickel NPs possessed excellent larvicidal parasitic potential. To the best of our knowledge, this is the very first report on larvicidal activity of blood-feeding parasites using synthesized Ni-NPs.

9.2.1.3 Use of Palladium Nanoparticles against Endoparasites

Palladium is a very precious metal, and palladium nanoparticles (Pd-NPs) are of particular interest in the field of medicine due to their affinity for hydrogen and catalytic properties. Pd-NPs possessing intermediate compounds have promising applications, like catalyst active membranes and sensors [88]. They were synthesized using a wide range of wet chemical methods like electrochemical, chemical reduction and sonochemical as well as ployol methods, and have been investigated for structure-controlled synthesis of NPs [89, 90].

In a study, Rajakumar et al. [91] reported the anti-plasmodial effectiveness of Pd-NPs. In another study, Minal and Parkash [92] reported the nano-efficacy of Pd-NPs against larvae of *Anopheles stephensi* (Liston). The results showed that larvicidal effectiveness of palladium nanoparticles increases with an increase in time. This can be useful in tackling mosquito-borne diseases and emerging insecticide resistance worldwide.

9.2.2 Metallic Nanoparticles against Ectoparasites

Kar et al. [93] carried out an in vitro study to check the anti-helmintic efficacy of gold nanoparticles against a model cestode (*Raillietina* spp.) that is an intestinal parasite of domestic fowl. This study proved the effectiveness of Au-NPs; however, further studies are required on the anthelmintic efficacy of Au-NPs. In another study, Roy et al. [94] used biopolymer functionalized gold particles against bovine parasite (*Setaria cervi*) and human filarial parasite (*Wuchereria bancrofti*). Au-NPs showed excellent anti-filarial activity against both bovine and human parasites along with induction of oxidative stress and apoptotic cell death mediated through mitochondria in filarial parasites. Furthermore, these NPs showed no toxic effects to the mammalian system.

Adeyemi and his colleagues [95] studied the anti-trypanosoma effects of silver and gold nanoparticles and their alloys. The findings of this study add support to the in vitro anti-parasitic activity of silver and gold NPs against *Trypanosoma congolense, Trypanosoma evansi* and *Trypanosoma brucei brucei*, and also reported that NPs might be effective against both intracellular and extracellular parasites and support future studies to investigate the effectiveness of silver and gold NPs at higher doses against *Trypanosoma* in vivo. In another study, Vazini [96] examined the effectiveness of gold nanoparticles to eliminate the *Leishmania major* both in vitro and in vivo. Therefore, these metal particles may be applied for the in vivo destruction of the promastigotes of *Leishmania major*. Moreover, this study provides a suitable alternative to the drugs used as remedial agents for the cutaneous leishmaniasis caused by *Leishmania major* and can be used without the adverse effects related to the use of chemical therapeutic agents such as glucantime. Furthermore, amphotericin B as a liposomal formulation of AmBisome is regarded as the first line of treatment for visceral leishmaniasis, caused by the *Leishmania donovani*. But, nephrotoxicity is very common due to aggregation of amphotericin and poor water solubility. Therefore, Kumar et al. [97] prepared a water-soluble covalent conjugate of Au-NPs with a polyene antibiotic amphotericin B for reduced cytotoxicity and improved anti-leishmanial effectiveness. Macrophages treated with amphotericin B along with Au-NPs showed an enhanced immunostimulatory Th1 (interferon-γ and IL-12) response in comparison with amphotericin B. Uptake of amphotericin B was approximately 5.5- and 3.7-fold higher for macrophages treated with Au-NPs–amphotericin within 1 and 2 hour of treatment, respectively. The ergosterol content in parasites treated with amphotericin with Au-NPs was nearly two times reduced as compared with amphotericin B-treated parasites. Further, amphotericin along with gold particles was significantly less hemolytic and cytotoxic than amphotericin B alone. Gold nanoparticles-based delivery of amphotericin B might be a cheaper, safer and better alternative than amphotericin B formulations alone.

Sazgarnia et al. [98] in his study, reported that the presence of Au-NPs during MW irradiation was found to be more lethal for amastigotes and promastigotes as compared to MW alone.

A common food and pet-borne disease associated with personality alteration and hallucinations often contracted by house pets is toxoplasmosis. The therapeutic agent was developed at the University of Technology Sydney, Australia, by treating gold nanoparticles with toxoplasmid-hunting antibodies. These gold carrying-antibodies have the ability to spread through the circulatory system and affix themselves to parasites in the blood. As soon as the Au-NPs are well distributed and broadly attached to the parasites, the laser light heats up the gold ultimately incinerating the parasites [99].

In another study, Bavand et al. [100] reported that gold nanoparticles at a concentration of 0.3 mg. ml^{-1} can be an effective formulation for killing *Giardia lamblia* cysts in vitro.

Silver nanoparticles are the most important commercial nanomaterials [101], as they possess unique characteristics, such as large surface area and their promising ability to produce reactive oxygen species (ROS), which help them to present effective anti-microbial activities [102–104]. Protists are good models of organisms for nanotoxicology because they have highly developed systems for the internalization of micro- and nano-scale particle sizes. [105]. The anti-protozoal activity of silver nanoparticles against the ciliates *Tetrahymena pyriformis* [106], *Tetrahymena thermophile* [104], trophonts and free-living stages of *Ichthyophthirius multifiliis* [107, 108], flagellates *Leishmania tropica* [109] and *L. amazonensis* [110] has been documented. But the toxicity of Ag-NPs can vary with structure, composition and size of the engineered NPs [107]. The exact mechanism of action of silver NPs in protists is still unknown, but involvement of outer membrane rupturing has been reported as in the cases of *I. multifiliis* and *Cryptosporidium parvum*. In a study, Saad et al. [80] reported a very important role for Ag and CuO nanoparticles in overcoming amoebiasis and cryptosporidium caused by *E. histolytica* and *C. parvum*, respectively.

Allahverdiyev et al. [111] used Ag-NPs on *Leishmania tropica* in vitro and demonstrated significant anti-leishmanial actions of Ag-NPs by inhibiting the metabolic activity and proliferation of promastigotes. Allahverdiyev et al. [111] also used Ag_2O and TiO_2 in another study on *Leishmania tropica* in vitro and demonstrated the anti-leishmanial effects of Ag_2O and TiO_2 nanoparticles.

Karimi et al. [113] used Ag-NPs against *Leishmania major* in vitro and concluded that Ag-NPs did not completely kill *Leishmania major* promastigotes when used singly, but when combined with direct current electricity had significant synergistic effects on the mortality of promastigotes. Khosarvi et al. [114] reported that IC50 for Ag-NPs solution on *Leishmania tropica* in vitro was significantly high (14.9 μg ml^{-1}). Jameii et al. [115] made a combined use of silver and selenium nanoparticles on *Leishmania major* in vivo and reported that Ag-NPs showed promising anti-leishmanial effects in vivo as compared to selenium nanoparticles. Gaafar et al. [116] reported the use of silver and chitosan on *Toxoplasma gondii* in vivo and showed that Ag-NPs used singly or combined with chitosan have significant anti-toxoplasma effects. Ponarulselvam et al. [117] reported the anti-plasmodial potential of Ag-NPs on *Plasmodium falciparum* in vitro. Gherbawy et al. [118] used Ag-NPs on *Fasciola* in vivo and in vitro, and reported that the percentage of non-hatching eggs when treated with triclabendazole drug was 69.67%. However, this percentage reached up to 89.67% when a combination of triclabendazole drug and Ag-NPs was used.

9.2.3 NANOPARTICLES OF GREEN ORIGIN AGAINST PARASITES

Parasitology and nanomedicines are facing a number of challenges that mostly deal with the paucity of the effectiveness of the preventive and therapeutic agents against mosquito-borne and other parasitic diseases. In this case, the use of invertebrate and plant extracts as stabilizing, capping and reducing agents for the production of nanoparticles is advantageous over the physical and chemical processes, as they are one-pot, cheap, do not require high energy, pressure or temperature and do not require the use of highly toxic chemical agents. Green fabricated nanoparticles have been successfully used against *Plasmodium* and mosquito-borne illnesses [130].

In recent years, a green synthesis approach of nanoparticles based on 12 green chemistry principles has emerged as an attractive alternative to traditional methods and includes the development and designing of NPs using renewable material, nontoxic chemicals, eco-friendly solvents and ultimately degradable waste products [131]. From a green chemistry perception, there are three important components

TABLE 9.1

Some Metals and Metal Oxide Nanoparticles along with Their Mechanism of Action against Parasites

Parasite	Metal NP Composition Size/Shape	Synthesis	Mechanism of Action	Reference
Schistosoma mansoni	Au-NPs, 10–15 nm, spherical	Chemical	Scavenge free radical	41
	Se-NPs	Chemical	Scavenge free radical	119
Trypanosoma brucei	Ag-NPs, 4–9 nm, spherical	Chemical	TbAK inhibitor	120
	Au-NPs, 7–22 nm, spherical	Chemical	TbAK inhibitor	120
Leishmania major	Se-NPs, 80–220 nm, spherical	*Bacillus sp.* MSh-1	DNA fragmentation	121
	Au-NPs, 40 nm	Chemical	ROS production	98
Leishmania donovani	Ag-NPs, 12.82 nm, spherical	*Euphorbia prostrate*	TR inhibitor	122
	TiO-NPs, 83.22 nm, circular and irregular shape			
Leishmania tropica/infantum	TiO-NPs, 90 nm, spherical	Chemical	ROS production	123
	Ag-NPs, rectangular			
Leishmania tropica	Ag-NPs, 10–40 nm, round shaped	Chemical	ROS production	79
Plasmodium berghei	Pd-NPs, 30–110 nm, mostly spherical	*Eclipta prostrate* leaf extract	N/A	91
	Au-NPs, 5–50 nm, polygonal	Marine actinobacteria	N/A	46
Plasmodium falciparum	Ag-NPs, 22–44 nm, triangular/hexagonal	Amylase enzyme	N/A	124
	Ag-NPs, 2–8 nm, spherical	Neem plant extract	N/A	125
	Ag-NPs, 35–65 nm, spherical	*Pteridium aquilinum* leaf extract	N/A	126
	Au-NPs, 29.2–43.8 nm, spherical and oval	*Couroupita guianensis* flower extract	N/A	127
	Ag-NPs, 5 nm	Tannic acid	PfThzK inhibitor	127
	Ag-NPs, 20–35nm, cubical	*Ulva lactula* seaweed extract	N/A	128
	Ag-NPs, 20–40 nm	*Codium tomentosum* seaweed extract	N/A	129

in the preparation of NPs: nontoxic reducing agent, harmless solvent medium and environmentally benign stabilization agents [132]. The selection of capping agent is another significant factor to passivate the surface of nanoparticles and shows a promising effect on the morphologies, size ranges and targeted applications [133].

Biosynthesis of nanoparticles has gained much attention over traditional methods as they are environment-friendly and involve the use of both multicellular and unicellular entities [134–151]. These biological entities importantly act as biological factories and offer a clean, nontoxic method of synthesizing nanoparticles with wide ranges of shapes, sizes, compositions and physiochemical properties [152]. Another important feature of the many biological entities is the ability to act as templates in the synthesis, assembly and organization of the nanometer scale materials to properly fabricate micro- and macro-scale structures. For instance, phage-based liposome assemblies have been used in targeted drug delivery processes [153–156], bacteriophages have been used to form micrometer and nanometer scale structures [157, 158], and viruses have been applied to assemble gold and iron oxide NPs to form microstructures [159]. Comparing the aforementioned biological entities and their potential activities to act as efficient biological factories, synthesis of nanoparticles using plants is a relatively advantageous and straightforward approach [160], as it does not need any special, complex and multi-step processes like isolation, culture preparation and maintenance. Therefore, plant-based nanomaterials are known to be the best material and have gained special attention due to ease, excess availability and biodegradability [161–164].

Recently, it has been reported that a wide range of biological synthesized nanoparticles (metals and their oxides) can be administered to inhibit the growth of different parasitic species. Table 9.2 contains some examples of biosynthesized NPs and their important properties.

9.3 EXAMPLES OF NON-METALLIC NANOMATERIALS AGAINST PARASITES

Liposomal encapsulation of the already resistant drugs for leishmaniasis have proven to be effective as treatment for visceral and cutaneous in animal models as well as in humans due to their elevated therapeutic index. These drugs include amphotericin B [179], atovaquone [180], quercetin and harmine [181], and miltefosine [182]. Similarly, the liposome drug delivery system has been extensively studied and investigated for the treatment of trypanosomiasis (*Trypanosoma cruzi and Trypanosoma brucei*). They do not colonize in the mononuclear phagocyte system (MPS) and become difficult for liposomes to target. In addition, liposomal encapsulation of anti-trypanosomal and other drugs have proven to be more effective due to improved potency and therapeutic index, including benzimidazole [183], etanizole [184], arsenic containing lipids [185] and lipid-based amphotericin B [186]. Nanoparticle-based delivery systems possess several characteristic features due to their tiny size and advanced nature, such as ameliorated bioavailability, enhanced pharmaceutical activity of active ingredients, extensive cell membrane permeability, improved stability by providing the nanoparticles protection from physical and chemical degradation, and improving

TABLE 9.2
Some Examples of Biosynthesized Parasitic Nanoparticles

Plant	Nanoparticle	Size (nm)	Morphology	Application	Reference
Catharanthus roseus	Ag	35–55	Cubic	Anti-plasmodial	165
Nerium oleander	Ag	20–35	Spherical, cubic	Anti-*Anopheles stephensi*	166
Teucrium stocksianum Boiss	Ag	10–40	Spherical	Anti-leishmanial	167
Mimosa pudica Gaertn	Ag	25–60	Spherical	*A. subpictus*, malaria vector, *R. microplus*, *C. quinquefasciatus*	168
Anisomeles indica	Ag	50–100	Spherical	*An. subpictus*, *Ae. Albopictus*, *Cx. Tritaeniorhynchus*	169
Azadirachta indica and *Cymbopogon citratus*	ZnO	15–25	Spherical (irregular)	*Hyalomma* ticks	170
Isatis tinctoria	Ag-NPs treated with Amphotericin B	10–20	Spherical	*Leishmania donovani*	171
Maytenus royleanus	Au	30	Hexagonal	*Leishmania tropica*	172
Dioscorea bulbifera	Au, Ag shell nanoparticles	15	Spherical	*Leishmania tropica*	173
Cymbopogon citratus	Au	20–50	Spherical	*Aedes aegypti*, *Aedes stephensi*	173
Callistemon citrinus	Au	1–50	Spherical (irregular)	Anti-plasmodial	173
Aegle marmelos correa	Ni	80–100	Cubic	Anti-mosquito larvicidal	174
Melia azedarach	NiO	21–35	Spherical	*Hyalomma dromedarii*	175
Lobelia leschenaultiana	ZnO	20–65	Spherical, hexagonal	*Rhipicephalus (Boophilus) microplus*	176
Eclipta prostrata	Pd	63 ± 1.4	Spherical	*Plasmodium berghei*	91
Citrus limon	Pd	2.3 ± 0.04	Spherical	*Anopheles stephensi*	92
Cocos nucifera	Ni	47	Cubic	*Aedes ageypti*	177
Tinospora cordifolia Miers	Pd	16	Spherical	*A. subpictus*, malaria vector, *C. Quinquefasciatus*	178

their therapeutic index and target specificity. Nanoparticulate delivery systems are categorized into polymeric nanoparticles, inorganic nanoparticles, nanocapsules, drug nanoparticles or nanosuspensions, solid lipid nanoparticles (SLNs), lipid drug conjugates (LDCs), dendrimers, polymeric micelles and nanogels. They all have their applications in the treatment of parasitic diseases.

Polymeric nanoparticles are colloidal particles solid in nature with size range of 1–1000 nm, obtained mostly from biodegradable polymers. Nanoparticles of primaquine [187], gelatin, albumin, glutaraldehyde and polyacrylamide [188] chitosan nanoparticles complexed with phosphorothioate anti-sense oligodeoxynucleotides [189], and fabricated halofantrine-loaded PEG-coated polylactic acid (PLA) nanocapsules for the treatment of malaria [190]. Primaquine-loaded polyalkylcyano acrylate (PACA) nanoparticles showed an anti-leishmaniasis effect on macrophages [191]. Pentamidine-loaded polymethyl methacrylate nanoparticles were experimented on *Leishmania*-infected macrophages and mice [192]. Poly (ε-caprolactone) nanoencapsulated amphotericin B proved to be two to three times more effective than its free anti-leishmanial agent [193]. 2′,6′-Dihydroxy-4′-methoxychalcone (DMC) obtained from *Piper aduncum* was studied as an anti-leishmanial agent after encapsulation in PLA nanoparticles. Nanoparticle encapsulation of other natural agents had anti-leishmaniasis activity, including quercetin, bacopasaponin C and arjunaglucoside [194].

Nifurtimox-loaded polyethylcyanoacrylate (PECA) nanoparticles have proven to be effective against *T. cruzi* (causative agent of Chagas disease) and showed anti-parasitic effects by autolysis of the cytoplasm, degeneration of kinetoplast and lyses of parasite membranes, with minimal or no adverse effect on host cells [195]. Allopurinol-encapsulated PECA nanoparticles are twice as effective against *T. cruzi* epimastigotes than its free form [196]. Stealth nanoparticles of DO870 (an investigational anti-trypanosomial agent) in the *T. cruzi*-infected experimental mouse model showed significant cure rates [197]. Daunomycin nanoparticles have also been studied against the blood forms of *T. brucei* causing human African trypanosomiasis [198].

Solid lipid nanoparticles (SLNs) are efficient, well-tolerated, low-cost drug delivery systems and possess the characteristics of both colloidal and nanoparticle-based drug delivery systems, like modulated drug release [199], physiologically acceptable mechanisms [200], target specificity and controlled drug release [201], with minimal disadvantages. They are microscopic particles of 1–1000 nm combined with lipids in nature to consolidate lipophilic and hydrophilic drugs in the lipid matrix. SLNs are used in the treatment of various endoparasitic diseases, like malaria, trypanosomiasis and leishmaniasis, by either as such, natural form or by small structural modifications like strong encapsulation [202].

Nanobiotechnology is now helping humankind in the rapid detection, treatment and control of arboviruses-borne, vector-borne and waterborne parasitic diseases [203]. Ticks are ectoparasites and serve as vectors for various bacterial, viral and protozoan diseases. Chemical- and plant-based acaricides have been used in the past, but all of them faced the same problem of the vector resistance. Banumathi et al. [204] have reviewed various nanoparticle-based acaricides used against the cattle tick *Rhipicephalus (Boophilus) microplus*, which serves as vector for *Anaplasma marginale* and *Babesia bigemina* causative agents of anaplasmosis and babesiosis, respectively. Plant-synthesized NPs have proven effective against arthropod pests of economic importance, such as moths [205], beetles [206], lice (e.g., *Pediculus humanus capitis*) [207], hard ticks (e.g., *Haemaphysalis bispinosa*) [208], louse flies (e.g., *Hippobosca maculata*) [209] and mosquitoes [210].

Nanoparticles of amorphous nanosilica obtained from various natural sources, like the shell wall of phytoplankton, epidermis of vegetables, burnt pretreated rice hulls, straw of thermoelectric plants and volcanic soil, act as insecticides [211]. Nanosilica (silicon nanoparticles, Si-NPs) act as pesticides and help in killing insects, pests, herbicides and animal ectoparasites [212]. Their mechanism of action is to damage the protective lipid coating of the insects leading to dehydration and controlled targeted release of the commercially available pesticides [213]. Some nanoparticles affect antioxidants and detoxification and other enzyme activities necessary for the insect survival causing oxidative stress and cell death.

9.4　CONCLUSIONS AND FUTURE OUTLOOK

Nanomaterials greatly alter the metabolic pathways at the cellular level in the target parasites. Many fascinating studies have reported promising efficacy of metallic and non-metallic nanomaterials against a variety of ecto- and endo-parasites. It is imperative to develop certain models for testing nanomaterials against invertebrates both in vitro and in vivo. Additionally, eco-health and safety should be the top priorities and must be ensured while reporting the efficient use of nanomaterials. The industry–academia–farmer linkages may be further strengthened to materialize the concept of commercial antiparasitic nanomaterials.

REFERENCES

1. Wang, Y., Yu, L., Kong, X., Sun, L. (2017). Application of nanodiagnostics in point-of-care tests for infectious diseases. *International Journal of Nanomedicine*, 12, 4789–4803.
2. Hikal, W.M. (2020). Parasitic contamination of drinking water and Egyptian standards for parasites in drinking water. *Open Journal of Ecology*, 10(1), 1–21.
3. Farhoudi, R. (2017). An overview on recent new nano-anti-parasitological findings and application. *Advances in Nano Research*, 5(1), 49–59.
4. World Health Organization. (2020). *World Malaria Report 2020: 20 Years of Global Progress and Challenges.*
5. Kotloff, K.L., Nataro, J.P., Blackwelder, W.C., Nasrin, D., Farag, T.H., Panchalingam, S., et al. (2013). Burden and aetiology of diarrhoeal disease in infants and young children in developing countries (the Global Enteric Multicenter Study, GEMS): A prospective, case-control study. *Lancet*, 382(9888), 209–222.
6. Momcilovic, S., Cantacessi, C., Arsić-Arsenijević, V., Otranto, D.S., Tasić-Otašević, S. (2019). Rapid diagnosis of parasitic diseases: Current scenario and future needs. *Clinical Microbiology and Infection*, 25(3), 290–309.
7. Fauci, A.S., Morens, D.M. (2012). The perpetual challenge of infectious diseases. *New England Journal of Medicine*, 366(5), 454–461.
8. Hauck, T.S., Giri, S., Gao, Y., Chan, W.C.W. (2010). Nanotechnology diagnostics for infectious diseases prevalent in developing countries. *Advanced Drug Delivery Reviews*, 62(4–5), 438–448.

9. Elechiguerra, J.L., Burt, J.L., Morones, J.R., Camacho-Bragado, A., Gao, X., Lara, H.H., Yacaman, M.J. (2005). Interaction of silver nanoparticles with HIV-1. *Journal of Nanobiotechnology*, 3, 6.

10. Huang, J., Li, Q., Sun, D., Lu, Y., Su, Y., Yang, X., et al. (2007). Biosynthesis of silver and gold nanoparticles by novel sundried *Cinnamomum camphora* leaf. *Nanotechnology*, 18(10), 105104.

11. Das, P., Chetia, B., Prasanth, R., Madhavan, J., Singaravelu, G., Benelli, G., Murugan, K. (2017). Green nanosynthesis and functionalization of gold nanoparticles as PTP 1B inhibitors. *Journal of Cluster Science*, 28(4), 2269–2277.

12. Murugan, K., Anitha, J., Suresh, U., Rajaganesh, R., Panneerselvam, C., Aziz, A.T., Tseng, L.C., Kalimuthu, K., Saleh Alsalhi, M., Devanesan, S., Nicoletti, M., Kumar, S.S., Benelli, G., Hwang, J.H. (2017). Chitosan-fabricated Ag nanoparticles and larvivorous fishes: A novel route to control the coastal malaria vector *Anopheles sundaicus*? *Hydrobiologia*, 797(1), 335–350.

13. Swain, P., Nayak, S.K., Sasmal, A., Behera, T., Barik, S.K., Swain, S.K., Mishra, S.S., Sen, A.K., Das, J.K., Jayasankar, P. (2014). Antimicrobial activity of metal based nanoparticles against microbes associated with diseases in aquaculture. *World Journal of Microbiology and Biotechnology*, 30(9), 2491–2502.

14. Li, Q.L., Mahendra, S., Lyon, D.Y., Brunet, L., Liga, M.V., Li, D., Alvarez, P.J.J. (2008). Antimicrobial nanomaterials for water disinfection and microbial control: Potential applications and implications. *Water Research*, 42(18), 4591–4602.

15. Rana, S., Kalaichelvan, P.T. (2011). Antibacterial activities of metal nanoparticles. *Advanced Biotechnology*, 11, 21–23.

16. Saleh, M., Kumar, G., Abdel-Baki, A.A., Al-Quraishy, S., El-Matbouli, M. (2016). In vitro antimicrosporidial activity of gold nanoparticles against *Heterosporis saurida*. *BMC Veterinary Research*, 12, 44.

17. Shaalan, M., Saleh, M., El-Mahdy, M., El-Matbouli, M. (2016). Recent progress in applications of nanoparticles in fish medicine: A review. *Nanomedicine: Nanotechnology, Biology, and Medicine*, 12(3), 701–710.

18. Alharbi, N.S., Bhakyaraj, K., Gopinath, K., Govindarajan, M., Kumuraguru, S., Mohan, S., Kaleeswarran, P., Kadaikunnan, S., Khaled, J.M., Benelli, G. (2017). Gum-mediated fabrication of eco-friendly gold nanoparticles promoting cell division and pollen germination in plant cells. *Journal of Cluster Science*, 28(1), 507–517.

19. Benelli, G., Pavela, R., Maggi, F., Petrelli, R., Nicoletti, M. (2017). Commentary: Making green pesticides greener? The potential of plant products for nanosynthesis and pest control. *Journal of Cluster Science*, 28(1), 3–10.

20. Borase, H.P., Patil, C.D., Suryawanshi, R.K., Koli, S.H., Mohite, B.V., Benelli, G., Patil, S.V. (2017). Mechanistic approach for fabrication of gold nanoparticles by Nitzschia diatom and their antibacterial activity. *Bioprocess and Biosystems Engineering*, 40(10), 1437–1446.

21. Gopinath, K., Kumaraguru, S., Bhakyaraj, K., Mohan, S., Venkatesh, K.S., Esakkirajan, M., Kaleeswarran, P.R., Naiyf, S.A., Kadaikunnan, S., Govindarajan, M., Benelli, G., Arumugam, A. (2016). Green synthesis of silver, gold and silver/gold bimetallic nanoparticles using the *Gloriosa superba* leaf extract and their antibacterial and antibiofilm activities. *Microbial Pathogenesis*, 101, 1–11.

22. Pavela, R., Murugan, K., Canale, A., Benelli, G. (2017). *Saponaria officinalis* synthesized silver nanocrystals as effective biopesticides and oviposition inhibitors against *Tetranychus urticae* Koch. *Industrial Crops and Products*, 97, 338–344.

23. Vijayakumar, S., Vaseeharan, B., Malaikozhundan, B., Gopi, N., Ekambaram, P., Pachaiappan, R., Velusamy, P., Murugan, K., Benelli, G., Kumar, R.S., Suriyanarayanamoorthy, M. (2017). Therapeutic effects of gold nanoparticles

synthesized using *Musa paradisiaca* peel extract against multiple antibiotic resistant *Enterococcus faecalis* biofilms and human lung cancer cells (A549). *Microbial Pathogenesis*, 102, 173–183.

24. Ebrahimi, K., Shiravand, M., Mahmoudvand, H. (2017). Biosynthesis of copper nanoparticles using aqueous extract of *Capparis spinosa* fruit and investigation of its antibacterial activity. *Marmara Pharmaceutical Journal*, 21(4), 866–8717.

25. Wolfram, J., Zhu, M., Yang, Y., Shen, J., Gentile, E., Paolino, D., Fresta, M., Nie, G., Chen, C., Shen, H., Ferrari, M., Zhao, Y. (2015). Safety of nanoparticles in medicine. *Current Drug Targets*, 16(14), 1671 –1681.

26. Samuel Singh, N., Phillips Singh, D. (2019). A review on major risk factors and current status of visceral leishmaniasis in North India A review on major risk factors and current status of visceral leishmaniasis in North India. *American Journal of Entomology*, 3(1), 6–14.

27. Akhlagh, A., Salehzadeh, A., Zahirnia, A.H., Davari, B. (2019). 10-year trends in epidemiology, diagnosis, and treatment of cutaneous leishmaniasis in Hamadan Province, West of Iran (2007–2016). *Frontiers in Public Health*, 7, 27.

28. Griensven, J. van, Diro, E. (2019). Visceral leishmaniasis: Recent advances in diagnostics and treatment regimens. *Infectious Disease Clinics of North America*, 33(1), 79 –99.

29. Chávez-Fumagalli, M.A., Ribeiro, T.G., Castilho, R.O., Fernandes, S.O., Cardoso, V.N., Coelho, C.S., Mendonça, D.V., Soto, M., Tavares, C.A., Faraco, A.A., Coelho, E.A. (2015). New delivery systems for amphotericin B applied to the improvement of leishmaniasis treatment. *Revista da Sociedade Brasileira de Medicina Tropical*, 48(3), 235 –242.

30. Pissuwan, D., Niidome, T., Cortie, M.B. (2011). The forthcoming applications of gold nanoparticles in drug and gene delivery systems. *Journal of Controlled Release*, 149(1), 65–71.

31. Suganya, P., Vaseeharan, B., Vijayakumar, S., Balan, B., Govindarajan, M., Alharbi, N.S., Kadaikunnan, S., Khaled, J.M., Benelli, G. (2017). Biopolymer zein-coated gold nanoparticles: Synthesis, antibacterial potential, toxicity and histopathological effects against the Zika virus vector *Aedes aegypti*. *Journal of Photochemistry and Photobiology, Part B: Biology*, 173, 404–411.

32. Tiwari, P.M., Vig, K., Dennis, V.A., Singh, S.R. (2011). Functionalized gold nanoparticles and their biomedical applications. *Nanomaterials*, 1(1), 31–63.

33. Zhou, Y., Kong, Y., Kundu, S., Cirillo, J.D., Liang, H. (2012). Antibacterial activities of gold and silver nanoparticles against *Escherichia coli* and bacillus Calmette-Guérin. *Journal of Nanobiotechnolology*, 10, 19.

34. Lima, E., Guerra, R., Lara, V., Guzmán, A. (2013). Gold nanoparticles as efficient antimicrobial agents for *Escherichia coli* and *Salmonella typhi*. *Chemistry Central Journal*, 7(1), 1–7.

35. Lolina, S., Narayanan, V. (2013). Antimicrobial and anticancer activity of gold nanoparticles synthesized from grapes fruit extract. *Chemical Science Transactions*, 2, 105–110.

36. Bratovcic, A. (2019). Different applications of nanomaterials and their impact on the environment. *International Journal of Material Science and Engineering*, 5(1), 1–7.

37. Lallawmawma, H., Sathishkumar, G., Sarathbabu, S., Ghatak, S., Sivaramakrishnan, S., Gurusubramanian, G., Kumar, N.S. (2015). Synthesis of silver and gold nanoparticles using *Jasminum nervosum* leaf extr*act and its larvicidal activity against filarial and arboviral vector Culex quinquefasciatus* Say (Diptera: Culicidae). *Environmental Science and Pollution Research International*, 22(22), 17753–17768.

38. Murugan, K., Benelli, G., Panneerselvam, C., Subramaniam, J., Jeyalalitha, T., Dinesh, D., Nicoletti, M., Hwang, J.S., Suresh, U., Madhiyazhagan, P. (2015). *Cymbopogon citratus*-synthesized gold nanoparticles boost the predation efficiency of copepod *Mesocyclops aspericornis* against malaria and dengue mosquitoes. *Experimental Parasitology*, 153, 129–138.

39. Benelli, G. (2016). Plant-mediated biosynthesis of nanoparticles as an emerging tool against mosquitoes of medical and veterinary importance: A review. *Parasitology Research*, 115(1), 23–34.

40. Subramaniam, J., Murugan, K., Panneerselvam, C., Kovendan, K., Madhiyazhagan, P., Dinesh, D., Kumar, P.M., Chandramohan, B., Suresh, U., Rajaganesh, R., Alsalhi, M.S., Devanesan, S., Nicoletti, M., Canale, A., Benelli, G. (2016). Multipurpose effectiveness of *Couroupita guianensis*-synthesized gold nanoparticles: High antiplasmodial potential, field efficacy against malaria vectors and synergy with *Aplocheilus lineatus* predators. *Environmental Science and Pollution Research International*, 23(8), 7543–7558.

41. Dkhil, M.A., Bauomy, A.A., Diab, M.S., Wahab, R., Delic, D., Al-Quraishy, S. (2015). Impact of gold nanoparticles on brain of mice infected with *Schistosoma mansoni*. *Parasitology Research*, 114(10), 3711–3719.

42. Dkhil, M.A., Nafady, D.A., Diab, M.S., Bauomy, A.A., Al-Quraishy, S. (2016). Nanoparticles against schistosomiasis. In: *Nanoparticles in the Fight against Parasites* (Editor Heinz Mehlhorn). Springer International Publishing, pp. 191–205.

43. Dkhil, M.A., Khalil, M.F., Diab, M.S., Bauomy, A.A., Al-Quraishy, S. (2017). Effect of gold nanoparticles on mice splenomegaly induced by schistosomiasis mansoni. *Saudi Journal of Biological Sciences*, 24(6), 1418–1423.

44. Kar, P.K., Murmu, S., Saha, S., Tandon, V., Acharya, K. (2014). Anthelmintic efficacy of gold nanoparticles derived from a phytopathogenic fungus, *Nigrospora oryzae*. *PLOS ONE*, 9(1), e84693.

45. Pissuwan, D., Valenzuela, S.M., Miller, C.M., Cortie, M.B. (2007). A golden bullet? Selective targeting of *Toxoplasma gondii* tachyzoites using antibody functionalized gold nanorods. *Nano Letters*, 7(12), 3808–3838.

46. Karthik, L., Kumar, G., Keswani, T., Bhattacharyya, A., Reddy, B.P., Rao, K.B. (2013). Marine actinobacterial mediated gold nanoparticles synthesis and their antimalarial activity. *Nanomedicine: Nanotechnology, Biology and Medicine*, 9(7), 951–960.

47. Adeyemi, O.S., Whiteley, C.G. (2014). Interaction of metal nanoparticles with recombinant arginine kinase from *Trypanosoma brucei*: Thermodynamic and spectrofluorimetric evaluation. *Biochimica et Biophysica Acta (BBA)-General Subjects*, 1840(1), 701–706.

48. Andreadou, M., Liandris, E., Gazouli, M., Taka, S., Antoniou, M., Theodoropoulos, G., Ikonomopoulos, J., Goutas, N., Vlachodimitropoulos, D., Kasampalidis, I., Ikonomopoulos, J. (2014). A novel non-amplification assay for the detection of *Leishmania* spp. in clinical samples using gold nanoparticles. *Journal of Microbiological Methods*, 96, 56–61.

49. Ahmad, A., Syed, F., Shah, A., Khan, Z., Tahir, K., Khan, A.U., Yuan, Q. (2015). Silver and gold nanoparticles from *Sargentodoxa cuneata*: Synthesis, characterization and antileishmanial activity. *RSC Advances*, 5(90), 73793–73806.

50. Ghosh, S., Jagtap, S., More, P., Shete, U.J., Maheshwari, N.O., Rao, S.J., et al. (2015). *Dioscorea bulbifera* mediated synthesis of novel Au core Ag shell nanoparticles with potent antibiofilm and antileishmanial activity. *Journal of Nanomaterials*, 16(1), 161.

51. Jebali, A., Kazemi, B. (2013). Nano-based antileishmanial agents: A toxicological study on nanoparticles for future treatment of cutaneous leishmaniasis. *Toxicology in Vitro*, 27(6), 1896–1904.

52. Thiruppathiraja, C., Kamatchiammal, S., Adaikkappan, P., Alagar, M. (2011). An advanced dual labeled gold nanoparticles probe to detect Cryptosporidium parvum using rapid immuno-dot blot assay. *Biosensors and Bioelectronics*, 26(11), 4624–4627.

53. Guirgis, B.S., Cunha, C.S., Gomes, I., Cavadas, M., Silva, I., Doria, G., Blatch, G.L., Baptista, P.V., Pereira, E., Azzazy, H.M., Mota, M.M., Prudêncio, M., Franco, R. (2012). Gold nanoparticle-based fluorescence immunoassay for malaria antigen detection. *Analytical and Bioanalytical Chemistry*, 402(3), 1019–1027.

54. Nash, M.A., Waitumbi, J.N., Hoffman, A.S., Yager, P., Stayton, P.S. (2012). Multiplexed enrichment and detection of malarial biomarkers using a stimuli-responsive iron oxide and gold nanoparticle reagent system. *ACS Nano*, 6(8), 6776–6785.

55. Jeon, W., Lee, S., Manjunatha, D.H., Ban, C. (2013). A colorimetric aptasensor for the diagnosis of malaria based on cationic polymers and gold nanoparticles. *Analytical Biochemistry*, 439(1), 11–16.

56. Rivas, L., de la Escosura-Muñiz, A., Serrano, L., Altet, L., Francino, O., Sánchez, A., Merkoçi, A. (2015). Triple lines gold nanoparticle-based lateral flow assay for enhanced and simultaneous detection of Leishmania DNA and endogenous control. *Nano Research*, 8(11), 3704–3714.

57. Sattarahmady, N., Movahedpour, A., Heli, H., Hatam, G.R. (2016). Gold nanoparticles-based biosensing of Leishmania major kDNA genome: Visual and spectrophotometric detections. *Sensors and Actuators. Part B: Chemical*, 235, 723–731.

58. Kumar, R., Ray, P.C., Datta, D., Bansal, G.P., Angov, E., Kumar, N. (2015). Nanovaccines for malaria using *Plasmodium falciparum* antigen Pfs25 attached gold nanoparticles. *Vaccine*, 33(39), 5064–5071.

59. Hikal, W.M., Bratovcic, A., Baeshen, R.S., Tkachenko, K.G., Said-Al Ahl, H.A. (2021). Nanobiotechnology for the detection and control of waterborne parasites. *Open Journal of Ecology*, 11(3), 203–223.

60. Sachs, J., Malaney, P. (2002). The economic and social burden of malaria. *Nature*, 415(6872), 680–685.

61. Benelli, G. (2015). Research in mosquito control: Current challenges for a brighter future. *Parasitology Research*, 114(8), 2801–2805.

62. Nájera, J.A., González-Silva, M., Alonso, P.L. (2011). Some lessons for the future from the global malaria eradication programme (1955–1969). *PLOS Medicine*, 8(1), e1000412, doi: 10.1371/journal.pmed.1000412.

63. Hempelmann, E., Krafts, K. (2013). Bad air, amulets and mosquitoes: 2,000 years of changing perspectives on malaria. *Malaria Journal*, 12, 232.

64. Hemingway, J., Shretta, R., Wells, T.N.C., Bell, D., Djimdé, A.A., Achee, N., Qi, G. (2016). Tools and strategies for malaria control and elimination: What do we need to achieve a grand convergence in malaria? *PLOS Biology*, 14(3), e1002380.

65. Benelli, G., Beier, J.C. (2017). Current vector control challenges in the fight against malaria. *Acta Tropica*, 174, 91–96.

66. Jensen, M., Mehlhorn, H. (2009). Seventy-five years of Resochin® in the fight against malaria. *Parasitology Research*, 105(3), 609–627.

67. Tu, Y. (2011). The discovery of artemisinin (qinghaosu) and gifts from Chinese medicine. *Nature Medicine*, 17(10), 1217–1220.

68. Vieites, M., Smircich, P., Buggeri, L., Marchan, E., Gomez-Barrio, A., Navarro, M., Garat, B., Gambino, D. (2009). Synthesis and characterization of a pyridine-2- thiol N- oxide gold(I) complex with potent antiproliferative effect against *Trypanosoma cruzi* and *Leishmania* sp. insight into its mechanism of action. *Journal of Inorganic Biochemistry*, 103(10), 1300–1306.

69. Angelucci, F., Sayed, A.A., Williams, D.L., Boumis, G., Brunori, M., Dimastrogiovanni, D., Miele, A.E., Pauly, F., Bellelli, A. (2009). Inhibition of *Schistosoma mansoni* thioredoxin-glutathione reductase by auranofin: Structural and kinetic aspects. *Journal of Biological Chemistry*, 284(42), 28977–28985.

70. Kuntz, A.N., Davioud-Charvet, E., Sayed, A.A., Califf, L.L., Dessolin, J., Arner, E.S., Williams, D.L. (2007). Thioredoxin glutathione reductase from *Schistosoma mansoni*: An essential parasite enzyme and a key drug target. *PLOS Medicine*, 4(6), e206.

71. Sannella, A.R., Casini, A., Gabbiani, C., Messori, L., Bilia, A.R., Vincieri, F.F., Majori, G., Severini, C. (2008). New uses for old drugs. Auranofin, a clinically established antiarthritic metallodrug. exhibits potent antimalarial effects in vitro: Mechanistic and pharmacological implications. *FEBS Letters*, 582(6), 844–847.

72. Ilari, A., Baiocco, P., Messori, L., Fiorillo, A., Boffi, A., Gramiccia, M., Di Muccio, T., Colotti, G. (2012). A gold-containing drug against parasitic polyamine metabolism: The X-ray structure of trypanothione reductase from Leishmania infantum in complex with auranofin reveals a dual mechanism of enzyme inhibition. *Amino Acids*, 42(2–3), 803–811.

73. Varela-Aramburu, S., Ghosh, C., Goerdeler, F., Priegue, P., Moscovitz, O., Seeberger, P.H. (2020). Targeting and inhibiting *Plasmodium falciparum* using ultra-small gold nanoparticles. *ACS Applied Materials and Interfaces*, 12(39), 43380–44338.

74. Benelli, G. (2018). Gold nanoparticles–against parasites and insect vectors. *Acta Tropica*, 178, 73–80.

75. Dkhil, M.A., Khalil, M.F., Diab, M.S.M., Bauomy, A.A., Santourlidis, S., Al-Shaebi, E.M., Al-Quraishy, S. (2019). Evaluation of nanoselenium and nanogold activities against murine intestinal schistosomiasis. *Saudi Journal of Biological Sciences*, 26(7), 1468–1472.

76. Franci, G., Falanga, A., Galdiero, S., Palomba, L., Rai, M., Morelli, G., Galdiero, M. (2015). Silver nanoparticles as potential antibacterial agents. *Molecules*, 20(5), 8856–8874.

77. Nafari, A., Cheraghipour, K., Sepahvand, M., Shahrokhi, G., Gabal, E., Mahmoudvand, H. (2020). Nanoparticles: New agents toward treatment of leishmaniasis. *Parasite Epidemiology and Control*, 10, e00156.

78. Lara, H.H., Ayala-Núnez, N.V., Turrent, L.D.C.I., Padilla, C.R. (2010). Bactericidal effect of silver nanoparticles against multidrug-resistant bacteria. *World Journal of Microbiology and Biotechnology*, 26(4), 615–621.

79. Allahverdiyev, A.M., Abamor, E.S., Bagirova, M., Ustundag, C., Kaya, C., Kaya, F., Rafailovich, M. (2011). Antileishmanial effect of silver nanoparticles and their enhanced antiparasitic activity underultraviolet light. *International Journal of Nanomedicine*, 6, 2705–2714.

80. Saad, H., Soliman, M.I., Azzam, A.M., Mostafa, B. (2015). Antiparasitic activity of silver and copper oxide nanoparticles against *Entamoeba histolytica* and *Cryptosporidium parvum* cysts. *Journal of the Egyptian Society of Parasitology*, 45(3), 593–602.

81. Hikal, W.M., Said-Al Ahl, H.A.H., Tkachenko, K.G. (2020). Present and future potential of antiparasitic activity of *Opuntia ficus-indica*. *Tropical Journal of Natural Product Research*, 4(10), 672–679.

82. Morilla, M.J., Romero, E.L. (2015). Nanomedicines against Chagas disease: An update on therapeutics, prophylaxis and diagnosis. *Nanomedicine*, 10(3), 465–481.

83. Baranwal, A., Chiranjivi, A.K., Kumar, A., Dubey, V.K., Chandra, P. (2018). Design of commercially comparable nanotherapeutic agent against human disease-causing parasite, Leishmania. *Scientific Reports*, 8(1), Article No. 8814.

84. Zhang, Q., Kusaka, Y., Zhu, X., Sato, K., Mo, Y., Kluz, T., Donaldson, K. (2003). Comparative toxicity of standard nickel and ultrafine nickel in lung after intratracheal instillation. *Journal of Occupational Health*, 45(1), 23–30.

85. Sivulka, D.J. (2005). Assessment of respiratory carcinogenicity associated with exposure to metallic nickel: A review. *Regulatory Toxicology and Pharmacology: RTP*, 43(2), 117–133.

86. Chandra, S., Kumar, A., Tomar, P.K. (2014). Synthesis of Ni nanoparticles and their characterizations. *Journal of Saudi Chemical Society*, 18(5), 437–442.

87. Rajakumar, G., Rahuman, A.A., Velayutham, K., Ramyadevi, J., Jeyasubramanian, K., Marikani, A., Elango, G., Kamaraj, C., Santhoshkumar, T., Marimuthu, S., Zahir, A.A., Bagavan, A., Jayaseelan, C., Kirthi, A.V., Iyappan, M., Siva, C. (2013). Novel and simple approach using synthesized nickel nanoparticles to control blood-sucking parasites. *Veterinary Parasitology*, 191(3–4), 332–339.

88. Fritsch, D., Kuhr, K., Mackenzie, K., Kopinke, F.D. (2003). Hydrodechlorination of chloroorganic compounds in ground water by palladium catalysts: Part 1. Development of polymer-based catalysts and membrane reactor tests. *Catalysis Today*, 82(1–4), 105–118.

89. Chen, W., Cai, W., Lei, Y., Zhang, L. (2001). A sonochemical approach to the confined synthesis of palladium nanoparticles in mesoporous silica. *Materials Letters*, 50(2–3), 53–56.

90. Korovchenko, P., Renken, A., Kiwi-Minsker, L. (2005). Microwave plasma assisted preparation of Pd-nanoparticles with controlled dispersion on woven activated carbon fibres. *Catalysis Today*, 102–103, 133–141.

91. Rajakumar, G., Rahuman, A.A., Chung, I.M., Kirthi, A.V., Marimuthu, S., Anbarasan, K. (2015). Antiplasmodial activity of eco-friendly synthesized palladium nanoparticles using *Eclipta prostrata* extract against *Plasmodium berghei* in Swiss albino mice. *Parasitology Research*, 114(4), 1397–1406.

92. Minal, S.P., Prakash, S. (2018). Characterization and nano-efficacy study of palladium nanoparticles against larvae of *Anopheles stephensi* (Liston). *International Journal of Advanced Engineering and Nano Technology*, 3(10), 1.

93. Kar, P.K., Murmu, S., Saha, S., Tandon, V., Acharya, K. (2014). Anthelmintic efficacy of gold nanoparticles derived from a phytopathogenic fungus, *Nigrospora oryzae*. *PLOS ONE*, 9(1), e84693.

94. Roy, P., Saha, S.K., Gayen, P., Chowdhury, P., Babu, S.P.S. (2018). Exploration of antifilarial activity of gold nanoparticle against human and bovine filarial parasites: A nanomedicinal mechanistic approach. *Colloids and Surfaces, Part B: Biointerfaces*, 161, 236–243.

95. Adeyemi, O.S., Molefe, N.I., Awakan, O.J., Nwonuma, C.O., Alejolowo, O.O., Olaolu, T., Maimako, R.F., Suganuma, K., Han, Y., Kato, K. (2018). Metal nanoparticles restrict the growth of protozoan parasites. *Artificial Cells, Nanomedicine, and Biotechnology*, 46(sup3), S86–S94.

96. Vazini, H. (2018). The in vitro and in vivo efficacy of gold nanoparticle in comparison to the glucantime as a therapeutic agent against L. Major. *Journal of Infectious Diseases and Therapy*, 6, 373.

97. Kumar, P., Shivam, P., Mandal, S., Prasanna, P., Kumar, S., Prasad, S.R., Kumar, A., Das, P., Ali, V., Singh, S.K., Mandal, D. (2019). Synthesis, characterization, and mechanistic studies of a gold nanoparticle-amphotericin B covalent conjugate with enhanced antileishmanial efficacy and reduced cytotoxicity. *International Journal of Nanomedicine*, 14, 6073–6101.

98. Sazgarnia, A., Taheri, A.R., Soudmand, S., Parizi, A.J., Rajabi, O., Darbandi, M.S. (2013). Antiparasitic effects of gold nanoparticles with microwave radiation on promastigotes and amastigotes of Leishmania major. *International Journal of Hyperthermia*, 29(1), 79–86.

99. Hikal, W.M., Bratovcic, A., Baeshen, R.S., Tkachenko, K.G., Said-Al Ahl, H.A. (2021). Nanobiotechnology for the detection and control of waterborne parasites. *Open Journal of Ecology*, 11(3), 203–223.

100. Bavand, Z., Gholami, S., Honary, S., Rahimi, E.B., Torabi, N., Barabadi, H. (2014). Effect of gold nanoparticles on Giardia lamblia cyst stage in in vitro. *Journal of Arak University of Medical Sciences*, 16, 27–37.

101. Valerio-García, R.C., Carbajal-Hernández, A.L., Martínez-Ruíz, E.B., Jarquín-Díaz, V.H., Haro-Pérez, C., Martínez-Jerónimo, F. (2017). Exposure to silver nanoparticles produces oxidative stress and affects macromolecular and metabolic biomarkers in the goodeid fish *Chapalichthys pardalis*. *Science of the Total Environment*, 583, 308–318.

102. Gherbawy, Y.A., Shalaby, I.M., El-sadek, M.S.A., Elhariry, H.M., Banaja, A.A. (2013). The anti-fasciolasis properties of silver nanoparticles produced by *Trichoderma harzianum* and their improvement of the anti-fasciolasis drug triclabendazole. *International Journal of Molecular Sciences*, 14(11), 21887–21898.

103. Juganson, K., Mortimer, M., Ivask, A., Pucciarelli, S., Miceli, C., Orupõld, K., Kahru, A. (2017). Mechanisms of toxic action of silver nanoparticles in the protozoan *Tetrahymena thermophila*: From gene expression to phenotypic events. *Environmental Pollution*, 225, 481–489.

104. Holbrook, R.D., Murphy, K.E., Morrow, J.B., Cole, K.D. (2008). Trophic transfer of nanoparticles in a simplified invertebrate food web. *Nature Nanotechnology*, 3(6), 352–355.

105. Shi, J.P., Ma, C.Y., Xu, B., Zhang, H.W., Yu, C.P. (2012). Effect of light on toxicity of nanosilver to *Tetrahymena pyriformis*. *Environmental Toxicology and Chemistry*, 31(7), 1630–1638.

106. Daniel, S.C.G.K., Sironmani, T.A., Dinakaran, S. (2016). Nano formulations as curative and protective agent for fish diseases: Studies on red spot and white spot diseases of ornamental gold fish *Carassius auratus*. *International Journal of Fisheries and Aquatic Studies*, 4, 255–261.

107. Saleh, M., Abdel-Baki, A.A., Dkhil, M.A., El-Matbouli, M., Al-Quraishy, S. (2017). Antiprotozoal effects of metal nanoparticles against *Ichthyophthirius multifiliis*. *Parasitology*, 144(13), 1802–1810.

108. Rossi-Bergmann, B., Pacienza-Lima, W., Marcato, P.D., De Conti, R., Durán, N. (2012). Therapeutic potential of biogenic silver nanoparticles in murine cutaneous leishmaniasis. *Journal of Nano Research*, 20, 89–97.

109. Griffitt, R.J., Luo, J., Gao, J., Bonzongo, J.C., Barber, D.S. (2008). Effects of particle composition and species on toxicity of metallic nanomaterials in aquatic organisms. *Environmental Toxicology and Chemistry: an International Journal*, 27(9), 1972–1978.

110. Cameron, P., Gaiser, B.K., Bhandari, B., Bartley, P.M., Katzer, F., Bridle, H. (2016). Silver nanoparticles decrease the viability of *Cryptosporidium parvum* oocysts. *Applied and Environmental Microbiology*, 82(2), 431–437.

111. Allahverdiyev, A.M., Abamor, E.S., Bagirova, M., Rafailovich, M. (2011). Antimicrobial effects of TiO_2 and Ag_2O nanoparticles against drug-resistant bacteria and *Leishmania* parasites. *Future Microbiology*, 6(8), 933–940.

113. Karimi, M., Dalimi, A., Jamei, F., Ghaffarifar, F., Dalimi, A. (2015). The killing effect of silver nanoparticles and direct electric current induction on Leishmania major promastigotes in vitro. *Modares Journal of Medical Sciences: Pathobiology*, 18, 87–96.

114. Khosravi, A., Sharifi, I., Barati, M., Zarean, M., Hakimi-Parizi, M. (2011). Antileishmanial effect of nanosilver solutions on *Leishmania tropica* promastigotes by in-vitro assay. *Zahedan Journal of Research in Medical Sciences*, 13, 8–12.

115. Jameii, F., Dalimi Asl, A., Karimi, M., Ghaffarifar, F. (2015). Healing effect comparison of selenium and silver nanoparticles on skin leishmanial lesions in mice. *Scientific Journal of Hamadan University of Medical Sciences*, 22, 217–223.

116. Gaafar, M., Mady, R., Diab, R., Shalaby, T.I. (2014). Chitosan and silver nanoparticles: Promising anti-toxoplasma agents. *Experimental Parasitology*, 143, 30–38.
117. Ponarulselvam, S., Panneerselvam, C., Murugan, K., Aarthi, N., Kalimuthu, K., Thangamani, S. (2012). Synthesis of silver nanoparticles using leaves of *Catharanthus roseus* Linn. G. Don and their antiplasmodial activities. *Asian Pacific Journal of Tropical Biomedicine*, 2(7), 574–580.
118. Gherbawy, Y.A., Shalaby, I.M., El-sadek, M.S.A., Elhariry, H.M., Banaja, A.A. (2013). The anti-Fasciolasis properties of silver nanoparticles produced by *Trichoderma harzianum* and their improvement of the anti-Fasciolasis drug triclabendazole. *International Journal of Molecular Sciences*, 14(11), 21887–21898.
119. Dkhil, M.A., Bauomy, A.A., Diab, M.S., Al-Quraishy, S. (2016). Protective role of selenium nanoparticles against *Schistosoma mansoni* induced hepatic injury in mice. *Biomedical Research*, 27, 214–219.
120. Adeyemi, O.S., Whiteley, C.G. (2013). Interaction of nanoparticles with arginine kinase from *Trypanosoma brucei*: Kinetic and mechanistic evaluation. *International Journal of Biological Macromolecules*, 62, 450–456.
121. Beheshti, N., Soflaei, S., Shakibaie, M., Yazdi, M.H., Ghaffarifar, F., Dalimi, A., Shahverdi, A.R. (2013). Efficacy of biogenic selenium nanoparticles against Leishmania major: In vitro and in vivo studies. *Journal of Trace Elements in Medicine and Biology*, 27(3), 203–207.
122. Zahir, A.A., Chauhan, I.S., Bagavan, A., Kamaraj, C., Elango, G., Shankar, J., Arjaria, N., Roopan, S.M., Rahuman, A.A., Singh, N. (2015). Green synthesis of silver and titanium dioxide nanoparticles using *Euphorbia prostrata* extract shows shift from apoptosis to G0/G1 arrest followed by necrotic cell death in *Leishmania donovani*. *Antimicrobial Agents and Chemotherapy*, 59(8), 4782–4799.
123. Allahverdiyev, A.M., Abamor, E.S., Bagirova, M., Baydar, S.Y., Ates, S.C., Kaya, F., Kaya, C., Rafailovich, M. (2013). Investigation of antileishmanial activities of Tio$_2$@Ag nanoparticles on biological properties of *L. tropica* and *L. infantum* parasites, in vitro. *Experimental Parasitology*, 135(1), 55–63.
124. Mishra, A., Kaushik, N.K., Sardar, M., Sahal, D. (2013). Evaluation of antiplasmodial activity of green synthesized silver nanoparticles. *Colloids and Surfaces, Part B: Biointerfaces*, 111, 713–718.
125. Shankar, S.S., Rai, A., Ahmad, A., Sastry, M. (2004). Rapid synthesis of Au, Ag, and bimetallic Au core-Ag shell nanoparticles using Neem (*Azadirachta indica*) leaf broth. *Journal of Colloid and Interface Science*, 275(2), 496–502.
126. Panneerselvam, C., Murugan, K., Roni, M., Aziz, A.T., Suresh, U., Rajaganesh, R., Madhiyazhagan, P., Subramaniam, J., Dinesh, D., Nicoletti, M., Higuchi, A., Alarfaj, A.A., Munusamy, M.A., Kumar, S., Desneux, N., Benelli, G. (2016). Fern-synthesized nanoparticles in the fight against malaria: LC/ MS analysis of *Pteridium aquilinum* leaf extract and biosynthesis of silver nanoparticles with high mosquitocidal and antiplasmodial activity. *Parasitology Research*, 115(3), 997–1013.
127. Yao, J., van Marwijk, J., Wilhelmi, B., Whiteley, C.G. (2015). Isolation, characterization, interaction of a thiazolekinase (*Plasmodium falciparum*) with silvernanoparticles. *International Journal of Biological Macromolecules*, 79, 644–653.
128. Murugan, K., Samidoss, C.M., Panneerselvam, C., Higuchi, A., Roni, M., Suresh, U., Chandramohan, B., Subramaniam, J., Madhiyazhagan, P., Dinesh, D., Rajaganesh, R., Alarfaj, A.A., Nicoletti, M., Kumar, S., Wei, H., Canale, A., Mehlhorn, H., Benelli, G. (2015). Seaweed-synthesized silver nanoparticles: An eco-friendly tool in the fight against Plasmodium falciparum and its vector *Anopheles stephensi*? *Parasitology Research*, 114(11), 4087–4097.

129. Murugan, K., Panneerselvam, C., Subramaniam, J., Madhiyazhagan, P., Hwang, J.S., Wang, L., Dinesh, D., Suresh, U., Roni, M., Higuchi, A., Nicoletti, M., Benelli, G. (2016). Eco-friendly drugs from the marine environment: Spongeweed-synthesized silver nanoparticles are highly effective on *Plasmodium falciparum* and its vector *Anopheles stephensi*, with little non-target effects on predatory copepods. *Environmental Science and Pollution Research International*, 23(16), 16671–16685.

130. Benelli, G. (2016). Green synthesized nanoparticles in the fight against mosquito-borne diseases and cancer—A brief review. *Enzyme and Microbial Technology*, 95, 58–68.

131. Virkutyte, J., Varma, R.S. (2011). Green synthesis of metal nanoparticles: Biodegradable polymers and enzymes in stabilization and surface functionalization. *Chemical Science*, 2(5), 837–846.

132. Raveendran, P., Fu, J., Wallen, S.L. (2003). Completely "green" synthesis and stabilization of metal nanoparticles. *Journal of the American Chemical Society*, 125(46), 13940–13941.

133. Annu, A.A., Ahmed, S. (2018) *Green Synthesis of Metal, Metal Oxide Nanoparticles, and Their Various Applications. Handbook of Ecomaterials.* Cham: Springer International Publishing, pp. 1–45.

134. Ahmad, A., Senapati, S., Khan, M.I., Kumar, R., Sastry, M. (2003). Extracellular biosynthesis of monodisperse gold nanoparticles by a novel extremophilic actinomycete, *Thermomonospora* sp. *Langmuir*, 19(8), 3550–3553.

135. Sastry, M., Ahmad, A., Khan, M.I., Kumar, R. (2003). Biosynthesis of metal nanoparticles using fungi and actinomycete. *Current Science*, 85, 162–170.

136. Roh, Y., Lauf, R.J., McMillan, A.D., Zhang, C., Rawn, C.J., Bai, J., Phelps, T.J. (2001). Microbial synthesis and the characterization of metal-substituted magnetites. *Solid State Communications*, 118(10), 529–534.

137. Lengke, M., Southam, G. (2006). Bioaccumulation of gold by sulphate-reducing bacteria cultured in the presence of gold (I)-thiosulfate complex. *Acta*, 70, 3646–3661.

138. Nair, B., Pradeep, T. (2002). Coalescence of nanoclusters and formation of submicron crystallites assisted by Lactobacillus strains. *Crystal Growth and Design*, 2(4), 293–298.

139. Joerger, T.K., Joerger, R., Olsson, E., Granqvist, C.G. (2001). Bacteria as workers in the living factor: Metal accumulating bacteria and their potential for materials science. *Trends in Biotechnology*, 19(1), 15–20.

140. Husseiny, M.I., El-Aziz, M.A., Badr, Y., Mahmoud, M.A. (2007). Biosynthesis of gold nanoparticles using *Pseudomonas aeruginosa*. *Spectrochimica Acta, Part A*, 67(3–4), 1003–1006.

141. Mukherjee, P., Ahmad, A., Mandal, D., Senapati, S., Sainkar, S.R., Khan, M.I., Parishcha, R., Aiayumar, P.V., Alam, M., Kumar, R., Sastry, M. (2001). Fungus-mediated synthesis of silver nanoparticles and their immobilization in the mycelia matrix: A novel biological approach to nanoparticles synthesis. *Nano Letters*, 1(10), 515–519.

142. Ahmad, A., Senapati, S., Khan, M.I., Kumar, R., Ramani, R., Srinivas, V., Sastry, M. (2003). Intracellular synthesis of gold nanoparticles by a novel alkalotolerant actinomycete Rhodococcus species. *Nanotechnology*, 14(7), 824–828.

143. Ahmad, A., Senapati, S., Khan, M.I., Kumar, R., Sastry, M. (2005). Extra-/intracellular, biosynthesis of gold nanoparticles by an alkalotolerant fungus, *Trichothecium* sp. *Journal of Biomedical Nanotechnology*, 1(1), 47–53.

144. Kuber, C., Souza, S.F. (2006). *Extracellular biosynthesis of silver nanoparticles using the fungus Aspergillus fumigates. Colloids and Surfaces, Part B: Biointerfaces*, 47(2), 160–164.

145. Philip, D. (2010). Green synthesis of gold and silver nanoparticles using *Hibiscus rosa sinensis*. *Physics and Engineering*, 42(5), 1417–1424.

146. Kumar, P., Singh, P., Kumari, K., Mozumdar, S., Chandra, R. (2011). A green approach for the synthesis of gold nanotriangles using aqueous leaf extract of *Callistemon viminalis*. *Materials Letters*, 65(4), 595–597.

147. Lee, S.W., Mao, C., Flynn, C., Belcher, A.M. (2002). Ordering of quantum dots using genetically engineered viruses. *Science*, 296(5569), 892–895.

148. Merzlyak, A., Lee, S.W. (2006). Phage as template for hybrid materials and mediators for nanomaterials synthesis. *Current Opinion in Chemical Biology*, 10(3), 246–252.

149. Dameron, C.T., Reeser, R.N., Mehra, R.K., Kortan, A.R., Carroll, P.J., Steigerwald, M.L., Brus, L.E., Winge, D.R. (1989). Biosynthesis of cadmium sulphide quantum semiconductor crystallites. *Nature*, 338(6216), 596–597.

150. Kowshik, M., Arhtaputre, S., Kharrazi, S., Vogel, W., Urban, J., Kulkarni, S.K., Paknikar, K.M. (2003). Extracellular synthesis of silver nanoparticles by a silver-tolerant yeast strain MKY3. *Nanotechnology*, 14(1), 95–100.

151. Gericke, M., Pinches, A. (2006). Biological synthesis of metal nanoparticles. *Hydrometallurgy*, 83(1–4), 132–140.

152. Mohanpuria, P., Rana, N.K., Yadav, S.K. (2008). Biosynthesis of nanoparticles: Technological concepts and future applications. *Journal of Nanoparticle Research*, 10(3), 507–517.

153. Ngweniform, P., Li, D., Mao, C. (2009). Self-assembly of drug-loaded liposomes on genetically engineered protein nanotubes: A potential anti-cancer drug delivery vector. *Soft Matter*, 5(5), 954–956.

154. Tang, S., Mao, C., Liu, Y., Kelly, D.Q., Banerjee, S.K. (2007). Protein-mediated nanocrystal assembly for flash memory fabrication. *IEEE Transactions on Electron Devices*, 54(3), 433–438.

155. Janardhananan, S.K., Narayan, S., Abbineni, G., Hayhurst, A., Mao, C. (2010). Architectonics of phage-liposome nanowebs as optimized photosensitizer vehicles for photodynamic cancer therapy. *Molecular Cancer Therapeutics*, 9(9), 2524–2535.

156. Ngweniform, P., Abbineni, G., Cao, B., Mao, C. (2009). Self-assembly of drug-loaded liposomes on genetically engineered target-recognizing M13 phage: A novel nanocarrier for targeted drug delivery. *Small*, 5(17), 1963–1969.

157. Courchesne, N.M.D., Klug, M., Chen, P.Y., Kooi, S.E., Yun, D.S., Hong, N., Fang, N.X., Belcher, A.M., Hammond, P.T. (2014). Assembly of a bacteriophage-based template for the organization of materials into nanoporous networks. *Advanced Materials*, 26(21), 3398–3404.

158. Cao, B., Xu, H., Mao, C. (2011). Transmission electron microscopy as a tool to image bioinorganic nanohybrids: The case of phage-gold nanocomposites. *Microscopy Research and Technique*, 74(7), 627–635.

159. Kale, A., Bao, Y., Zhou, Z., Prevelige, P.E., Gupta, A. (2013). Directed self-assembly of CdS quantum dots on bacteriophage P22 coat protein templates. *Nanotechnology*, 24(4), 045603.

160. Khan, A.A., Fox, E.K., Gorzny, M.L., Nikulina, E., Brougham, D.F., Wege, C., Bittner, A.M. (2013). pH control of the electrostatic binding of gold and iron oxide nanoparticles to tobacco mosaic virus. *Langmuir*, 29(7), 2094–2098.

161. Iravani, S. (2011). Green synthesis of metal nanoparticles using plants. *Green Chemistry*, 13(10), 2638–2650.

162. Thakkar, K.N., Mhatre, S.S., Parikh, R.Y. (2010). Biological synthesis of metallic nanoparticles. *Nanotechnol. Biol. med. Nanomed.*, 6(2), 257–262.

163. Swami, A., Selvakannan, P.R., Pasricha, R., Sastry, M. (2004). One-step synthesis of ordered two dimensional assemblies of silver nanoparticles by the spontaneous reduction of silver ions by pentadecylphenol Langmuir monolayers. *Journal of Physical Chemistry. Part B, Condensed Matter, Materials, Surfaces, Interfaces and Biophysical*, 108(50), 19269–19275.

164. Jha, A.K., Prasad, K., Kulkarni, A.R. (2009). Plant system: Nature's nanofactory. *Colloids and Surfaces, Part B: Biointerfaces*, 73(2), 219–223.

165. Bar, H., Bhui, D.K., Sahoo, G.P., Sarkar, P., de Sankar, P., Misra, A. (2009). Green synthesis of silver nanoparticles using latex of *Jatropha curcas*. *Colloids and Surfaces. Part A: Physicochemical and Engineering Aspects*, 339(1–3), 134–139.

166. Roni, M., Murugan, K., Panneerselvam, C., Subramaniam, J., Hwang, J.-S. (2013). Evaluation of leaf aqueous extract and synthesized silver nanoparticles using nerium oleander against *Anopheles stephensi* (Diptera: Culicidae). *Parasitology Research*, 112(3), 981–990.

167. Ullah, I., Cosar, G., Abamor, E.S., Bagirova, M., Shinwari, Z.K., Allahverdiyev, A.M. (2018). Comparative study on the antileishmanial activities of chemically and biologically synthesized silver nanoparticles (AgNPs). *3 Biotech*, 8(2), 1–8.

168. Marimuthu, S., Rahuman, A.A., Rajakumar, G., Santhoshkumar, T., Kirthi, A.V., Jayaseelan, C., Bagavan, A., Zahir, A.A., Elango, G., Kamaraj, C. (2011). Evaluation of green synthesized silver nanoparticles against parasites. *Parasitology Research*, 108(6), 1541–1549.

169. Govindarajan, M., Rajeswary, M., Veerakumar, K., Muthukumaran, U., Hoti, S.L., Benelli, G. (2016). Green synthesis and characterization of silver nanoparticles fabricated using *Anisomeles indica*: Mosquitocidal potential against malaria, dengue and Japanese encephalitis vectors. *Experimental Parasitology*, 161, 40–47.

170. Zaheer, T., Imran, M., Pal, K., Sajid, M.S., Abbas, R.Z., Aqib, A.I., Hanif, M.A., Khan, S.R., Khan, M.K., Sindhu, ZuD., Rahman, S. (2021). Synthesis, characterization and acaricidal activity of green-mediated ZnO nanoparticles against Hyalomma ticks. *Journal of Molecular Structure*, 1227, 129652.

171. Slepicka, P., Elashnikov, R., Ulbrich, P., Staszek, M., Kolská, Z., Svorcik, V. (2015). Stabilization of sputtered gold and silver nanoparticles in PEG colloid solutions. *Journal of Nanoparticle Research*, 17(1), 11.

172. Slepicka, P., Mala, Z., Rimpelova, S., Svorcik, V. (2016). Antibacterial properties of modified biodegradable PHB non-woven fabric. *Materials Science and Engineering C – Materials for Biological Applications*, 65, 364–368.

173. Rotimi, L., Ojemaye, M.O., Okoh, O.O., Sadimenko, A., Okoh, A.I. (2019). Synthesis, characterization, antimalarial, antitrypanocidal and antimicrobial properties of gold nanoparticle. *Green Chemistry Letters and Reviews*, 12(1), 61–68.

174. Angajala, G., Ramya, R., Subashini, R. (2014). In-vitro anti-inflammatory and mosquito larvicidal efficacy of nickel nanoparticles phytofabricated from aqueous leaf extracts of *Aegle marmelos* Correa. *Acta Tropica*, 135, 19–26.

175. Abdel-Ghany, H.S., Abdel-Shafy, S., Abuowarda, M.M., El-Khateeb, R.M., Hoballah, E., Hammam, A.M.M., Fahmy, M.M. (2021). In vitro acaricidal activity of green synthesized nickel oxide nanoparticles against the camel tick, *Hyalomma dromedarii* (Ixodidae), and its toxicity on Swiss albino mice. *Experimental and Applied Acarology*, 83(4), 611–633.

176. Banumathi, B., Malaikozhundan, B., Vaseeharan, B. (2016). Invitro acaricidal activity of ethnoveterinary plants and green synthesis of zinc oxide nanoparticles against Rhipicephalus (Boophilus) microplus. *Veterinary Parasitology*, 216, 93–100.

177. Elango, G., Roopan, S.M., Dhamodaran, K.I., Elumalai, K., Al-Dhabi, N.A., Arasu, M.V. (2016). Spectroscopic investigation of biosynthesized nickel nanoparticles and its larvicidal, pesticidal activities. *Journal of Photochemistry and Photobiology, Part B: Biology*, 162, 162–167.

178. Jayaseelan, C., Gandhi, P.R., Rajasree, S.R.R., Suman, T.Y., Mary, R.R. (2018). Toxicity studies of nanofabricated palladium against filariasis and malaria vectors. *Environmental Science and Pollution Research International*, 25(1), 324–332.

179. Amato, V., Rabello, A., Rotondo-Silva, A., Kono, A., Maldonado, T.P., Alves, I.C., Floeter-Winter, L.M., Neto, V.A., Shikanai-Yasuda, M.A. (2004). Successful treatment of cutaneous leishmaniasis with lipid formulations of amphotericin B in two immuno-compromised patients. *Acta Tropica*, 92(2), 127–132.

180. Cauchetier, E., Paul, M., Rivollet, D., Fessi, H., Astier, A., Deniau, M. (2000). Therapeutic evaluation of free and liposome-encapsulated atovaquone in the treatment of murine leishmaniasis. *International Journal of Parasitology*, 30(6), 777–783.

181. Sarkar, S., Mandal, S., Sinha, J., Mukhopadhyay, S., Das, N., Basu, M.K. (2002). Quercetin: Critical evaluation as an antileishmanial agent in vivo in hamsters using different vesicular delivery modes. *Journal of Drug Targeting*, 10(8), 573–578.

182. Papagiannaros, A., Bories, C., Demetzos, C., Loiseau, P.M. (2005). Antileishmanial and trypanocidal activities of new miltefosine liposomal formulations. *Biomedicine and Pharmacotherapy*, 59(10), 545–550.

183. Morilla, M.J., Montanari, J.A., Prieto, M.J., Lopez, M.O., Petray, P.B., Romero, E.L. (2004). Intravenous liposomal benznidazole as trypanocidal agent: Increasing drug delivery to liver is not enough. *International Journal of Pharmacy*, 278, 311–318.

184. Morilla, M., Montanari, J., Frank, F., Malchiodi, E., Corral, R., Petray, P., Romero, E.L. (2005). Etanidazole in pH-sensitive liposomes: Design, characterization, and in vitro/in vivo anti-*Trypanosoma cruzi* activity. *Journal of Controlled Release*, 103(3), 599–607.

185. Antimisiaris, S., Ioannou, P.V., Loiseau, P.M. (2003). In-vitro antileishmanial and try-panocidal activities of arsonoliposomes and preliminary in-vivo distribution in Balb/c mice. *Journal of Pharmacy and Pharmacology*, 55(5), 647–652.

186. Yardley, V., Croft, S.L. (1999). In vitro and in vivo activity of amphotericin B-lipid formulations against experimental Trypanosoma cruzi infections. *American Journal of Tropical Medicine and Hygiene*, 61(2), 193–197.

187. Mbela, T.K.M., Poupaert, J.H., Dumont, P. (1992). Poly(diethylmethylidene malo-nate) nanoparticles as primaquine delivery system to liver. *International Journal of Pharmacy*, 79(1–3), 29–38.

188. Labhasetwar, V.D., Dorle, A.K. (1990). Nanoparticles-A colloidal drug delivery system for primaquine and metronidazole. *Journal of Controlled Release*, 12(2), 113–119.

189. Föger, F., Noonpakdee, W., Loretz, B., Joojuntr, S., Salvenmoser, W., Thaler, M., Bernkop-Schnürch, A. (2006). Inhibition of malarial topoisomerase II in *Plasmodium falciparum* by antisense nanoparticles. *International Journal of Pharmacy*, 319(1–2), 139–146.

190. Mosqueira, V.C.F., Bories, C., Legrand, P., Devissaguet, J.P., Barratt, G. (2004). Efficacy and pharmacokinetics of intravenous nanocapsule formulations of halofantrine in *Plasmodium berghei*-infected mice. *Antimicrobial Agents and Chemotherapy*, 48(4), 1222–1228.

191. Gaspar, R., Opperdoes, F. R., Préat, V., Roland, M. (1992). Drug targeting with polyalkylcya-noacrylate nanoparticles: in vitro activity of primaquine-loaded nanoparticles against intra-cellular Leishmania donovani. *Annals of Tropical Medicine & Parasitology*, 86(1), 41–49.

192. Durand, R., Rivollet, D., Houin, R., Astier, A., Deniau, M. (1997). Activity of pent-amidine-loaded methacrylate nanoparticles against Leishmania infantum in a mouse model. *International Journal of Parasitology*, 27(11), 1361–1367.

193. Espuelas, M., Loiseau, P.M., Bories, C., Barratt, G., Irache, J.M. (2002). In vitro anti-leishmanial activity of amphotericin B loaded in poly (ε-caprolactone) nanospheres. *Journal of Drug Targeting*, 10(8), 593–599.

194. Sinha, J., Das, N., Medda, S., Garai, S., Mahato, S.B., Basu, M.K. (2002). Bacopasaponin C: Critical evaluation of antileishmanial properties in various delivery modes. *Drug Delivery*, 9(1), 55–62.

195. Sanchez, G., Cuellar, D., Zulantay, I., Gajardo, M., González-Martin, G.-M. (2002). Cytotoxicity and trypanocidal activity of nifurtimox encapsulated in ethylcyanoacrylate nanoparticles. *Biological Research*, 35(1), 39–45.

196. Gonzalez-Martin, G., Merino, I., Osuna, A. (2000). Allopurinol encapsulated in polycyanoacrylate nanoparticles as potential lysosomatropic carrier: preparation and trypanocidal activity. *European Journal of Pharmaceutics and Biopharmaceutics*, 49(2), 137–142.

197. Molina, J., Gref, R., Brener, Z., Rodrigues Júnior JM (2001). Cure of experimental Chagas' disease by the bis-triazole DO870 incorporated into 'stealth' polyethyleneglycol–polylactide nanospheres. *Journal of Antimicrobial Chemotherapy*, 47(1), 101–104.

198. Flaig, R., Wink, M., Fricker, G. (2005). Ktenate nanoparticles (bdellosomes): A novel strategy for delivering drugs to parasites or tumors. *Journal of Drug Delivery Science and Technology*, 15(1), 59–63.

199. Mehnert, W., Mäder, K. (2001). Solid lipid nanoparticles: Production, characterization and applications. *Advanced Drug Delivery Reviews*, 47, 165–196.

200. Müller, R.H., Mäder, K., Gohla, S. (2000). Solid lipid nanoparticles (SLN) for controlled drug delivery—A review of the state of the art. *European Journal of Pharmaceutics and Biopharmaceutics*, 50, 161–178.

201. Goeppert, T., Müller, R.H. (2005). Polysorbate-stabilized solid lipid nanoparticles as colloidal carriers for intravenous targeting of drugs to the brain: Comparison of plasma protein adsorption patterns. *Journal of Drug Targeting*, 13, 179–187.

202. Date, A.A., Joshi, M.D., Patravale, V.B. (2007). Parasitic diseases: Liposomes and polymeric nanoparticles versus lipid nanoparticles. *Advanced Drug Delivery Reviews*, 59(6), 505–521.

203. Hikal, W.M.-A., Bratovcic, A., Baeshen, R.S., Tkachenko, K.G., Said-Al Ahl, H.A.H. (2021). Nanobiotechnology for the detection and control of waterborne parasites. *Open Journal of Ecology*, 11(3), 203–223.

204. Banumathi, B.V., Vaseeharan, B., Rajasekar, P., Prabhu, N.M., Ramasamy, P., Murugan, K., Canale, A., Benelli, G. (2017). Exploitation of chemical, herbal and nanoformulated acaricides to control the cattle tick, Rhipicephalus (Boophilus) microplus–a review. *Veterinary Parasitology*, 244, 102–110.

205. Roni, M., Murugan, K., Panneerselvam, C., Subramaniam, J., Nicoletti, M., Madhiyazhagan, P., Dinesh, D., Suresh, U., Khater, H.F., Wei, H., Canale, A., Alarfaj, A.A., Munusamy, M.A., Higuchi, A., Benelli, G. (2015). Characterization and biotoxicity of *Hypnea musciformis*-synthesized silver nanoparticles as potential eco-friendly control tool against *Aedes aegypti* and *Plutella xylostella*. *Ecotoxicology and Environment Safety*, 121, 31–38.

206. Abduz Zahir, A., Bagavan, A., Kamaraj, C., Elango, G., Rahuman, A.A. (2012). Efficacy of plant-mediated synthesized silver nanoparticles against *Sitophilus oryzae*. *Journal of Biopesticides*, 5(Supplementary), 95–102.

207. Jayaseelan, C., Rahuman, A.A., Rajakumar, G., Vishnu Kirthi, A., Santhoshkumar, T., Marimuthu, S., Bagavan, A., Kamaraj, C., Zahir, A.A., Elango, G. (2011). Synthesis of pediculocidal and larvicidal silver nanoparticles by leaf extract from heartleaf moonseed plant, *Tinospora cordifolia* Miers. *Parasitology Research*, 109(1), 185–194.

208. Abduz Zahir, A., Rahuman, A.A. (2012). Evaluation of different extracts and synthesised silver nanoparticles from leaves of *Euphorbia prostrata* against *Haemaphysalis bispinosa* and *Hippobosca maculata*. *Veterinary Parasitology*, 187(3–4), 511–520.

209. Jayaseelan, C., Rahuman, A.A., Rajakumar, G., Santhoshkumar, T., Kirthi, A.V., Marimuthu, S., Bagavan, A., Kamaraj, C., Zahir, A.A., Elango, G., Velayutham, K., Rao, K.V., Karthik, L., Raveendran, S. (2012). Efficacy of plant-mediated synthesized silver nanoparticles against hematophagous parasites. *Parasitology Research*, 111(2), 921–933.

210. Benelli, G. (2016). Green synthesized nanoparticles in the fight against mosquito-borne diseases and cancer—A brief review. *Enzyme and Microbial Technology*, 95, 58–68.

211. Athanassiou, C.G., Kavallieratos, N.G., Benelli, G., Losic, D., Usha Rani, P., Desneux, N. (2018). Nanoparticles for pest control: Current status and future perspectives. *Journal of Pest Science*, 91(1), 1–15.

212. Ulrichs, C., Mewis, I., Goswami, A. (2005). Crop diversification aiming nutritional security in West Bengal: Biotechnology of stinging capsules in nature's water-blooms. *Ann. Tech. Issue of State Agri. Technologists Service Assoc., Ann. Arbor*, 1–18.

213. Rastogi, A.T., Tripathi, D.K., Yadav, S., Chauhan, D.K., Živčák, M., Ghorbanpour, M., El-Sheery, N.I., Brestic, M. (2019). Application of silicon nanoparticles in agriculture. *3 Biotech*, 9(3), 1–11.

10 Toxicity of Metal Oxide Nanoparticles in Freshwater Fish

Sana Aziz and Sajid Abdullah

CONTENTS

10.1 Introduction 257
10.2 Factors Affecting Toxicity of Nanoparticles 259
    10.2.1 Particle Size and Shape 259
    10.2.2 Particle Surface and Its Modification 261
    10.2.3 Chemical Composition 262
    10.2.4 Solubility and Particle Dissolution 262
    10.2.5 pH of the System 262
10.3 Toxicological Mechanism of Engineered Nanoparticles 263
    10.3.1 Overproduction of Reactive Oxygen Species 263
    10.3.2 Release of Metallic Ions 265
    10.3.3 Particle Cell Interaction, Uptake, and Bioaccumulation of
            Nanoparticles 265
10.4 Systemic Toxicity of Nanoparticles to Fish 267
10.5 Conclusion and Future Prospects 267
References 273

10.1 INTRODUCTION

Particles that have diameters in the 1–100 nm range are known as nanoparticles (NPs). They have unique characteristics that are different from their bulk forms and their different types can be distinguished. NPs are found in soils, natural waters, and volcanic dust naturally and can be released both accidentally and intentionally into the environment (Figure 10.1). Engineered metal-based NPs, an important class of nanomaterials, are made up of one metal (e.g., Ag, and Zn NPs), metal oxides (e.g., TiO_2 and ZnO-NPs), or complex of respective metals. Nowadays, metal oxide nanoparticles (MeO-NPs) are produced on a large-scale in industries due to wide applications in telecommunication, electronics, transportation, imaging, mechanical engineering, pollution remediation, biological and chemical sensors, and therapeutic and diagnostics devices (Buzea et al., 2007) (Figure 10.2), which increase their discharge into the aquatic environment and increase concerns about the environmental risk to aquatic animals (Mueller

DOI: 10.1201/9781003126256-10

257

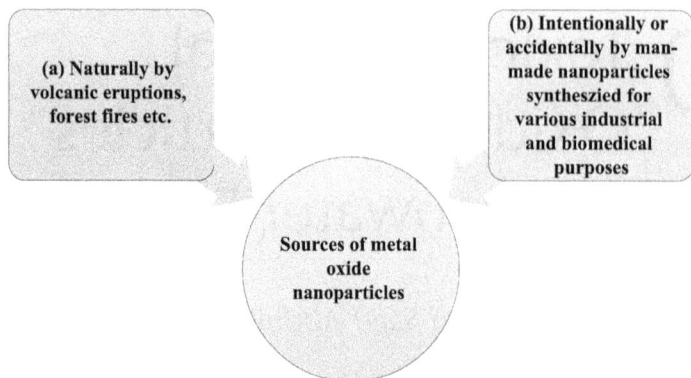

(a) Naturally by volcanic eruptions, forest fires etc.

(b) Intentionally or accidentally by man-made nanoparticles syntheszied for various industrial and biomedical purposes

Sources of metal oxide nanoparticles

FIGURE 10.1 Sources of metal oxide nanoparticles in the environment. There are two main sources that release nanoparticles: (a) Naturally – they enter through volcanic eruptions, forest fires, etc. (b) Intentionally or accidentally – man-made nanoparticles released during synthesis for various industrial and biomedical purposes.

et al., 2012; Theron et al., 2008). Information on the fundamental interaction of NPs with both biotic and abiotic factors is deficient, and presently there are few reliable techniques that can evaluate the toxicity of nanomaterials (Arora et al., 2012; Handy et al., 2008). The toxicity of metallic substances in the conventional soluble form is comparatively well noted and the main aim of this chapter is to study our present conceptual models of metal oxide nanoparticles toxicity in fish to check their toxic effects. The different concentrations of heavy metal nanoparticles (i.e., CuO and Ag nanoparticles) can also lead into assimilation in aquatic animals or fish that are at the end of the aquatic food chain and disturb the food chain condition (Torre et al., 2013) by entering into human beings by food inducing lethal and sublethal diseases (Al-Yousuf et al., 2000). Fish health can give a clear indication of the health status of the food chain due to their position in the aquatic food chain. So, understanding the effects of metal oxide nanoparticles on fish is therefore a crucial prospect to study as a whole aquatic environment. After the entrance of MeO-NPs into the aquatic environment, they can remain as individual particles, soluble in water, release ions and reactive oxygen species after light exposure, and start homo- and heteroaggregation that cause accumulation of NPs on the bottom surface that may be an essential path of toxic impacts. After sedimentation, MeO-NPs can interact with various constituents within the sediment, for example, sulfidation can change the bioavailability of particles. NPs can attach to the outer surface of algal or bacterial cells that trigger their toxicity. Attachment onto the surface can be a main driver for toxicity caused by non-ion-releasing engineered NPs (Oberdorster et al., 2006; Dabrunz et al., 2011). The adsorption of dissolved organic matter onto the surface of NPs can modify their surface charge and suppress aggregation or agglomeration and enhance their movement in the ecosystem (Figure 10.3).

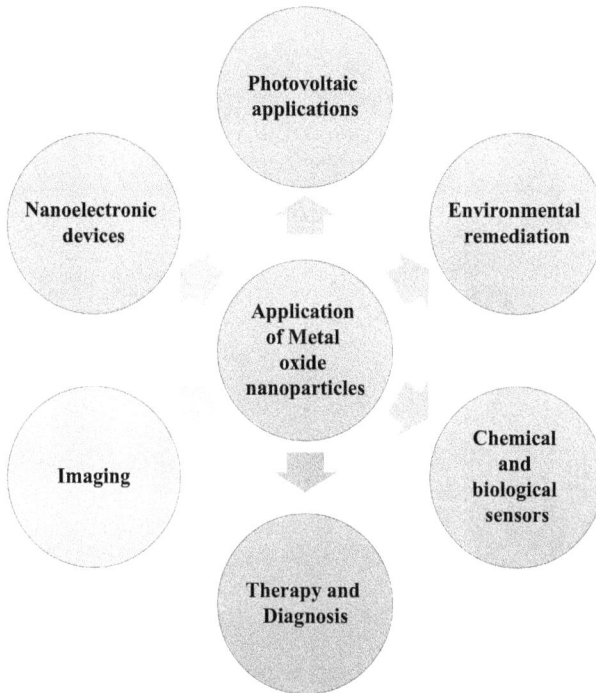

FIGURE 10.2 Applications of metal oxide nanoparticles. Metal oxide nanoparticles show broad photovoltaic applications and are used as environmental remediations, chemical and biological sensors, therapy and diagnosis, imaging, and nanoelectronic devices.

10.2 FACTORS AFFECTING TOXICITY OF NANOPARTICLES

The toxicity of NPs varies with their shape, size, dissolution, chemical surroundings, surface, and material composition (Figure 10.4). Water chemistry, viz., pH, salinity, temperature, ionic strength, and divalent ions also are important factors that determine the exact chemistry and properties of NPs (Shaw and Handy, 2011). For example, silver NPs that entered into aquatic environments with a low-level of ionic contents could be predicted to stay as tiny agglomerates of particles (Farmen et al., 2012).

10.2.1 PARTICLE SIZE AND SHAPE

The size of metal-based nanoparticles affects the interactions of NPs with biological systems of the body, and many biological NP-relevant mechanisms like cellular intake, rate of releasing ions, and particle processing ability depend on it (Aillon et al., 2009; Huang et al., 2017). As the size of particle decreases, the surface–volume ratio increases. Therefore, nanomaterials cause more harmful effects than their bulk types due to their reduced size, which helps in deep cell penetration and high accumulation (Lovern and Klaper, 2006; Mironava et al., 2010). Kaya et al. (2015) compared the

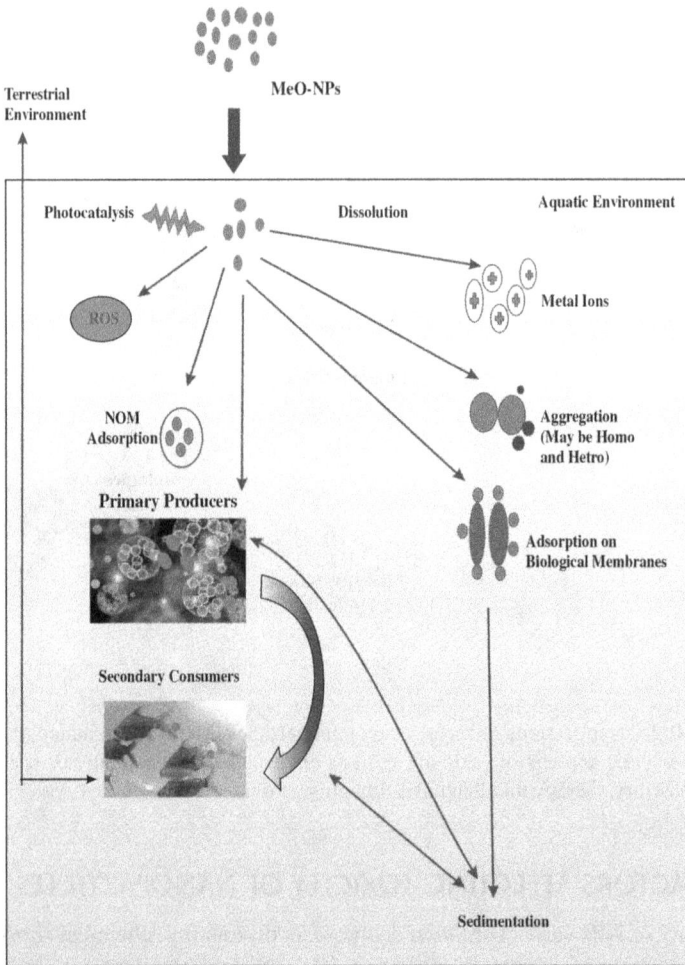

FIGURE 10.3 Schematic presentation of fate and transfer of metal oxide nanoparticles in environment. Nanoparticles can stay as individual particles, dissolve in water as metallic ions, start homo- and heteroaggregations, adsorb on biological membranes, deposit by sedimentation and transformation processes, directly adsorb on natural organic matter (NOM), start formation of reactive oxygen species (ROS) after photocatalysis, adsorb on primary producers, and enter into ultimately secondary consumers like fish through ingestion of primary or secondary producers or sediments, or enter into terrestrial environment via consumption of secondary consumers like fish.

toxicity and accumulation pattern of tilapia after 14-day exposure to two different doses (1 and 10 mg/L) of large and small zinc oxide nanoparticles (ZnO-NPs). Both sizes of NPs showed significant accumulation in the organs. Bioaccumulation for the large NPs was significantly lower than smaller NPs under similar exposure. As particle size increases, the amount of uptake decreases by the membrane of cells (Wang

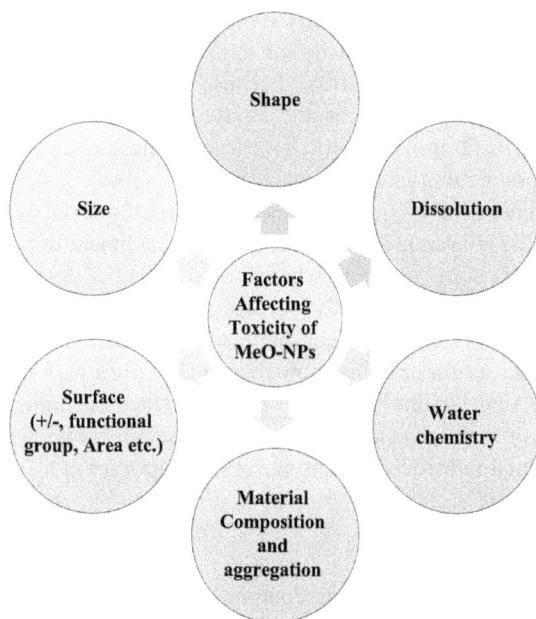

FIGURE 10.4 Schematic illustration of the factors affecting toxicity of nanoparticles in biological systems. The main factors that affect toxicity include particle size, shape, dissolution, water chemistry, material composition, and aggregation behavior and surface (+/–, functional group and area, etc.).

et al., 2010). Copper oxide nanoparticles (CuO-NPs) produce more DNA damage than other small-sized particles and nanoscale zinc oxide and titanium dioxide (Ahamed et al., 2010; Nel et al., 2009). According to this, macrophages easily get rid of larger NPs than smaller ones, which eludes this defense mechanism well (Van Der Zande et al., 2012). NPs having sizes > 50 nm via injections link all body tissues quicker and cause more toxicity (Ajdary et al., 2018). NPs having sizes < 6 nm cannot be eliminated and accumulated in specific target body organs (Albanese et al., 2012). Moreover, NPs having various morphologies like tubes, fibers, spheres, and planes can have various levels of dispersion, aggregation, and dissolution of ions in the medium that affect level of toxicity (Raza et al., 2016; Nam and An, 2019), because shapes may influence their endocytosis and toxicokinetic processes. Spherical NPs of the same size show quicker internalization through endocytosis than rodlike NPs because they require less membrane-covering time (Geng et al., 2007).

10.2.2 Particle Surface and Its Modification

The fundamental interaction of MeO-NPs with cells and their dissolution ability depend on the nature of the NPs' surface (Risom et al., 2005). The surface coating of NPs can modify their chemical, optical, electrical, and magnetic properties, which influence their cytotoxic attributes by influencing pharmacokinetics, biodistribution, storage, and toxicity (Ospina et al., 2015). As compared to negative charge or neutral NPs, positively charged NPs are easily taken up by biological systems (Slowing et al., 2006; Thorek and Tsourkas, 2008), because membranes of cells have a slight negative charge that affects electrostatic attraction (Jin et al., 2009; Wang et al., 2008). For small NPs (2 nm), the positive charge on the surface can disrupt the cellular membrane potential after influxing calcium ions into cells, ultimately inhibiting the development of cells. For large NPs (4–20 nm), the surface charge stimulates the buildup of lipid (bilayers) (Wang et al., 2008). Attachment of negatively charged NPs to a lipid bilayer starts gelation, whereas attachment of positively charged NPs cause fluidity. The chemistry of NPs' surface seems to find the protein type that will adsorb onto the surface and affect the level of interaction (Lynch and Dawson, 2008). For example, serum proteins show their interactions with negatively charged DNA-coated NPs that facilitate cellular uptake. Surface chemical modification by functionalization of surface is an essential scheme used in biomedical applications to increase stability and reduce the toxicity level (Chompoosor et al., 2010).

10.2.3 CHEMICAL COMPOSITION

The chemistry of NPs is another main component contributing to the effect the cellular interactions. Regarding the chemical composition of NPs, Griffitt et al. (2008) determined different levels of toxicity in zebrafish, algae, and daphnids species for both copper and silver NPs with respect to TiO (titanium oxide). Additionally, the aggregation state of NPs must also be taken into consideration, but aggregation behavior also depends on the material type, size, and surface load. It has been suggested that higher dose rates of NPs caused more aggregation and ultimately reduced toxicity (Gatoo et al., 2014). Liao et al. (2011) reported that graphene oxide and graphene toxicity depend on the aggregation behavior and interaction mode with the cells.

10.2.4 SOLUBILITY AND PARTICLE DISSOLUTION

Intercellular dissolution is a causal factor affecting the NPs elicited cytotoxicity (Studer et al., 2010), because when NPs are in a biological system or a medium, the ions generated from dissolution of the NPs can also affect the level of toxicity. Franklin et al. (2007) determined that ions of Zn^{+2} discharged from the dissolved form of ZnO-NPs caused disturbance of cell homeostasis, damage in lysosomes and mitochondria, and ultimately cell death in organisms. Later on, Xia et al. (2008) also reported that soluble ions from ZnO-NPs–induced cytotoxicity. Dissolution of Ni ions from nanoscale nickel induced cell death through the stimulating intrinsic apoptotic pathway in mitochondria (Faisal et al., 2013).

10.2.5 pH OF THE SYSTEM

The change in pH affects aggregation and the ion dissolution of metal oxide NPs. Bian et al. (2011) reported that ZnO-NP dissolution increased with decreasing pH from 11 to 1. Furthermore, large size aggregates of ZnO-NPs were determined at pH 9, when repulsion interactions decrease between NPs that can favor fast sedimentation of NPs. He et al. (2012) reported that nanometals can stimulate free radicals via Fenton-type reactions, and the free radical generation is dependent on the pH of the medium. The rate of free radical generation was in the following order: pH 4.6 < 3.6 < 1.2. At pH 7.4, there was no evidential synthesis of free radicals.Wang et al. (2012) observed that both maghemite nano-iron oxide and hematite nano-iron oxide elicited hydroxyl radicals, with the order pH 7.2 < 4.2 < 1.2. They determined that a lower pH level assists Fenton-type reactions to synthesize hydroxyl radicals.

10.3 TOXICOLOGICAL MECHANISM OF ENGINEERED NANOPARTICLES

Various studies have reported that exposure of NPs of transition metals can affect aquatic organisms, such as fish, at the cellular level. Metal-based NPs can cause toxicity mainly by three defined toxic mechanisms (Brunner et al., 2006). First, toxic substances may develop after surface interactions of particles with media, e.g., reactive oxygen species. Second, NPs may liberate harmful ionic substances into media, e.g., Cu^+ ions from particles of copper. Third, NPs may directly interrupt the biological target areas, e.g., interaction of NPs with DNA or cell membranes (Ma et al., 2013). Many mechanisms have been proposed to describe the toxicity of ZnO-NPs, such as binding with cell membranes (Ma et al., 2013), apoptosis and necrosis (Buerki-Thurnherr et al., 2013), initiation of reactive oxygen species (ROS) in cells (Aziz et al., 2020) cardiorespiratory toxicity (Bessemer et al., 2015), and developmental toxicity (Choi et al., 2016).

10.3.1 OVERPRODUCTION OF REACTIVE OXYGEN SPECIES

ROS and free radical production are the direct mechanisms of NP toxicity. They may be due to oxidative stress and resultant damage to membranes, proteins, and DNA. ROS are the ions of oxygen [superoxide (O_2^-), singlet oxygen] or any oxygen-holding radicals (hydroxyl, etc.). Reaction products [e.g.. hydrogen peroxide (H_2O_2)] of ROS are progressively constituted as signaling intermediates that can bring molecular responses. The key components that are involved in ROS production by NPs include (i) pro-oxidant functional groups on the NP surface, (ii) active oxidation–reduction cycling on the NP surface, and (iii) interactions between particles and cells (Risom et al., 2005; Knaapen et al., 2004). Both the shape and size of NPs as well as the experimental conditions influence ROS production (Shvedova et al., 2012). The intracellular ROS increased in embryos of zebrafish exposed to ZnO-NPs and enforced some toxic effects. The light absorption by chromophoric groups can start synthesis of ROS, e.g., singlet oxygen and O_2, which may increase the toxicity of the ZnO nanoparticles or chemically modify them in different forms. The metal-settled nanoparticles like copper can be easily adapted with different ligands on the surface like 16-mercaptohexadecanoic acid, and showed variable oxidation responsiveness on the surface that ends in ROS production through Fenton-type reactions, which finally excite the defensive systems of the body (antioxidant) (Heinlaan et al., 2008; Shi et al., 2012; Wang et al., 2014). The NPs have been documented to effect transcriptional factors, intracellular calcium concentration, and production of cytokines by production of free radicals. The interactions of NPs with the cell may produce pro-oxidant effects by the intracellular ROS synthesis regarding mitochondrial respiration and stimulation of NADPH-like systems (Manke et al., 2013). To get over the intense ROS production, cells activate either an injurious or defensive response raising a chain of harmful biological responses. Phagocytic cells such as neutrophils of the immune system also induce significant ROS (Fadeel and Kagan, 2003). The oxidative outburst of phagocytic cells is related to many of the physicochemical attributes of NPs. ROS causes lipid peroxidation and oxidative DNA damage by strand breaks in DNA, DNA-protein cross links, and alkali-labile sites (Kawanishi et al., 2002; Shi et al., 2004), and appear as carcinogens (Knaapen et al., 2004) (Figure 10.5). Testing of genotoxicity is important for the risk evaluation of NPs. OH• is a highly effective radical and known to interact with all constituents of DNA through synthesis of DNA adduct (8-hydroxy-2'-deoxyguanosine), which is a signal of DNA lesions (Pilger and Rüdiger, 2006; Valavanidis et al., 2009). Both in vivo (Inoue et al., 2006) and in vitro (Eblin et al., 2006) studies showed that NPs can cause mutagenesis. A recent study compared ROS-mediated genotoxicity of metal oxide NPs (Cu, Fe, Ti, and Ag) by DNA damage and micronuclei (Song et al., 2012). The concept of redox homeostasis implicates signaling cascades such as NF-κB, HIF-1,

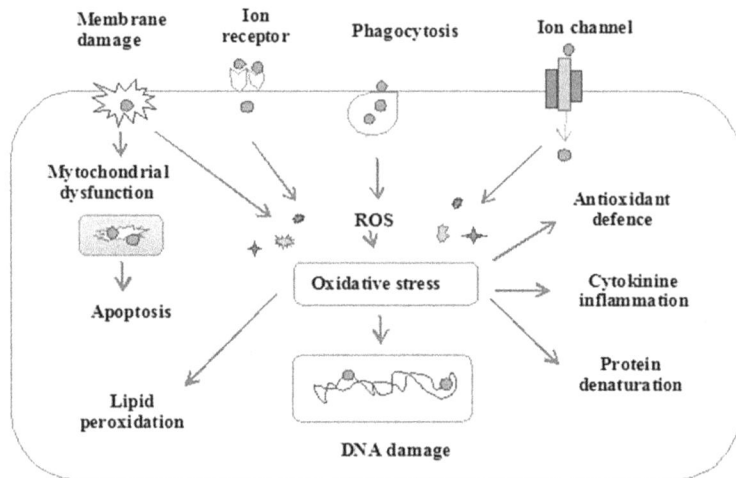

FIGURE 10.5 Possible uptake and pathways for toxicity induced by metal oxide nanoparticles. Nanoparticles can enter the eukaryotic cells through membrane damage, ion receptors, phagocytosis, and ion channels. After entrance, excessive ROS generation causes oxidative stress that can stimulate an antioxidant defense system, cytokine inflammation through cells signaling pathways, protein denaturation, DNA damage by strand breaks, and lipid peroxidation. Mitochondrial dysfunction by metal oxide nanoparticles can lead to apoptosis and cell death.

MAPK, and PI3K, which control metastasis, proliferation, growth of cells, survival, apoptosis, and inflammation (Poljak-Blazi et al., 2010).

10.3.2 RELEASE OF METALLIC IONS

Release of metallic ions is considered to be one of the important mechanisms in the toxicities related to metal-based NPs. The biological mechanism of action of newly synthesized and aged NPs is importantly varied due to the variable quantity of releasing metallic ions (Kittler et al., 2010), because the liberation of ions mainly depends on the physicochemical parameters of the medium. Midander et al. (2009) documented that CuO-NPs can release Cu^{2+} and that soluble amount was different in different media, but different cytotoxicological studies demonstrated that the nanoscale CuO caused toxicity, and their released amount in Cu^{+2} form only accounted for a small part of these toxic effects. Chevallet et al. (2016) demonstrated that released Zn ions from ZnO-NPs are the main mechanisms of toxicity to organisms that interrupt cell homeostasis. Dissolution of NPs and the resultant release of metallic ions can raise ROS generation (Knaapen et al., 2004). When exposure extended to 24 h, released Cu^{2+} from CuO-NPs induced a high percentage of DNA damage (Wang et al., 2012). It is considered that both dissolved ions and metal oxide NPs will give in to cytotoxicity after exposure to metal oxide NPs. Shaw and Handy (2011) compared the influence of NPs on fish caused by metal ions and nanometals. They demonstrated that many nanometals showed more short-term toxicity to some species of fish than their solubilized forms. For example, the values of 48-h LC_{50} are 0.71 and 1.78 mgL^{-1} for nano and soluble forms of copper for *Danio rerio*, respectively. Nanotoxicity linked with ZnO is mainly due to the liberation of Zn ions (Boran and Ulutas, 2016). It has now been accepted that nanometals do have harmful effects on fish, and their risk is different from the conventional dissolved forms.

10.3.3 PARTICLE CELL INTERACTION, UPTAKE, AND BIOACCUMULATION OF NANOPARTICLES

The uptake potency of nanomaterials by fish is one of the most crucial factors in evaluating the nanoparticle toxicity. Therefore, the accumulation level of metals is used as an indicator in different toxicological studies (Birungi et al., 2007). There are some basic conflicts relevant to bioavailability and uptake of NPs, because the behavior and chemical composition of nanometals concerns dynamical aspects of aggregation/agglomeration theory, rather than traditionally used equilibrium models for free metallic ions. Regarding the uptake mechanism, researchers have proposed that NPs can pass the cell membrane through membrane damage and diffusion, and uptake specifically through phagocytosis and endocytosis (Figure 10.5). Studies from research work exhibited that bioaccumulation of metallic substances in tissues/organs is primarily dependent on the exposure period and water concentrations of metals and also on several other environmental components such as salinity, pH, oxygen concentration, temperature, hardness, alkalinity, and dissolved organic carbon (DOC) of water (Linbo et al., 2009). The uptake of nanoparticles by aquatic organisms depends on various elements, such as size and shape of NPs, species, route of exposure, organs, exposure duration, environmental situations, and exposure concentration (Sajid et al., 2015). For metal oxide nanoparticles, both the blood capillaries and lymphatic tissues absorb these particles and transport them to other organs where they start accumulation. An inflammation induction test of vascular endothelial cells pointed out that entrance of NPs into vessels can cause toxicity in cells (Chang et al., 2012). Due to variable preferences in attraction of ligands (imidazole, sulfhydryl group, and amino residues) of proteins with metals, final concentration of metals in fish varies (Tamas et al., 2014). Under the stress of metals, different living organisms synthesize a particular kind of protein, metallothionein, which helps in detoxification of the metallic ion by attaching and accumulating it as a component. Metallothioneins are proteins rich in cysteine having low molecular weight that can selectively attach to heavy metals (Thirumoorthy et al., 2007). Metal-binding proteins generally attach to primary trace elements, such as Cu and Zn (Papagiannis et al., 2004) and xenobiotic (cadmium, lead, mercury, nickel) metals, binding them through specific motifs that disturb the internal homeostasis of the body. The extent of metallothioneins has been projected to be a symbol of exposure to heavy metals in carps (Roy et al., 2011). Engineered nanoparticles or metal ion toxicity can possibly result from adhesion to the body or uptake via ingestion and respiration (Klaine et al., 2008). Once Zn, and Cu are absorbed, metals bind to the protein carrier albumin in blood plasma. Benthic and pelagic fish can take up the aggregates of nanoparticles that sink into the bottom, and filter feeding organisms can take particles from the water column. Higher organisms may also uptake metal oxide NPs, and their biological fate depends on the equilibrium state of four main processes: absorption, biodistribution, metabolism, and elimination. Different organs accumulate different concentrations of nanoparticles. The variation among different tissues in accumulating metallic nanoparticles is related to their metabolic activities and physiological roles in maintaining homeostasis and their metal regulative potential (Tunçsoy and Erdem, 2014). Kaya et al. (2015) investigated significant assimilation of Zn in the gill tissues of tilapia after exposure to both small and large nanoscale ZnO. The trend of Zn assimilation changed with the enhanced time period and the tissue type. In other toxicological research done with nanoscale CuO, TiO_2, and Ag, it was also demonstrated that nanoparticles were taken from the aquatic system by the respiratory organ (gills) (Scown et al., 2010; Zhao et al., 2011; Ates et al., 2015). Bioaccumulation in gills might be due to the trapping of zinc oxide NPs in the mucus layer of gills and throughout the membrane of cells (Hao et al., 2013). Zn^{2+} had been taken up by the tissues of gills through Ca^{+2} transmission channels and lateral Ca-ATPase placed in the mitochondria, ion-moving chloride cells (Bury et al., 2003). After absorption of metals through gills, they are transported to other tissues, like the liver and muscle via blood circulation and diffusion (Wang et al., 2004). The liver plays an essential role in absorption, elimination, and detoxification of metallic substances (Marijic and Raspor, 2006), and metallothioneins are synthesized mainly in the liver; therefore, metals could amass in this tissue in higher amounts. The trophic transfer of NPs is now attracting lots of research interest. The central knowledge gaps present in our understanding of how MeO-NPs transfer is influenced by the interior dispersion in prey and the physiology of the predator's digestive system. In addition, NP relationships with sediments may be a main process that ends in the transportation of entire particles within the aquatic food webs. Different aquatic species need to be examined in order to clear up this issue. The attainable biomagnifications of NPs in aquatic food webs may eventually drive the possible risk to human beings.

10.4 SYSTEMIC TOXICITY OF NANOPARTICLES TO FISH

The widespread and increasing synthesis and use of engineered MeO-NPs increases the potency of their release into the aquatic environment that is raising concerns regarding nanotoxicity. Many people show positive attitudes toward the biological effects of MeO-NPs, although some critical voices are seemingly louder regarding possible environmental risks caused by metal oxide NPs. Many research papers have supplied much of this information, exhibiting the variable toxicities that MeO-NPs can generate. Many metal oxide NPs can drive lethal impacts, which are dramatic, and many other effects, including hepatotoxicity, pulmonary toxicity, immunotoxicity, neurotoxicity, renal toxicity, and testis damage, are also reported in animals. In view of freshwater nanotoxicity using fish models, effects including immunotoxicity, developmental toxicity, genotoxicity, and disruption of lipid metabolism are mostly discussed, and various toxicities from various NPs are present in Table 10.1. Xiong et al. (2011) suggested that nano ZnO and TiO_2 caused significantly high SOD activity in the gut and liver of zebrafish, declaring a tissue-specific response. Data on metallic NPs have shown that nanomaterials can induce oxidative stress, without considerable accumulation of metals in the affected organs/tissues (e.g., high TBARS levels in trout gills, brain, and intestine after TiO_2-NPs exposure; Federici et al., 2007). Zhao et al. (2013) also reported DNA damage after ZnO-NPs exposure in zebrafish embryos by single-cell gel electrophoresis. The significant alteration in DNA damage was found statistically in gill tissues of common carp after exposure to CuO-NPs (Chelomin et al., 2017).

10.5 CONCLUSION AND FUTURE PROSPECTS

In this chapter, we systematically summarize the toxicity of metal oxide nanoparticles to freshwater fish. Many influential factors, such as the particle size and shape, surface chemistry and its modification, and solubility affect the nanotoxicity. The overproduction of ROS, release of metallic ions, bioavailability, exposure route, biodistribution, and particle cell interaction affect the toxicity in the target organs. However, the number of knowledge gaps required to fill in nanotoxicity studies, especially with respect to characterization of NPs and exposure media, action mechanism of NPs in complex exposure media, and the long-term effects as a result of environmentally relevant concentration exposure. Future ecotoxicological research of MeO-NPs should focus on the chronic effects of NPs at low exposure rates as well as the accumulation and trophic transportation of NPs in freshwater environments to fill the knowledge gaps. Measurement of the fluxes in the aquatic environment is

TABLE 10.1

Summary of Studies Carried Out to Study the Toxicity of Metal Oxide Nanoparticles to Freshwater Fish

Metal Oxide Nanoparticle	Model Organism	Toxic Effects	Concentration	Exposure Time	Reference
Copper oxide nanoparticles (CuO-NPs)	Danio rerio	Developmental toxicity	0.1, 0.5, 5, and 50 mg/L	120 hpf*	Xu et al., 2017
	Cyprinus carpio	Lesions, hyperplasia, Aurism, cell inflation and edema	10, 40, and 80 mg/L	6 weeks	Sahraei et al., 2018
	Oncorhynchus mykis	No mortality, fluctuations in hematological parameters	1, 5, 20, and 100 ppm	96 hours	Khabbazi et al., 2015
	Cyprinus carpio	Apoptosis and oxidative stress	0.1, 0.2, 0.5, and 1 mg/L	7 days	Naeemia et al., 2020
	Caspian trout (Salmo trutta caspius)	Hematological and metabolic enzymes	10% of 96-h LC$_{50}$	Acute, 96 hours;chronic, 28 days	Kaviani et al., 2019
	Oreochromis niloticus	Increase in SOD, CAT, GPX	10, 50, and 100 µg/L	1, 7, and 15 days	Tunçsoy and Erdem, 2018
	Guppy (Poecilia reticulata)	Histopathological damages	20 µg/l	10 days	Mansouri et al., 2015
	Ceriodaphnia silvestrii and Hyphessobrycon eques	Cellular hyperplasiaOverproduction of reactive oxygen species Necrosis and apoptosis	Acute (0.7, 10, 13, 16, 19 µg/L); chronic (0.5, 1, 2, 4, 8, 10 µg/L)	Acute, 24 hours;chronic, 8 days	Mansano et al., 2018
	African catfish (C. gariepinus)	No side effects	1, 3, 5, 7, and 10 mg/kg	7 weeks	Onuegbu et al., 2018
	Cyprinus carpio	Inhibition of cholinesterase activity, free Cu2+ ions in the body of fish	10, 50, 100, 200, 300, 500, and 1000 mg/l; 100 mg/ml	96 hours; 30 days	Zhao et al., 2011
	Nile tilapia (Oreochromis niloticus)	LC$_{50}$/96 h of copper oxide particles = 150 mg/l, sublethal caused oxidative stress and biochemical alteration	15 and 7.5 mg/L	Acute and chronic	Abdel-Khalek et al., 2015
	Zebrafish	96-h LC$_{50}$ = 2.25mg/L, ROS	0.0625, 0.125, 0.25, 0.5, 1.0, 2.0, 4.0, and 8.0 mg/L	96 hours	Hou et al., 2018

(Continued)

TABLE 10.1 (CONTINUED)

Summary of Studies Carried Out to Study the Toxicity of Metal Oxide Nanoparticles to Freshwater Fish

Metal Oxide Nanoparticle	Model Organism	Toxic Effects	Concentration	Exposure Time	Reference
	Zebrafish embryos	Hepatotoxicity and neurotoxicity	50, 25, 12.5, 6.25, or 1 mg/L	4 hpf	Sun et al., 2016
	Goldfish (*Carassius auratus*)	Oxidative stress in dose-dependent manner	1 and 10 μg mL^{-1}	21 days	Ates et al., 2015
	Poecilia reticulata	Less reproductive stress, more parturition time, and mortality	0.5–45 and 0.5–10 mg/L	96 hours	Vajargah et al., 2020
	Zebrafish	Drastic effects on the morphology and physiology	0.01, 0.1, 1.0, 10.0, and 50 μg/ml	96 hpf	Kaur et al., 2019
Zinc oxide nanoparticles	*Oreochromis niloticus*	Histopathological changes	1 and 2 mgL^{-1}	15 days	Alkaladi et al., 2014
	Zebrafish embryo	Mortality of embryos retardation in the embryo hatching, reduced development	1, 5, 10, 25, 50, and 100 mg/L	96 hpf	Bai et al., 2010
	Zebrafish	Delayed hatching and inhibited hatching, induction of metallothionein	0.2, 1, and 5 mg/L	96 hpf	Brun et al., 2014
	Catostomus commersonii	Impact cardiorespiratory function	1.0 mg L^{-1}	25 hours	Bessemer et al., 2015
	Cyprinus carpio	Decrease of GSH and increase of LPO	0.5, 5.0, and 50.0 mg/L	14 days	Hao and Chen, 2012
	Zebrafish	96-h LC$_{50}$ = 48.2 mg/L, overproduction of ROS, DNA damage	1.0, 2.0, 4.0, 8.0, 16.0, 32.0, 64.0, and 128.0 mg/L	96 hours	Hou et al., 2018
	Goldfish (*Carassius auratus*)	Oxidative stress and lipid peroxidation in dose-dependent manner	1 and 10 μg mL^{-1}	21 days	Ates et al., 2015
	Cyprinus carpio	Hinder kidney and liver function in fish	50 and 500 mg kg^{-1}	6 weeks	Chupani et al., 2018
	Oncorhynchus mykiss	Hinder cytochrome P450 metabolic processes	300 or 1000 mg ZnO-NPs/kg feed	10 days	Connolly et al., 2016
	Oncorhynchus mykiss	Structural damages to gills	500 μg/L and 0.05 μg/L	4 days	Mansouri et al., 2018
	Danio rerio	Oxidative stress and damage	5 mg/L	96 hours	Xiong et al., 2011
	Common carp (*Cyprinus carpio*)	DNA damage on the exposure time and concentration	2.5, 6.25, and 10 mg/L	14 days	Nikdehghan et al., 2018

(*Continued*)

TABLE 10.1 (CONTINUED)

Summary of Studies Carried Out to Study the Toxicity of Metal Oxide Nanoparticles to Freshwater Fish

Metal Oxide Nanoparticle	Model Organism	Toxic Effects	Concentration	Exposure Time	Reference
	Tilapia (*Oreochromis niloticus*)	Oxidative stress and lipid peroxidation	1 and 10 mg/L	1.4 days	Kaya et al., 2015
	Channa punctatus	Lipid peroxidation (TBARS) and reduced glutathione (GSH)	–	–	Banerjee et al., 2014
	Danio rerio	96-h LC_{50} of 4.92 mg/L and caused oxidative stress	–	96 hours	Xiong et al., 2011
	Nile tilapia (*Oreochromis niloticus*) and redbelly tilapia (*Tilapia zilli*)	ZnO-NPs (500 µg/l) stimulated the brain antioxidant system, but inhibited it at high concentration	500 and 2000 µg/l	15 days, aqueous exposure	Saddick et al., 2017
	Zebrafish (*Danio rerio*) embryos	ZnO-NPs retarded development of nervous and vascular system	1, 5, 10, 20, 50, and 100 ppm	48, 72, and 96 hours, aqueous exposure	Kteeba et al., 2018
	Zebrafish (*Danio rerio*) embryos	96-h LC_{50} = 1.79 mg/L	–	96 hours	Zhu et al., 2008
	Zebra fish (*Danio rerio*)	96-h LC_{50} = 3.969 mg/L	–	96 hours	Yu et al., 2011
	Larval zebrafish	DNA damage	0, 0.2, 1, 2, 4, and 6 mg/L	96 hours	Boran and Ulutas, 2016
	Zebrafish	Life stage dependent toxicity. LC_{50} = 3.5–9.1 mg/L	–	24 hours	Wehmas et al., 2015
	Oreochromis mossambicus	Oxidative stress, induction of antioxidant defense, DNA damage	0.5, 1.0, and 1.5 mg/L	14 days	Shahzad et al., 2018
	Oreochromis mossambicus	Tissue damaging effect and oxidative stress	70 ppm, 80 ppm, 90 ppm, and 100 ppm	12 days	Amutha and Subramanian, 2009
	Cyprinus carpio	Bioaccumulation. severe histopathological changes	50 mg/L	30 days	Hao et al., 2013
	Nile tilapia	Oxidative stress	1 and 2 mg/L	5 days	Abdelazim et al., 2018
	Labeo rohita	96-h LC_{50} = 31.15, oxidative stress	1/3 and 1/5 of 96 hour LC_{50}	80 days	Aziz et al., 2020
	Zebrafish (*Danio rerio*)	No significant uptake of Zn in fish tissues up to exposure 5 mg/l	50, 500, or 5000 µg/L	–	Johnston et al., 2010

(Continued)

TABLE 10.1 (CONTINUED)
Summary of Studies Carried Out to Study the Toxicity of Metal Oxide Nanoparticles to Freshwater Fish

Metal Oxide Nanoparticle	Model Organism	Toxic Effects	Concentration	Exposure Time	Reference
Titanium oxide/dioxide nanoparticles	Rainbow trout (*Oncorhynchus mykiss*)	Gill pathologies, decrease in Na⁺K⁺-ATPase activity	0.1, 0.5, or 1.0 mg/L	14 days	Federici et al., 2007
	Zebrafish (*Danio rerio*)	Brain damage	5, 10, 20, and 40 µg/L	45 days	Sheng et al., 2016
	Cyprinus carpio	Oxidative stress and lipid peroxidation	100 and 200 mg/L	20 days	Hoa et al., 2009
	Prochilodus lineatus	Changes in hematology, red blood cell (RBC) genotoxicity/mutagenicity, liver function after acute and subchronic exposure	0, 1, 5, 10, and 50 mg/L	Acute and 14 days subchronic exposure	Carmo et al., 2019
	Zebrafish (*Danio rerio*) larvae	Induced Parkinson's disease-like symptoms and faster embryos hatching	For 96 hours post fertilization	0, 0.1, 1, 10 µg/ml	Hu et al., 2017
	Zebrafish (*Danio rerio*) and rainbow trout (*Oncorhynchus mykiss*)	Difficult penetration into the brain	500 and 5000 µg/L	24 hours and 14 days	Johnston et al., 2010
	Cyprinus carpio	Induced GST and CAT levels and histopathological changes	5, 10, 20, 40, and 80 mg/L	48 hours	Lee et al., 2012
	Danio rerio	TiO₂ caused mortality and growth inhibition	1.0, 2.0, 4.0, 5.0, and 7.0 mg/L	6 months	Chen et al., 2011a
	Larval zebrafish (*Danio rerio*)	Larval swimming parameters significantly affected	0.1, 0.5, 1, 5, and 10 mg/L	120 hpf	Chen et al., 2011b
	Danio rerio	Oxidative stress and damage	50 mg/L	96 hours	Xiong et al., 2011
	Carassius auratus and *Danio rerio*	Liver alterations	–	–	Diniz et al., 2013
	Rainbow trout (*Oncorhynchus mykiss*)	Reduced velocities of sperm cells after 10 mg/L and an increase in SOD and TGSH levels	0.01, 0.1, 0.5, 1, 10, and 50 mg/L	3 hours	Özgür et al., 2018
Cerium oxide	*Danio rerio*	96 h LC₅₀ =124.5 mg/L and oxidative stress	–	96 hours	Xiong et al., 2011
	Zebrafish (*Danio rerio*)	Genotoxic response in zebrafish (after acute exposure)	10, 20, and 50 µg/l	24, 72, and 120 hours	Bhagat et al., 2019
	Rainbow trout (*Oncorhynchus mykiss*)	Oxidative stress (GST activity was increased in gills, increase in CAT activity of livers), alterations in the gills and liver tissues in dose-dependent manner	0.1, 0.01, and 0.001 µg/L	28 days	Correia et al., 2019

(Continued)

TABLE 10.1 (CONTINUED)

Summary of Studies Carried Out to Study the Toxicity of Metal Oxide Nanoparticles to Freshwater Fish

Metal Oxide Nanoparticle	Model Organism	Toxic Effects	Concentration	Exposure Time	Reference
	Zebrafish (*Danio rerio*)	Did not show much of the effects	0.01, 0.1, 1.0, 10, and 50 µg/ml	96 hpf	Kaur et al., 2019
Aluminum oxide	*Oreochromis mossambicus*	Intense damages and architectural loss target tissues	120, 150, and 180 ppm	96 hours	Murali et al., 2018
	Oreochromis mossambicus	Median lethal concentrations was 140 mg/L, adverse toxic effects	10, 20, 30, 40, 50, 60, and 70 mg/L	96 hours	Vidya and Chitra, 2017
Silicon dioxide	*Labeo rohita*	Dose and time-dependent alterations in hematological parameters	1, 5, and 25 mg/L	96 hours	Priya et al., 2015
	Oreochromis niloticus	Upregulated expression of genes in the gills and hepatic tissues	20, 40, and 100 mg/L	3 weeks	Abdel-Latif et al., 2021
	Oreochromis mossambicus	Dose-dependent elevation of urea and creatinine levels, which indicated a possible damage to the renal tissues of the fishes	60, 100, and 140 ppm	96 hours	Athif et al., 2020
	Oreochromis mossambicus	Median lethal concentrations = 120 mg/L	5, 25, 50, 75, 100, 125, 150, and 175 mg/L	96 hours	Vidya and Chitra, 2017
	Rainbow trout sperm cells	Caused oxidative stress and significant reduction in sperms velocities	1, 10, 25, 50, and 100 mg/L	24 hours	Özgür et al., 2019
Iron oxide	Rainbow trout spermatozoon (*Oncorhynchus mykiss*)	Effect antioxidant system and kinematics of spermatozoon	(50, 100, 200, 400, and 800 mg/L	24 hours	Özgür et al., 2018
	Labeo rohita	Affect certain hematological, ionoregulatory and gill Na⁺/K⁺-ATPase activity	500 mg/L	25 days	Remya et al., 2015
	Zebrafish (*Danio rerio*)	Developmental toxicity	100, 50, 10, 5, 1, 0.5, and 0.1 mg/L	168 hours	Zhu et al., 2012

Hpf, hours post fertilization.

required to make the relevant toxicological work more significant. Collaboration is needed from researchers of biological, environmental, toxicological, and chemical sciences for more extensive studies on nanotoxicology. In addition, standard testing techniques are needed to make toxicological studies more comparable.

REFERENCES

Ahamed, M., Siddiqui, M. A., Akhtar, M. J., Ahmad, I., Pant, A. B., & Alhadlaq, H. A. (2010). Genotoxic potential of copper oxide nanoparticles in human lung epithelial cells. *Biochemical and Biophysical Research Communications*, 396(2), 578–583.

Aillon, K. L., Xie, Y., El-Gendy, N., Berkland, C. J., & Forrest, M. L. (2009). Effects of nanomaterial physicochemical properties on in vivo toxicity. *Advanced Drug Delivery Reviews*, 61(6), 457–466.

Ajdary, M., Moosavi, M. A., Rahmati, M., Falahati, M., Mahboubi, M., Mandegary, A., Jangjoo, S., Mohammadinejad, R., & Varma, R. S. (2018). Health concerns of various nanoparticles: A review of their in vitro and in vivo toxicity. *Nanomaterials*, 8(9), 634–661.

Albanese, A., Tang, P. S., & Chan, W. C. (2012). The effect of nanoparticle size, shape, and surface chemistry on biological systems. *Annual Review of Biomedical Engineering*, 14, 1–16.

Alkaladi, A., Afifi, M., Mosleh, Y. Y., & Abu-Zinada, O. (2014). Ultra structure alteration of sublethal concentrations of zinc oxide nanoparticals on Nile Tilapia (*Oreochromis niloticus*) and the protective effects of vitamins C and E. *Life Science Journal*, 11(10), 257–262.

Al-Yousuf, M. H., El-Shahawi, M. S., & Al-Ghais, S. M. (2000). Trace metals in liver, skin and muscle of *Lethrinus lentjan* fish species in relation to body length and sex. *Science of the Total Environment*, 256(2–3), 87–94.

Arora, S., Rajwade, J. M., & Paknikar, K. M. (2012). Nanotoxicology and in vitro studies: The need of the hour. *Toxicology and Applied Pharmacology*, 258(2), 151–165.

Ates, M., Arslan, Z., Demir, V., Daniels, J., & Farah, I. O. (2015). Accumulation and toxicity of CuO and ZnO nanoparticles through waterborne and dietary exposure of goldfish (*Carassius auratus*). *Environmental Toxicology*, 30(1), 119–128.

Amutha, C., & Subramanian, P. (2009). Tissue damaging effect of zinc oxide nanoparticle on *Oreochromis mossambicus. Biochemical and Cellular Archives*, 9, 235–239.

Aziz, S., Abdullah, S., Abbas, K., & Zia, M. A. (2020). Effects of engineered zinc oxide nanoparticles on freshwater fish, *Labeo rohita*: Characterization of ZnO nanoparticles, acute toxicity and oxidative stress. *Pakistan Veterinary Journal*, 40(4), 479–483.

Bai, W., Zhang, Z., Tian, W., He, X., Ma, Y., Zhao, Y., & Chai, Z. (2010). Toxicity of zinc oxide nanoparticles to zebrafish embryo: A physicochemical study of toxicity mechanism. *Journal of Nanoparticle Research*, 12(5), 1645–1654.

Abdel-Khalek, A., Kadry, M. A. M., Badran, S. R.,Marie, M. A. S. (2015). Comparative toxicity of copper oxide bulk and nanoparticles in Nile Tilapia; *Oreochromis niloticus*. Biochemical oxidative stress. *The Journal of Basic & Applied Zoology*, 72: 43–57.

Abdel-Latif, H. M. R., Shukry, M., El-Euony, O. I., Soliman, M. M., Noreldin, A. E., Ghetas, H. A., Dawood M. A. O., & Khallaf, M. A. (2021). Hazardous effects of SiO2 nanoparticles on liver and kidney functions, Histopathology characteristics, and transcriptomic responses in Nile tilapia (*Oreochromis niloticus*) Juveniles. *Biology*, 10, 183.

Athif, P., Murali, M., Suganthi, P., Bukhari, A. S., Mohammed, H. E. S., Basu, H., & Singhal, R. K. (2020). Alterations in renal markers of tilapia fish exposed to silicon dioxide nanoparticle. *Uttar Pradesh Journal of Zoology*, 41, 48–55.

Abdelazim, A. M., Saadeldin, I. M., Swelum, A. A. A., Afifi, M. M., & Alkaladi, A. (2018). Oxidative stress in the muscles of the fish Nile tilapia caused by zinc oxide nanoparticles and its modulation by vitamins C and E. *Oxidative Medicine and Cellular Longevity*, 2018, 1–9.

Bessemer, R. A., Butler, K. M. A., Tunnah, L., Callaghan, N. I., Rundle, A., Currie, S., Dieni, C. A., & MacCormack, T. J. (2015). Cardiorespiratory toxicity of environmentally relevant zinc oxide nanoparticles in the freshwater fish *Catostomus commersonii*. *Nanotoxicology*, 9(7), 861–870.

Bian, S. W., Mudunkotuwa, I. A., Rupasinghe, T., & Grassian, V. H. (2011). Aggregation and dissolution of 4 nm ZnO nanoparticles in aqueous environments: influence of pH, ionic strength, size, and adsorption of humic acid. *Langmuir*, 27(10), 6059–6068.

Birungi, Z., Masola, B., Zaranyika, M. F., Naigaga, I., & Marshall, B. (2007). Active biomonitoring of trace heavy metals using fish (*Oreochromis niloticus*) as bioindicator species. The case of Nakivubo wetland along Lake Victoria. *Physics and Chemistry of the Earth, Parts A/B/C*, 32(15–18), 1350–1358.

Boran, H., & Ulutas, G. (2016). Genotoxic effects and gene expression changes in larval zebrafish after exposure to $ZnCl_2$ and ZnO nanoparticles. *Diseases of Aquatic Organisms*, 117(3), 205–214.

Brun, N. R., Lenz, M., Wehrli, B., & Fent, K. (2014). Comparative effects of zinc oxide nanoparticles and dissolved zinc on zebrafish embryos and eleuthero-embryos: Importance of zinc ions. *Science of the Total Environment*, 476–477, 657–666.

Brunner, T. J., Wick, P., Manser, P., Spohn, P., Grass, R. N., Limbach, L. K., Bruinink, A., & Stark, W. J. (2006). In vitro cytotoxicity of oxide nanoparticles: comparison to asbestos, silica, and the effect of particle solubility. *Environmental Science & Technology*, 40(14), 4374–4381.

Buerki-Thurnherr, T., Xiao, L., Diener, L., Arslan, O., Hirsch, C., Maeder-Althaus, X., Grieder, K., Wampfler, B., Mathur, S., Wick, P., & Krug, H. F. (2013). In vitro mechanistic study towards a better understanding of ZnO nanoparticle toxicity. *Nanotoxicology*, 7(4), 402–416.

Bury, N. R., Walker, P. A., & Glover, C. N. (2003). Nutritive metal uptake in teleost fish. *Journal of Experimental Biology*, 206(1), 11–23.

Banerjee, P., Das, D., Mitra, P., Sinha, M., Dey, S., & Chakrabarti, S. (2014). Solar photocatalytic treatment of wastewater with zinc oxide nanoparticles and its ecotoxicological impact on *Channa punctatus* a freshwater fish. *Journal of Materials and Environmental Sciences*, 5, 1206–1213.

Buzea, C., Pacheco, I. I., & Robbie, K. (2007). Nanomaterials and nanoparticles: Sources and toxicity. *Biointerphases*, 2(4), MR17–MR71.

Bhagat, J., Greeshma, S. S., & Shyama, S. K. (2019). Genotoxicity of cerium oxide nanoparticle in zebrafish and green mussel *Perna Viridis* using alkaline comet assay. *International Journal of Health and Life Sciences*, 4, 118–127.

Correia, A. T., Rebelo, D., Marquesa, J., & Nunes, B. (2019). Eects of the chronic exposure to cerium dioxide nanoparticles in *Oncorhynchus mykiss*: Assessment of oxidative stress, neurotoxicity and histological alterations. *Environmental Toxicology and Pharmacology*, 68, 27–36.

Chang, Y. N., Zhang, M., Xia, L., Zhang, J., & Xing, G. (2012). The toxic effects and mechanisms of CuO and ZnO nanoparticles. *Materials*, 5(12), 2850–2871.

Chelomin, V. P., Slobodskova, V. V., Zakhartsev, M., & Kukla, S. (2017). Genotoxic potential of copper oxide nanoparticles in the bivalve mollusk *Mytilus trossulus*. *Journal of Ocean University of China*, 16(2), 339–345.

Chen, J., Dong, X., Xin, Y., & Zhao, M. (2011a). Effects of titanium dioxide nanoparticles on growth and some histological parameters of zebrafish (*Danio rerio*) after a long-term exposure. *Aquatic Toxicology*, 101(3–4), 493–499.

Chen, T. H., Lin, C. Y., & Tseng, M. C. (2011b). Behavioral effects of titanium dioxide nanoparticles on larval zebrafish (*Danio rerio*). *Marine Pollution Bulletin*, 63, 303–308.

Chevallet, M., Gallet, B., Fuchs, A., Jouneau, P., Um, K., Mintz, E., & Michaud-Soret, I. (2016). Metal homeostasis disruption and mitochondrial dysfunction in hepatocytes exposed to sub-toxic doses of zinc oxide nanoparticles. *Nanoscale*, 8, 18495–18506.

Choi, J. S., Kim, R. O., Yoon, S., & Kim, W. K. (2016). Developmental toxicity of zinc oxide nanoparticles to zebrafish (*Danio rerio*): A transcriptomic analysis. *PLoS One*, 11(8), e0160763.

Chompoosor, A., Saha, K., Ghosh, P. S., Macarthy, D. J., Miranda, O. R., Zhu, Z. J., Arcaro, K. F., & Rotello, V. M. (2010). The role of surface functionality on acute cytotoxicity, ROS generation and DNA damage by cationic gold nanoparticles. *Small*, 6(20), 2246–2249.

Chupani, L., Niksirat, H., Velíšek, J., Stará, A., Hradilová, Š., Kolařík, J., Panáček, A., & Zusková, E. (2018). Chronic dietary toxicity of zinc oxide nanoparticles in common carp (*Cyprinus carpio L.*): Tissue accumulation and physiological responses. *Ecotoxicology and Environmental Safety*, 147, 110–116.

Carmo, T. L. L., Siqueira, P. R., Azevedo, V. C., Tavares, D., Pesenti, E. C., Cestari, M. M., Martinez C. B. R., & Fernandes, M. N. (2019). Overview of the toxic effects of titanium dioxide nanoparticles in blood, liver, muscles, and brain of a Neotropical detritivorous fish. *Environmental Toxicology*, 34, 457–468.

Connolly, M., Fernández, M., Conde, E., Torrent, F., Navas, J. M., & Fernández-Cruz, M. L. (2016). Tissue distribution of zinc and subtle oxidative stress effects after dietary administration of ZnO nanoparticles to rainbow trout. *Science of the Total Environment*, 551, 334–343.

Diniz, M. S., de-Matos, A. P. A., Lourenço. J., Castro, L., Isabel, P., Mendonca, E., & Picado, A. (2013). Liver alterations in two freshwater fish species (*Carassius auratus* and *Danio rerio*) following exposure to different TiO₂ nanoparticle concentrations. *Microscopy and Microanalysis*, 19, 1131–1140.

Dabrunz, A., Duester, L., Prasse, C., Seitz, F., Rosenfeldt, R., Schilde, C., Schaumann, G. E., & Schulz, R. (2011). Biological surface coating and molting inhibition as mechanisms of TiO₂ nanoparticle toxicity in *Daphnia magna. PloS One*, 6(5), e20112.

Eblin, K. E., Bowen, M. E., Cromey, D. W., Bredfeldt, T. G., Mash, E. A., Lau, S. S., & Gandolfi, A. J. (2006). Arsenite and monomethylarsonous acid generate oxidative stress response in human bladder cell culture. *Toxicology and Applied Pharmacology*, 217(1), 7–14.

Fadeel, B., & Kagan, V. E. (2003). Apoptosis and macrophage clearance of neutrophils: Regulation by reactive oxygen species. *Redox Report*, 8(3), 143–150.

Faisal, M., Saquib, Q., Alatar, A. A., Al-Khedhairy, A. A., Hegazy, A. K., & Musarrat, J. (2013). Phytotoxic hazards of NiO-nanoparticles in tomato: a study on mechanism of cell death. *Journal of Hazardous Materials*. 318, 250–251.

Farmen, E., Mikkelsen, H. N., Evensen, Ø., Einset, J., Heier, L. S., Rosseland, B. O., Salbu, B., Tollefsen, K. E., & Oughton, D. H. (2012). Acute and sub-lethal effects in juvenile Atlantic salmon exposed to low µg/L concentrations of Ag nanoparticles. *Aquatic Toxicology*, 108, 78–84.

Federici, G., Shaw, B. J., & Handy, R. D. (2007). Toxicity of titanium dioxide nanoparticles to rainbow trout (*Oncorhynchus mykiss*): gill injury, oxidative stress, and other physiological effects. *Aquatic Toxicology*, 84(4), 415–430.

Franklin, N. M., Rogers, N. J., Apte, S. C., Batley, G. E., Gadd, G. E., & Casey, P. S. (2007). Comparative toxicity of nanoparticulate ZnO, bulk ZnO, and ZnCl₂ to a freshwater microalga (*Pseudokirchneriella subcapitata*): The importance of particle solubility. *Environmental Science & Technology*, 41(24), 8484–8490.

Gatoo, M. A., Naseem, S., Arfat, M. Y., Mahmood Dar, A., Qasim, K., & Zubair, S. (2014). Physicochemical properties of nanomaterials: implication in associated toxic manifestations. *BioMed Research International*, 2014, 498420.

Geng, Y. A. N., Dalhaimer, P., Cai, S., Tsai, R., Tewari, M., Minko, T., & Discher, D. E. (2007). Shape effects of filaments versus spherical particles in flow and drug delivery. *Nature Nanotechnology*, 2(4), 249–255.

Griffitt, R. J., Luo, J., Gao, J., Bonzongo, J. C., & Barber, D. S. (2008). Effects of particle composition and species on toxicity of metallic nanomaterials in aquatic organisms. *Environmental Toxicology and Chemistry: An International Journal*, 27(9), 1972–1978.

Handy, R. D., Henry, T. B., Scown, T. M., Johnston, B. D., & Tyler, C. R. (2008). Manufactured nanoparticles: their uptake and effects on fish-a mechanistic analysis. *Ecotoxicology*, 17(5), 396–409.

Hao, L., & Chen, L. (2012). Oxidative stress responses in different organs of carp (*Cyprinus carpio*) with exposure to ZnO nanoparticles. *Ecotoxicology and Environmental Safety*, 80, 103–110.

Hao, L., Chen, L., Hao, J., & Zhong, N. (2013). Bioaccumulation and sub-acute toxicity of zinc oxide nanoparticles in juvenile carp (*Cyprinus carpio*): A comparative study with its bulk counterparts. *Ecotoxicology and Environmental Safety*, 91, 52–60.

Hou, J., Liu, H., Wang, L., Duan, L., Li, S., & Wang, X. (2018). molecular toxicity of metal oxide nanoparticles in *Danio rerio*. *Environmental Science and Technology*, 52, 7996–8004.

He, W., Zhou, Y. T., Wamer, W. G., Boudreau, M. D., & Yin, J. J. (2012). Mechanisms of the pH dependent generation of hydroxyl radicals and oxygen induced by Ag nanoparticles. *Biomaterials*, 33(30), 7547–7555.

Hoa, L., Wang, Z., & Xing, B. (2009). Effect of sub-acute exposure to TiO$_2$ nanoparticles on oxidative stress and histopathological changes in juvenile carp (*Cyprinus carpio*). *Journal of Environmental Science*, 21, 1459–1466.

Heinlaan, M., Ivask, A., Blinova, I., Dubourguier, H. C., & Kahru, A. (2008). Toxicity of nanosized and bulk ZnO, CuO and TiO$_2$ to bacteria *Vibrio fischeri* and crustaceans *Daphnia magna* and *Thamnocephalus platyurus*. *Chemosphere*, 71(7), 1308–1316.

Hu, Q., Guo, F., Zhao, F., & Fu, Z. (2017). Effects of titanium dioxide nanoparticles exposure on parkinsonism in zebrafish larvae and PC12. *Chemosphere*, 173, 373–379.

Inoue, K. I., Takano, H., Yanagisawa, R., Hirano, S., Sakurai, M., Shimada, A., & Yoshikawa, T. (2006). Effects of airway exposure to nanoparticles on lung inflammation induced by bacterial endotoxin in mice. *Environmental Health Perspectives*, 114(9), 1325–1330.

Jin, H., Heller, D. A., Sharma, R., & Strano, M. S. (2009). Size-dependent cellular uptake and expulsion of single-walled carbon nanotubes: Single particle tracking and a generic uptake model for nanoparticles. *ACS Nano*, 3(1), 149–158.

Johnston, B. D., Scown, T. M., Moger, J., Cumberland, S. A., Baalousha, M., Linge, K., van Aerle, R., Jarvis, K., Lead, J. R., & Tyler, C. R. (2010). Bioavailability of nanoscale metal oxides TiO$_2$, CeO$_2$, and ZnO to fish. *Environmental Science & Technology*, 44(3), 1144–1151.

Kawanishi, S., Hiraku, Y., Murata, M., & Oikawa, S. (2002). The role of metals in site-specific DNA damage with reference to carcinogenesis. *Free Radical Biology and Medicine*, 32(9), 822–832.

Kaya, H., Aydın, F., Gürkan, M., Yılmaz, S., Ates, M., Demir, V., & Arslan, Z. (2015). Effects of zinc oxide nanoparticles on bioaccumulation and oxidative stress in different organs of tilapia (*Oreochromis niloticus*). *Environmental Toxicology and Pharmacology*, 40(3), 936–947.

Kittler, S., Greulich, C., Diendorf, J., Koller, M., & Epple, M. (2010). Toxicity of silver nanoparticles increases during storage because of slow dissolution under release of silver ions. *Chemistry of Materials*, 22(16), 4548–4554.

Klaine, S. J., Alvarez, P. J., Batley, G. E., Fernandes, T. F., Handy, R. D., Lyon, D. Y., Mahendra, S., Mclaughlin, M. J., & Lead, J. R. (2008). Nanomaterials in the environment: Behavior, fate, bioavailability, and effects. *Environmental Toxicology and Chemistry: An International Journal*, 27(9), 1825–1851.

Kaur, J., Khatri, M., & Puri, S. (2019). Toxicological evaluation of metal oxide nanoparticles and mixed exposures at low doses using zebra fish and THP1 cell line. *Environmental Toxicology*, 34(4), 375–387.

Knaapen, A. M., Borm, P. J., Albrecht, C., & Schins, R. P. (2004). Inhaled particles and lung cancer. Part A: Mechanisms. *International Journal of Cancer*, 109(6), 799–809.

Kteeba, S. M., El-Ghobashy, A. E., El-Adawi, H. I., El-Rayis, O. A., Sreevidya, V. S., Guo, L., & Svoboda, K. R. (2018). Exposure to ZnO nanoparticles alters neuronal and vascular development in zebrafish: Acute and transgenerational effects mitigated with dissolved organic matter. *Environmental Pollution*, 242, 433–448.

Khabbazi, M., Harsij, M., Hedayati, S. A. A., Gholipoor, H., Gerami, M. H., & Farsani, H.G. (2015). Effect of CuO nanoparticles on some hematological indices of rainbow trout *Oncorhynchus mykiss* and their potential toxicity. *Nanomedicine*, 2, 67–73.

Kaviani, E. F., Naeemi, A. S., & Salehzadeh, A. (2019). Influence of Copper Oxide nanoparticle on hematology and plasma biochemistry of Caspian Trout (*Salmo trutta caspius*), following acute and chronic exposure. *Pollution*, 5, 225–234.

Liao, K. H., Lin, Y. S., Macosko, C. W., & Haynes, C. L. (2011). Cytotoxicity of graphene oxide and graphene in human erythrocytes and skin fibroblasts. *ACS Applied Materials & Interfaces*, 3(7), 2607–2615.

Linbo, T. L., Baldwin, D. H., McIntyre, J. K., & Scholz, N. L. (2009). Effects of water hardness, alkalinity, and dissolved organic carbon on the toxicity of copper to the lateral line of developing fish. *Environmental Toxicology and Chemistry: An International Journal*, 28(7), 1455–1461.

Lovern, S. B., & Klaper, R. (2006). *Daphnia magna* mortality when exposed to titanium dioxide and fullerene (C_{60}) nanoparticles. *Environmental Toxicology and Chemistry: An International Journal*, 25(4), 1132–1137.

Lynch, I., & Dawson, K. A. (2008). Protein-nanoparticle interactions. *Nano Today*, 3(1–2), 40–47.

Lee, B. C., Kim, K. T., Cho, J. G., Lee, J. W., Ryu, T. K., Yoon, J. H., Lee, S. H., Duong, C. N., Eom, I. G. C., Kim, P. J., & Choi, K. H. (2012). Oxidative stress in juvenile common carp (*Cyprinus carpio*) exposed to TiO_2 nanoparticles. *Molecular and Cellular Toxicology*, 8, 357–366.

Vidya, P., & Chitra, K. C. (2017). Assessment of acute toxicity (LC_{50}-96 h) of aluminium oxide, silicon dioxide and titanium dioxide nanoparticles on the freshwater fish, *Oreochromis mossambicus*. *International Journal of Fisheries and Aquatic Studies*, 5, 327–332.

Mansouri, B., Johari, S. A., Azadi, N. A., & Sarkheil, M. (2018). Effects of waterborne ZnO nanoparticles and Zn2+ ions on the gills of rainbow trout (*Oncorhynchus mykiss*): Bioaccumulation, histopathological and ultrastructural changes. *Turkish Journal of Fish and Aquatic Science*, 18, 739–746.

Mansouri, B., Rahmani, R., Azadi, N. A., Davari, B. Johari, S.A., & Sobhani, P. (2015). Effect of waterborne copper oxide nanoparticles and copper ions on guppy (*Poecilia reticulata*): Bioaccumulation and histopathology. *Journal of Advances in Environmental Health Research*, 3, 215–223.

Ma, H., Williams, P. L., & Diamond, S. A. (2013). Ecotoxity of manufactured ZnO nanoparticles–a review. *Environmental Pollution*, 172, 76–85.

Manke, A., Wang, L., & Rojanasakul, Y. (2013). Mechanisms of nanoparticle-induced oxidative stress and toxicity. *BioMed Research International*, 2013, 1–15.

Mansano, A. S., Souza, J. P., Cancino-Bernardi, J., Venturini, F. P., Marangoni, V. S., & Zucolotto, V. (2018). Toxicity of copper oxide nanoparticles to neotropical species *Ceriodaphnia silvestrii* and *Hyphessobrycon eques*. *Environmental Pollution*, 243, 723–733.

Marijic, V. F., & Raspor, B. (2006). Age-and tissue-dependent metallothionein and cyto-solic metal distribution in a native Mediterranean fish, *Mullus barbatus*, from the Eastern Adriatic Sea. *Comparative Biochemistry and Physiology Part C: Toxicology & Pharmacology*, 143(4), 382–387.

Midander, K., Cronholm, P., Karlsson, H. L., Elihn, K., Möller, L., Leygraf, C., & Wallinder, I. O. (2009). Surface characteristics, copper release, and toxicity of nano-and microm-eter-sized copper and copper (II) oxide particles: A cross-disciplinary study. *Small*, 5(3), 389–399.

Mironava, T., Hadjiargyrou, M., Simon, M., Jurukovski, V., & Rafailovich, M. H. (2010). Gold nanoparticles cellular toxicity and recovery: Effect of size, concentration and exposure time. *Nanotoxicology*, 4(1), 120–137.

Mueller, N. C., Braun, J., Bruns, J., Černík, M., Rissing, P., Rickerby, D., & Nowack, B. (2012). Application of nanoscale zero valent iron (NZVI) for groundwater remediation in Europe. *Environmental Science and Pollution Research*, 19(2), 550–558.

Murali, M., Athif, P., Suganthi, P., Bukhari, A. S., Mohammed, H. E. S., Basu, H. & Singhal, R. K. (2018). Toxicological effect of Al_2O_3 nanoparticles on histoarchitecture of the fresh-water fish *Oreochromis mossambicus*. *Environmental Toxicology and Pharmacology*, 59, 74–81.

Naeemia, A. S., Elmib, F., Vaezic, G., & Ghorbankhahd, M. (2020). Copper oxide nanopar-ticles induce oxidative stress mediated apoptosis in carp (*Cyprinus carpio*) larva. *Gene Reports*, 19, 100676.

Nam, S. H., & An, Y. J. (2019). Size-and shape-dependent toxicity of silver nanomaterials in green alga *Chlorococcum infusionum*. *Ecotoxicology and Environmental Safety*, 168, 388–393.

Nel, A. E., Mädler, L., Velegol, D., Xia, T., Hoek, E. M., Somasundaran, P., Klaessig, F., Castranova, V., & Thompson, M. (2009). Understanding biophysicochemical interac-tions at the nano–bio interface. *Nature Materials*, 8(7), 543–557.

Nikdehghan, N., Kashiri, H., & Hedayati, A. A. (2018). CuO nanoparticles-induced micronu-clei and DNA damage in *Cyprinus carpio*. *ACCL Bioflux*, 11, 925–936.

Oberdorster, E., Zhu, S., Blickley, T. M., McClellan-Green, P., & Haasch, M. L. (2006). Ecotoxicology of carbon-based engineered nanoparticles: effects of fullerene (C_{60}) on aquatic organisms. *Carbon*, 44(6), 1112–1120.

Özgür, M. E., Balcıoğlu, S., Ulu, A., Özcan, I., Okumuş, F., Köytepe, S. & Ateş, B. (2018). The in vitro toxicity analysis of titanium dioxide (TiO_2) nanoparticles on kinematics and biochemical quality of rainbow trout sperm cells. *Environmental Toxicology and Pharmacology*, 62, 11–19.

Özgür, M. E., Ulu, A., Özcan, I., Balcıoğlu, S., Ateş, B., & Köytepe, S. (2019). Investigation of toxic effects of amorphous SiO_2 nanoparticles on motility and oxidative stress markers in rainbow trout sperm cells. *Environmental Science and Pollution Research International*, 26(15), 15641–15652.

Özgür, M. E., Ulu, A., Balcıoğlu, S., Özcan, I., Köytepe, S., & Ates, B. (2018). The toxicity assessment of Iron Oxide (Fe3O4) nanoparticles on physical and biochemical quality of rainbow trout spermatozoon. *Toxics*, 6, 62.

Ospina, S. P., Favi, P. M., Gao, M., Johana Sepúlveda Arango, L., Morales, M., Pavon, J. J., & Webster, T. J. (2015). Shape and surface effects on the cytotoxicity of nanoparticles: Gold nanospheres versus gold nanostars. *Journal of Biomedical Materials Research Part A*, 103(11), 3449–3462.

Onuegbu, C. Aggarwal, A., & Singh, N. B. (2018). ZnO nanoparticles as feed supplement on growth performance of cultured African catfish fingerlings. *Journal of Scientific and Industrial Research*, 77(4), 213–218.

Papagiannis, I., Kagalou, I., Leonardos, J., Petridis, D., & Kalfakakou, V. (2004). Copper and zinc in four freshwater fish species from Lake Pamvotis (Greece). *Environment international*, 30(3), 357–362.

Pilger, A., & Rüdiger, H. W. (2006). 8-Hydroxy-2′-deoxyguanosine as a marker of oxidative DNA damage related to occupational and environmental exposures. *International Archives of Occupational and Environmental Health*, 80(1), 1–15.

Priya, K. K., Ramesh, M., Saravanan, M., & Ponpandian, N. (2015). Ecological risk assessment of silicon dioxide nanoparticles in a fresh- water fish *Labeo rohita*: Hematology, ionoregulation and gill Na+/K+ ATPase activity. *Ecotoxicology and Environmental Safety*, 120, 295–302.

Poljak-Blazi, M., Jaganjac, M., & Žarković, N. (2010). Cell oxidative stress: Risk of metal nanoparticles. Sattler KD, editor,*Handbook of Nanophysics: Nanomedicine and Nanorobotics*, CRC Press, New York, USA, 1–17.

Remya, A.S., Ramesh, M., Saravanann M., Poopal, R. K., Barathi, S., Naaraj, D. (2015). Iron oxide nanoparticles to an Indian major carp, *Labeo rohita*: Impacts on hematology, iono regulation and gill Na+/K+ ATPase activity. *Journal of King Saud University- Science,* 27(2),151–160.

Raza, M. A., Kanwal, Z., Rauf, A., Sabri, A. N., Riaz, S., & Naseem, S. (2016). Size-and shape-dependent antibacterial studies of silver nanoparticles synthesized by wet chemical routes. *Nanomaterials*, 6(4), 74–88.

Risom, L., Møller, P., & Loft, S. (2005). Oxidative stress-induced DNA damage by particulate air pollution. *Mutation Research/Fundamental and Molecular Mechanisms of Mutagenesis*, 592(1–2), 119–137.

Roy, U. S., Chattopadhyay, B., Datta, S., & Mukhopadhyay, S. K. (2011). Metallothionein as a biomarker to assess the effects of pollution on Indian major carp species from wastewater-fed fishponds of East Calcutta wetlands (a Ramsar Site). *Environmental Research, Engineering and Management*, 58(4), 10–17.

Sun, Y., Zhang, G., He, Z., Wang, Y., Cui, J., & Li, Y. (2016). Effects of copper oxide nanoparticles on developing zebrafish embryos and larvae. *International Journal of Nanomedicine*, 11, 905–918.

Saddick, S., Afifi, M., & Zinada, O. A. A. (2017). Effect of Zinc nanoparticles on oxidative stress-related genes and antioxidant enzymes activity in the brain of *Oreochromis niloticus* and *Tilapia zillii*. *Saudi Journal of Biological Sciences*, 24(7), 1672–1678.

Sahraei, H., Hoseini, S. A., Hedayati, S. A., & Ghorbani, R. (2018). Gill histopathological changes of common carp (*Cyprinus carpio*) during exposure to sub-lethal concentrations of copper oxide nanoparticles. *Journal of Environmental Science and Technology*, 20(1), 93–103.

Sajid, M., Ilyas, M., Basheer, C., Tariq, M., Daud, M., Baig, N., & Shehzad, F. (2015). Impact of nanoparticles on human and environment: Review of toxicity factors, exposures, control strategies, and future prospects. *Environmental Science and Pollution Research*, 22(6), 4122–4143.

Scown, T. M., Van-Aerle, R., & Tyler, C. R. (2010). Do engineered nanoparticles pose a significant threat to the aquatic environment? *Critical Reviews in Toxicology*, 40(7), 653–670.

Shahzad, K., Khan, M. N., Jabeen, F., Kosour, N., Chaudhry, A. S., Sohail, M., & Ahmad, N. (2018). Toxicity of zinc oxide nanoparticles (ZnO-NPs) in tilapia (*Oreochromis mossambicus*): Tissue accumulation, oxidative stress, histopathology and genotoxicity. *International Journal of Environmental Science and Technology*, 16(4), 1973–1984.

Shaw, B. J., & Handy, R. D. (2011). Physiological effects of nanoparticles on fish: A comparison of nanometals versus metal ions. *Environment International*, 37(6), 1083–1097.

Sheng, L., Wang, L., Su, M., Zhao, X., Hu, R., Yu, X., Hong, J., Liu, D., Xu, B., Zhu, Y., Wang, H., & Hong, F. (2016). Mechanism of TiO_2 nanoparticle-induced neurotoxicity in zebrafish (*Danio rerio*). *Environmental Toxicology*, 31(2), 163–175.

Shi, H., Hudson, L. G., & Liu, K. J. (2004). Oxidative stress and apoptosis in metal ion-induced carcinogenesis. *Free Radical Biology and Medicine*, 37(5), 582–593.

Shi, M., Kwon, H. S., Peng, Z., Elder, A., & Yang, H. (2012). Effects of surface chemistry on the generation of reactive oxygen species by copper nanoparticles. *ACS Nano*, 6(3), 2157–2164.

Shvedova, A. A., Pietroiusti, A., Fadeel, B., & Kagan, V. E. (2012). Mechanisms of carbon nanotube-induced toxicity: focus on oxidative stress. *Toxicology and Applied Pharmacology*, 261(2), 121–133.

Slowing, I., Trewyn, B. G., & Lin, V. S. Y. (2006). Effect of surface functionalization of MCM-41-type mesoporous silica nanoparticles on the endocytosis by human cancer cells. *Journal of the American Chemical Society*, 128(46), 14792–14793.

Song, M. F., Li, Y. S., Kasai, H., & Kawai, K. (2012). Metal nanoparticle-induced micronuclei and oxidative DNA damage in mice. *Journal of Clinical Biochemistry and Nutrition*, 50(3), 211–216.

Studer, A. M., Limbach, L. K., Van Duc, L., Krumeich, F., Athanassiou, E. K., Gerber, L. C., Moch, H., & Stark, W. J. (2010). Nanoparticle cytotoxicity depends on intracellular solubility: Comparison of stabilized copper metal and degradable copper oxide nanoparticles. *Toxicology Letters*, 197(3), 169–174.

Tamas, M. J., Sharma, S. K., Ibstedt, S., Jacobson, T., & Christen, P. (2014). Heavy metals and metalloids as a cause for protein misfolding and aggregation. *Biomolecules*, 4(1), 252–267.

Theron, J., Walker, J. A., & Cloete, T. E. (2008). Nanotechnology and water treatment: Applications and emerging opportunities. *Critical Reviews in Microbiology*, 34(1), 43–69.

Thirumoorthy, N., Kumar, K. M., Sundar, A. S., Panayappan, L., & Chatterjee, M. (2007). Metallothionein: an overview. *World Journal of Gastroenterology*, 13(7), 993–996.

Thorek, D. L. J., & Tsourkas, A. (2008). Size, charge and concentration dependent uptake of iron oxide particles by non-phagocytic cells. *Biomaterials*, 29(26), 3583–3590.

Torre, A., Trischitta, F., & Faggio, C. (2013). Effect of $CdCl_2$ on regulatory volume decrease (RVD) in *Mytilus galloprovincialis* digestive cells. *Toxicology in Vitro*, 27(4), 1260–1266.

Tuncsoy, M., & Erdem, C. (2014). Accumulation of copper, zinc and cadmium in liver, gill and muscle tissues of *Oreochromis niloticus* exposed to these metals separately and in mixture. *Fresenius Environmental Bulletin*, 23(5), 1143–1149.

Tunçsoy, M., & Erdem, C. (2018). Copper accumulation in tissues of *Oreochromis niloticus* exposed to copper oxide nanoparticles and copper sulphate with their effect on antioxidant enzyme activities in liver. *Water, Air, & Soil Pollution*, 229(8), 1–10.

Vajargah, M. F., Yalsuyi, A. M., Sattari, M., Prokić, M. D., & Faggio, C. (2020). Effects of copper oxide nanoparticles (CuO-NPs) on parturition time, survival rate and reproductive success of guppy fish, *Poecilia reticulata*. *Journal of Cluster Science*, 31(2), 499–506.

Valavanidis, A., Vlachogianni, T., & Fiotakis, C. (2009). 8-hydroxy-2′-deoxyguanosine (8-OHdG): A critical biomarker of oxidative stress and carcinogenesis. *Journal of Environmental Science and Health Part C*, 27(2), 120–139.

Van Der Zande, M., Vandebriel, R. J., Van Doren, E., Kramer, E., Herrera Rivera, Z., Serrano-Rojero, C. S., Gremmer, E. R., Mast, J., Peters, R. J., Hollman, P. C., Hendriksen, P. J., Marvin, H. J., Peijnenburg, A. A., & Bouwmeester, H. (2012). Distribution, elimination, and toxicity of silver nanoparticles and silver ions in rats after 28-day oral exposure. *ACS Nano*, 6(8), 7427–7442.

Wang, B., Yin, J.-J., Zhou, X., et al. (2012). Physicochemical origin for free radical generation of iron oxide nanoparticles in biomicroenvironment: Catalytic activities mediated by surface chemical states. *J Phys Chem C*;117:383e92.

Wang, B., Zhang, L., Bae, S. C., & Granick, S. (2008). Nanoparticle-induced surface reconstruction of phospholipid membranes. *Proceedings of the National Academy of Sciences, USA*, 105(47), 18171–18175.

Wang, C., Wang, X. M., Xu, B. Q., Zhao, J. C., Mai, B. X., Sheng, G. Y., Peng, P. A., & Fu, J. M. (2004). Enhanced photocatalytic performance of nanosized coupled ZnO/SnO$_2$ photocatalysts for methyl orange degradation. *Journal of Photochemistry and Photobiology A: Chemistry*, 168(1–2), 47–52.

Wang, J., Tian, S., Petros, R. A., Napier, M. E., & DeSimone, J. M. (2010). The complex role of multivalency in nanoparticles targeting the transferrin receptor for cancer therapies. *Journal of the American Chemical Society*, 132(32), 11306–11313.

Wang, T., Long, X., Cheng, Y., Liu, Z., & Yan, S. (2014). The potential toxicity of copper nanoparticles and copper sulphate on juvenile *Epinephelus coioides*. *Aquatic Toxicology*, 152, 96–104.

Wehmas, L. C., Anders, C., Chess J., Punnoose, A., Pereira, C. B., Greenwood, J. A. Tanguay, R. L. (2015). Comparative metal oxide nanoparticle toxicity using embryonic zebrafish. *Toxicology Reports*, 2, 702–715.

Xia, T., Kovochich, M., Liong, M., Madler, L., Gilbert, B., Shi, H., Yeh, J. I., Zink, J. I., & Nel, A. E. (2008). Comparison of the mechanism of toxicity of zinc oxide and cerium oxide nanoparticles based on dissolution and oxidative stress properties. *ACS Nano*, 2(10), 2121–2134.

Xiong, D., Fang, T., Yu, L., Sima, X., & Zhu, W. (2011). Effects of nano-scale TiO$_2$, ZnO and their bulk counterparts on zebrafish: Acute toxicity, oxidative stress and oxidative damage. *Science of the Total Environment*, 409(8), 1444–1452.

Xu, J., Zhang, Q., Li, X., Zhan, S., Wang, L., & Chen, D. (2017). The effects of copper oxide nanoparticles on dorsoventral patterning, convergent extension, and neural and cardiac development of zebrafish. *Aquatic Toxicology*, 188, 130–137.

Yu, L. P., Fang, T., Xiong, D.-W., Zhu, W.-T., & Sima, X. F. (2011). Comparative toxicity of nano-ZnO and bulk ZnO suspensions to zebrafish and the effects of sedimentation, ·OH production and particle dissolution in distilled water. *Journal of Environmental Monitoring*, 13, 1975–1982.

Zhao, J., Wang, Z., Liu, X., Xie, X., Zhang, K., & Xing, B. (2011). Distribution of CuO nanoparticles in juvenile carp (*Cyprinus carpio*) and their potential toxicity. *Journal of Hazardous Materials*, 197, 304–310.

Zhu, X., Zhu, L, Duan, Z., Qi, R., Li, Y., & Lang, Y. (2008). Comparative toxicity of several metal oxide nanoparticle aqueous suspensions to Zebrafish (*Danio rerio*) early developmental stage. *Journal of Environmental Science and Health, Part A*. Feb 15; 43(3), 278–284.

Zhu, X., Tian, S., & Cai, Z. (2012). Toxicity assessment of Iron Oxide nanoparticles in Zebrafish (*Danio rerio*) early life stages. *PLoS ONE*, 7(9), e46286.

Zhao, X., Wang, S., Wu, Y., You, H., & Lv, L. (2013). Acute ZnO nanoparticles exposure induces developmental toxicity, oxidative stress and DNA damage in embryo-larval zebrafish. *Aquatic Toxicology*, 136, 49–59.

11 The Future of Nanoscience: Where To, Where From?

Anum Haleem, Tean Zaheer, Rao Zahid Abbas, Sadia Muneer, Kaushik Pal, Afrah Nawaz and Alisha Tahir

CONTENTS

11.1 Introduction to Preparation of Nanomaterials...284
    11.1.1 Background of Nanomaterials...284
        11.1.1.1 Prehistory 284
        11.1.1.2 Stone Age 285
        11.1.1.3 Bronze Age...285
        11.1.1.4 Iron Age..285
        11.1.1.5 Modern Material Sciences 285
    11.1.2 Metals..285
    11.1.3 Polymers..285
    11.1.4 Composites 286
11.2 Methods to Synthesize Nanomaterials 286
    11.2.1 Bottom-Up Method 286
        11.2.1.1 Gas Phase Method..286
        11.2.1.2 Microwave Method...287
        11.2.1.3 Hydrothermal and Solvothermal Methods.....................287
        11.2.1.4 Spray Pyrolysis Method 288
        11.2.1.5 Sonochemical Method..288
        11.2.1.6 Sol-Gel Method 288
        11.2.1.7 Biosynthesis Method 289
    11.2.2 Top-Down Method 289
        11.2.2.1 Sputtering Method..290
        11.2.2.2 Thermal Decomposition Method 290
        11.2.2.3 Lithography Process...290
11.3 Principal Applications of Nanomaterials 290
    11.3.1 Application in Electronics..290
    11.3.2 Applications in Transportation and Automation 291
    11.3.3 Applications in Robotics 291
    11.3.4 Applications in Biological Science..292

DOI: 10.1201/9781003126256-11

        11.3.5   Applications in Environmental Science 292
        11.3.6   Applications as Day-to-Day Nanoproducts 294
        11.3.7   Applications in Aquaculture.. 294
                    11.3.7.1   Applications in Vaccines for Aquaculture 295
                    11.3.7.2   Applications in Water Treatment................................... 296
11.4   Conclusions.. 297
References.. 298

11.1   INTRODUCTION TO PREPARATION OF NANOMATERIALS

Many techniques have been established to manufacture and fabricate novel materials with better controlled size, shape, more functionalities and low cost. At nanoscale the nanoparticles show a characteristic chemical, biological and physical properties that are totally changed from their single atoms, molecules or bulk materials [1]. This is due to the large surface area-to-volume ratio, increased in mechanical strength, high stability or reactivity, spatial confinement etc., and because of these properties nanoparticles are used in different fields [2]. The synthesis and control of materials in the nanometer range can develop new properties in the material and device characteristics in extraordinary ways [3]. Nanocrystalline materials can also be treated as nanocomposites with the grain internal part as the matrix and the grain boundary as the reinforcement phase. Nanomaterials have unique characteristics, such as high chemical bioactivity and reactivity, tissue or organ penetration capability and good bioavailability. There is a variety of physical and chemical methods that are extensively used for the synthesis of nanomaterials with desired characteristics [4, 5, 6].

11.1.1   BACKGROUND OF NANOMATERIALS

Materials are so important to societal development that our civilization is often named after a particular material (e.g., Bronze Age and Iron Age). Essential materials at the foundation of our present civilization are far too divers to name just a few. Electronics, photonics and switchable sensors, nevertheless, capture the advancing technologies and industries that affect every aspect of our daily life such as, computing, analysis in healthcare and communication. Innovation in photonics, electronics and switchable sensors open a new era in catalyzing revolutionary changes and transforming the platform that technologies are built on [7]. The study of materials was aided by the convergence of chemistry and physics, and as a result, the interdisciplinary study of materials science emerged from the combination of these experiments. Following is a history of industrial-scale materials.

11.1.1.1   Prehistory

Many cultures leave behind their materials, which anthropologists use as a record of their existence. These materials help anthropologists to distinguish among people. The heating and molding of copper with hammer began about the year 5000 BCE. Melting and casting began about the year 4000 BCE. The first use of bronze alloy was around 3000 BCE [8].

11.1.1.2 Stone Age

The people of the Stone Age were limited to the rocks that were locally available and on the rocks that they could get by trading. The stones were mostly used for tools, jewelry and weapons. The first tool used in the Paleolithic Age is called the Oldowan. In the Neolithic Age, tools for farming were discovered [9].

11.1.1.3 Bronze Age

By chance, or through experiment, the addition of copper led to increased hardness in the form of a new metal alloy, called bronze [10]. Early civilizations used copper for different purposes.

11.1.1.4 Iron Age

In 1200 BCE, the use of iron was at its peak. People mostly used iron for manufacturing tools and weapons. Other relevant ages are antiquity, Middle Ages, early modern period and Silicon Age.

11.1.1.5 Modern Material Sciences

In the early 20th century, most universities had metallurgy and ceramics departments. Northwestern University founded the first materials science department in 1955. Ever since, this branch of science has seen revolutionary transformation, reshaping the outlook of research, academia and industry toward nanoscience [1]. In materials chemistry, the industrial-scale materials most often used are metals, polymers, ceramics and composites.

11.1.2 METALS

Metals are mostly solid at room temperature, except mercury, which is liquid at room temperature, and exist in the form of crystal lattice. They are held together by strong metallic bonds. Metals are grouped into ferrous and nonferrous materials. They can further be classified as plain and alloy metals. Plain metals are those materials that have a base material as the major constituent and also contain the traces of other constituents as impurities. Alloys are the materials that contain more than one base material and also the major constituent of other components, and their presence also affects the specific properties of metal [11]. Metals are mostly used in the form of wrought metals and castings.

11.1.3 POLYMERS

Polymers are materials that have a high number of repeating molecules called monomers. They form chains [12]. The first synthetic polymer was Bakelite, which was discovered by Leo Baekeland in 1907. Polymers are solid materials and belong in the category of organic compounds. The simplest organic molecules combine together to form a long chain hydrocarbon by a process called polymerization. A polymer must contain simple organic molecules. As a result of polymerization, the newly formed molecule may have different physical properties as compared to the derivative molecule. These molecules are formed by manipulating a set of conditions and have the ability to retain their shape. The types of polymers are thermoplastic, thermosets and elastomers [13].

11.1.4 COMPOSITES

Composites are materials that consist of more than two materials blended together by adhesive or mechanical bonding. They pursue different properties after blending. They are called engineering materials because they are designed for the production of desired quality and use. They are made with metals, plastics, ceramics and fibers. They mostly have fillers and a base or matrix. Fillers are stiff and possess high strength, whereas a base has low strength. Fillers are the load carriers and the base behave as a holder of that load [14]. Composites are used in the formation of aircrafts, bicycles, electrical products and industrial instruments. Composites provide a high strength-to-weight ratio, are more resistant to fatigue loss than steel or aluminum, provide high wear tolerance and can be formulated to have high hardness and damping. Based on the form of the filler, composites can be divided into three types: fiber-reinforced, particulates and laminates. Another method of categorizing matrix composites is to use the matrix material type: silicone, metal or ceramic matrix composites.

11.2 METHODS TO SYNTHESIZE NANOMATERIALS

In general, bottom-up and top-down are the two main methods used for the production of nanomaterials. With bottom-up, the material synthesis starts from the atomic level. With top-down, the size reduction is from bulk assemblies down to nano levels.

11.2.1 BOTTOM-UP METHOD

The bottom-up method is also called the building-up approach because nanomaterial is formed from relatively smaller substances, such as atom by atom, molecule by molecule or cluster by cluster. Nanomaterials can be synthesized by reduction of materials up to the atomic level, enhanced with the self-assembly procedure. During self-assembly, the physical forces applied at the nano level are used to combine units into a larger stable structure [15]. Sol-gel, spinning, chemical vapor deposition (CVD), physical vapor deposition (PVD), pyrolysis, sonochemical, microwave and biosynthesis are the most commonly used bottom-up methods for nanoparticle production. The bottom-up approach plays an important role in the fabrication and processing of nanostructures and nanomaterials.

11.2.1.1 Gas Phase Method

Gas phase methods are widely used for the synthesis of thin films. The gas phase technique is divided into two parts: chemical vapor deposition and physical vapor deposition [16]. CVD is a substrate-based method in which gaseous molecules transform into solid material in the form of nanomaterials on the substrate surface. In this method one or more of the precursor's compounds are placed in a vapor phase to the reaction chamber. These precursor molecules decompose on a heated substrate after mixing in the reaction chamber. The substrate is employed in the CVD chamber at 1200°C and 10^{-3} Torr vacuum pressure, where the precursors are diffused into the substrate at growth temperature followed by the rapid quenching of the substrate. CVD can coat large parts in a short period of time. PVD is another thin film deposition method in which films are formed from the gas phase but deprived of a chemical changing from precursor to product [17].

Gas phase is a versatile nanofabrication method for fabrication of various nano-materials, including nanocomposites, and controlled instantaneous deposition of numerous materials, including metal, ceramics, semiconductors, insulators and polymers. The advantages of gas phase methods are that they produce highly pure, uniform, hard and strong nanoparticles. The drawbacks of these methods are they require special apparatus and produce toxic byproducts.

11.2.1.2 Microwave Method

The microwave method is a very neat and simple method in which electromagnetic radiations of low energy and high frequency (900–2450 MHz) are used to synthe-size nanostructure materials. Microwave radiation produces heat by interacting with the solvent molecule, and the reorientation of their electric dipoles creates friction between them. The microwave method has been used for the production of various materials, such as carbides, complex oxides, nitrides, silicides, zeolites and apatites [18]. The microwave method has successfully produced CuO nanoparticles with con-trolled size distributions [19]. Synthesis of silver nanoparticles by the microwave method has been reported [20]. Copper, silver and bimetallic Cu/Ag has been syn-thesized by this method [21]. When the microwave handling cycle is increased, the particle size decreases. This preparation method is efficient and reliable. It requires a low synthesis temperature, reduced reaction time, low power depletion, and gener-ates small yet uniformly sized particles.

11.2.1.3 Hydrothermal and Solvothermal Methods

The hydrothermal method is a heterogeneous chemical synthesis in which the chem-ical reaction is carried out in a closed sealed vessel called an autoclave under highly optimum conditions (temperature higher than 100°C and pressure above 1 atm) in the presence of water as a solvent. The autoclave is made of steel and has Teflon lin-ing. In this process, a suitable solution is placed in an autoclave and it is sealed. The container is then placed in an oven at a high temperature and pressure. The sample obtained is then filtered, washed and annealed at a suitable temperature to make the nanostructure into a fine-powdered form. For example, titanium dioxide synthesized by the hydrothermal method has a diverse crystal structure and numerous morpholo-gies, depending on the hydrothermal condition, such as pH, time, temperature and concentration, in the reaction intermediate [22].

The hydrothermal method is widely used for the synthesis of smaller particles in ceramic industries for better quality and growing larger crystal. The synthesis of the nanoparticle depends on the solubility of precursor materials in hot water at high pressure. This method is also used for the synthesis of nanotubes, nanowires and nanorods. This process is overall best reported. The hydrothermal route has been employed to synthesize CuO and SnO_2 NPs [23, 24].

The solvothermal method follows similar route to the hydrothermal method. The main difference between the hydrothermal and solvothermal methods is the solvent that is used. The solvothermal method is used mainly for the synthesis of NPs. The drawback of this method is that it needs expensive autoclaves, high temperature and the impossibility of observing the crystal during its growth.

11.2.1.4 Spray Pyrolysis Method

Spray pyrolysis is a solution-based method in which synthesis of nanoparticles is carried out by a spinning disc reactor (SDR). An SDR is comprised of a spraying unit, a liquid feeding unit and rotating disc inside a reaction chamber in which temperature or other physical parameters can be controlled. The vessel is filled with inert gases to remove unwanted gas or chemical reactions [25]. The disc revolves at different speeds to spray precursor molecules on the heated substrate surface. The spinning causes the atoms or molecules to fuse into each other, which are precipitated, collected and dried to form a chemical compound [26]. This technique depends on several parameters, such as the liquid flow rate, disc rotation speed, liquid–precursor ratio, location of feed and disc surface. These operating parameters determine the nanoparticle's features [27]. The benefits of this technique are that it is simple, has a low cost and does not require high-quality substrate or chemicals. ZnO nanoparticles have been synthesized by the spray pyrolysis method [28].

11.2.1.5 Sonochemical Method

Sonochemical reduction is the most popular physical method used for the synthesis of nanoparticles. A chemical reaction occurs by using ultrasonic waves. Acoustic cavitation is the main phenomenon that produces a sonochemical effect in solutions. Cavitation is basically the production, growth and then collapse of bubbles in liquid, and produces high pressure, intense heating and liquid jets steams. In this method, an ultrasonic bath is used where a solution of metal oxide and solvent is stirred for two hours under normal temperature. The resulting solution is then shifted to an ultrasonic bath where this solution is irradiated with ultrasonic waves for a specific temperature and time. The obtained solution is then centrifuged and dried at a specific temperature to obtain the desired material. Although this method is used for obtaining a narrow size and pure samples, the availability of a large-scale ultrasonic reactor is less reliable. Zinc sulfide nanoparticles have been synthesized by the sonochemical method with small size distribution [29, 30].

11.2.1.6 Sol-Gel Method

The sol-gel method is a method in which the colloidal liquid (sol) is converted into a semisolid (gel) form. Sol-gel synthesis is used to synthesize material with a variety of shapes, such as porous structures, thin fibers, dense powders and thin films. Sol is a colloidal (the dispersed phase is so small that only Van der Waals forces and surface charges are present) or molecular suspension of solid particles of ions in a solvent A gel. A semirigid mass formed as the solvent from the sol evaporates and the particles or ions left behind continue to link together in a continuous 3D network. The chemistry of this method is based on hydrolysis and polycondensation [31]. In this process, particles from a colloidal suspension are settled into a preexisting surface resulting in ceramic material. The fabrication of materials in the sol-gel method starts with a solution containing metal compounds, such as metal chlorides, and a metal alkoxide as a precursor source. Water is used as a hydrolysis agent, alcohol is used as a solvent, and an acid or base is used as a catalyst. Hydrolysis and polycondensation of precursors occur near room temperature giving rise to sol, in which polymers or fine particles are dispersed [32, 33].

The sol particles interact with each other through hydrogen bonding or Van der Waals forces. This procedure is used for the construction of an integrated network that is gel from the colloidal particle or chemical solution (sol). Further reaction bonds the ions together, solidifying a sol into a wet gel that also retains water and solvents. Usually, different shapes are formed when sol is slowly transformed toward the liquid-like phase gel. After the drying, process vaporization of water and solvents occurs and gel is formed. To increase mechanical properties and for polycondensation, the gel is further treated thermally through calcination. SnO_2 nanoparticles and silica have been prepared by the sol-gel method [34]. This technique has several advantages, such as high purity, good homogeneity, ability to control the particle or grain size, and proficiency in monitoring the textural and surface properties at low temperature of nanomaterials.

11.2.1.7   Biosynthesis Method

Biosynthesis is a green and eco-friendly method for the production of nanoparticles that are nontoxic and biodegradable [35]. Different natural sources, such as leaves, roots, peel or former parts of the plant, are reprocessed as a reducing or a capping agent in the synthesis of nanomaterials. The biosynthesized nanoparticles have unique and superior properties that find their way in biomedical applications. Nanoparticles synthesized by this method are more stable and the rate of synthesis is faster than in any other case. The biosynthesis method has many benefits, such as its simplicity, inexpensive, high stability of sample, less time expenditure, harmless byproducts and large-scale production. Synthesis of SnO_2 nanoparticle by *Persia americana* seed [36], Ag nanoparticle by *Mentha longifolia* [37] and ZnO by *Cymbopogon citratus* leaf [38] has been reported.

11.2.2   Top-Down Method

The top-down method is also called the destructive approach and included in the typical solid-state treating of materials. This method begins with the bulk material, then makes it smaller. Physical procedures like crushing, milling or grinding are used for breaking up larger particles. Top-down method includes thermal decomposition, sputtering and lithography. The drawback of the top-down approach is the defect creation, crystallographic damage and imperfection of the surface structure. Such imperfection would have a significant effect on the physical and surface properties of nanostructures and nanomaterials [39, 40].

11.2.2.1 Sputtering Method

The sputtering process involves the deposition of nanoparticles on a surface by ejecting atoms or clusters from it by colliding with ions. Sputtering is usually a deposition of a thin layer of nanoparticles followed by annealing. Sputtering is usually characterized by the deposition of a thin film of nanoparticles accompanied by annealing. A highly focused beam of inert gas, such as helium or argon, is used for ejection. The shape and size of the nanoparticles depends on the thickness of the layer, temperature, duration of annealing and substrate type. Currently, for the synthesis of nanomaterials, different types of sputtering methods, such as ion-assisted deposition, ion beam, reactive, high-target utilization, high-power impulse magnetron and gas flow, are used [41, 42].

11.2.2.2 Thermal Decomposition Method

Thermal decomposition is an endothermic chemical decomposition method in which chemical bonds in the compound are broken by heat. The decomposition temperature is a specific temperature at which an element chemically decomposes and is responsible for controlling the quality of the nanomaterials. The nanoparticles are manufactured by decomposing the metal at specific temperatures undergoing a chemical reaction producing secondary products [43].

11.2.2.3 Lithography Process

Nanolithography is the study of fabricating nanometric scale structures with a minimum of one dimension in the size range of 1 to 100 nm. There are numerous nanolithographic methods used for the synthesis of nanoparticles, including optical, electron-beam, multiphoton, nanoimprint and scanning probe lithography. Normally, lithography is the procedure of printing a desired shape or structure on a light-sensitive material that selectively eliminates a portion of material to create the required shape and structure. Colloidal lithography is also used for synthesis of various nanostructure materials based on processes in which colloidal crystals are used as masks for etching and deposition of various nanostructures on planar and nonplanar substrates with low-cost and high area fabrication of materials. The main benefits of using this method are the production from a single nanoparticle to a cluster with preferred shape and size. The drawbacks are that it requires complex apparatus and needs to be cost effective [44, 45].

11.3 PRINCIPAL APPLICATIONS OF NANOMATERIALS

Materials that are the billionth part of the meter and like a grain in size are called nanomaterial or nanocrystalline materials [46]. Because of advantageous chemical, physical and mechanical properties, nanomaterials have vast applications. Some of them are described in the following sections.

11.3.1 Application in Electronics

In reference to electronics, nanotechnology has given the focus on miniaturization, which involves in the reduction of sizes of resistors, capacitors and transistors. This reduction in size leads to the development of fast microprocessors and this speed enhances the speed of computation. Transistors have become smaller in size with the help of nanotechnology. At the start of the 21st century, the size of transistors was 130 to 250 nm, which decreased to 14 nm in 2014. By 2015 the first 7 nm transistors were created by IBM. In 2016, Lawrence Berkeley National Lab successfully manufactured a 1 nm transistor. Because of this development, we can assume that computer memory will soon fit on a single chip [47].

With the use of magnetic random access memory (MRAM) computers are able to boot automatically. MRAM is generated by using a nanometer-scale magnetic tunnel junction and it enhances resume play features. A nanometer-scale magnetic tunnel junction is also very helpful in the advancement of memory chips for thumb drives and android phones, ultraresponsive hearing devices, production of the antimicrobial coating for keyboards and production of variable displays for e-book users [48].

11.3.2 APPLICATIONS IN TRANSPORTATION AND AUTOMATION

Nanotechnology plays a significant role in the transportation infrastructure. Development in the production of efficient, fast and smart automobiles, spacecraft and aircraft is a contribution of nanoparticles and nanocomposites [49]. A significant development has been the reduction in weight of jets. With the use of nanocomposites and nanoparticles, the size of jets has reduced almost 20% since their original weight. By this reduction of weight, the consumption of the fuel has also reduced 15%. A report of a trial performed by the National Aeronautical and Space Administration (NASA) showed that with the use of nanotechnology and advance nanocomposites in the manufacturing of spacecraft doubled the strength of jets and the net weight of a launch vehicles reduced 63%. These improvements saves the energy that is used for launching spacecraft into orbit, leading to the reduction of costs, increased reliability, time savings and further innovative ideas in the field of automation and material sciences [50].

Nanoengineering of different metals and elements has led to improvements in strength, moldable, ductile, resistance and durability of materials, and as a result the costs of automation and manufacturing have been reduced. This improvement is a reason of the increasing demand of high-power batteries, temperature-controlling materials, and highly efficient sensors and vehicles [51]. However, industry support for use of nanomaterials in automobiles is lacking. This arena of nanotechnology applications therefore remains underexplored.

11.3.3 APPLICATIONS IN ROBOTICS

Nanorobotics is the branch of nanotechnology that deals with the creation of machines and robot under the scale of nanometer (10^{-9} m). Nanomaterials are specially used in the creation of nanobots or tiny robots [52]. Nanobots are used in the fields of medicine and research, biohazard defense and hematology. Recently, nanobots made it possible to separate and isolate internal cell organelles and by combining these organelles with nanoparticles, scientists are able to study intracellular reactions like cell membrane disruption and endocytosis [53].

Nanobots are also used as a delivery tool for oxygen with the help of respirocytes. The size of this nanodevice is tantamount to 1 micron. These nanobots are designed to carry 236 times more oxygen than erythrocytes in the blood. This technology provides the solution for blood transfusions and made the process more efficient and easier [54].

The pharmaceutical field got ample benefits after the invention of nanobots. With the discovery of nanobots, scientists are able to diagnose cancer cells and any plaque formation inside arteries, which leads to atherosclerosis (heart disease). Nanobots can be inside the human body and move inside the human circulatory system. They can detect minute changes inside the human body and make perfectly correct diagnoses [55]. Scientist can use these nanobots in the cure of diseases. For example, in 2017, nanobots were able to diagnose cancer cells and drill inside them. This leads to the destruction of the cancer tumor and death of these harmful cells. This method is useful in the removal of cancer cells without operation and long-term chemotherapy treatment. Nanobots are also serving as an antigen for the human immune system and also as a delivery tool for stimulation in the body [56]. Nanomaterials have also been deployed in the tools of artificial intelligence aimed for better healthcare monitoring and applications.

11.3.4 APPLICATIONS IN BIOLOGICAL SCIENCE

Nanotechnology and biotechnology are highly emerging fields in science in 21st century. By combining these two fields we get nanobiotechnology and this field is helpful in providing the solution of many problems regarding health and life, and from a business point of view. They both serve humanity in many ways and provide improvements in the fields of gene and drug delivery, florescent biological labels, probing of genetic material, tissue culturing, detection of pathogens, hyperthermia isolation of biological molecules, formation of protein chips, dentistry, neurosurgery, resolving cardiac diseases, medical lab technology, MRI enhancement, phagokinetic studies, commercial exploration and so on [57].

Nanotechnology and biotechnology are in their early stages, but with their rate of progress, they will soon become pivotal for human life and survival [58]. We could cure many diseases that weren't possible until this nanotech revolution. It seems that diseases that are not curable today could in the future be treated with advancements and research in nanotechnology and biotechnology (Table 11.1).

11.3.5 APPLICATIONS IN ENVIRONMENTAL SCIENCE

The quality of the ecosphere is an intricate combination of determinants and is crucial to the survival and replication of all living species [69]. Nanotechnology has devised smart solutions to many environmental issues. Carbon emissions can be prevented by deploying nanotech-based smart solutions. Similarly, nanobased smart grids are being foreseen as clean and sustainable energy solutions [70]. Nanoparticles have been shown to remediate and control air contaminants, improving air quality [71]. Efficient methods for water supply expansion and cost-effective quality enhancement

TABLE 11.1

Overview of Nanoparticles in Medical Science

Core Element	Size of Nanoparticle	Target/Organism	Application	Domain of Application	Reference
Al (aluminum)	60 um (length) 5–500 nm (radius)	Biomolecules	Cell separation and probing	Separation and diagnosis	[59]
C (carbon)	90 nm	Tumors (mice model)	Detection and destruction of cancer cells	Cancer therapy	[60]
Magnetic	11.9 nm (diameter)	Tumors (mice model)	Primary and metastatic cancer	Cancer immunotherapy	[61]
C (Carbon)	Nanotubes	Clot or plague in arteries	Removing atherosclerosis and cardiovascular plaques	Surgery	[62]
Au (Aurum)	Gold nano-spike biosensor	SARS-CoV-2 antibodies	Identifying DNA sequence (viral detection)	Detection and diagnosis	[63]
P (phosphorus)	3.1 ± 1.8 nm (average lateral size)	Cancer	Photothermal treatment	Treatment	[64]
Si-NW (silicon nanowires)	Si-NW = 20–100 nm (diameter) Decorated with Ag and Cu	Bacteria (*E. coli*)	Antibacterial	Antimicrobial therapy	[65]
Fe (iron oxide)	3 nm diameter	Toxicity analysis	MRI	Diagnostic tool and safety evaluation	[66]
Ni (nickel–nanolipoprotein)	23.2 nm (diameter)	West Nile virus	Increased potency and efficacy	Vaccine potentiation	[67]
Zr-MOF (zirconium–metal organic framework)	120 nm	Hyperphosphatemia	Phosphate sensing and binding	Sensing and decrement of phosphates	[68]

can be achieved by nanotechnology [72]. Based on the potential of nanomaterials in air, soil and water quality enhancement, it could be forecasted that the nanorevolution will make its way to sustainable remediation and environment protection.

11.3.6 APPLICATIONS AS DAY-TO-DAY NANOPRODUCTS

Nanotechnology has provided solutions to large-scale problems. From heavy industries to the submolecular engineering, nanotechnology has efficiently provided sustainable options. In the recent COVID-19 global pandemic, nanotechnology has shown huge success in efficiently containing, detecting, preventing and treating the virus [73]. Hydroxyapatite-based products have been extensively utilized for oral health in humans. One of the commercial nanobased hydroxyapatite products was found to be safe for human use, without altering normal microflora [74]. Nanomaterials are being used in cosmetics/makeup, antiaging products and hair care and skin care products, and silver nanomaterials are the most widely utilized [75]. Another contemporary approach of higher consumer appeal is the use of products based on green nanoparticles [76]. Fortified fruit juices, multivitamin formulas/functional drinks, mineral water and kitchenware is being marketed by various industries worldwide [77, 78].

Personal protective equipment (PPE) has gained considerable value amidst the current pandemic. The concern of developing PPE that caters to the need of protection against smaller microorganisms has been addressed largely by the combination use of nanotechnology. CuO-nanoparticle-based masks have been shown to not only entrap but also kill viruses up to 99.9% [79]. There's also an extensive list of nanobased commercial product use in almost all areas. This spread of nanomaterial applications demonstrates the nature of multidisciplinary research and collaboration among research and development groups worldwide.

11.3.7 APPLICATIONS IN AQUACULTURE

In recent years, aquaculture has become one of the fastest growing food industries by supplying high nutritional food, incomes and employment throughout the globe. However, the aquaculture industry also faces some problems with optimizing production rates some of which were found to be nutrition, disease and water pollution, which impact the cultured organisms and environment as well. So, aquaculture is using some innovative technologies like metallic nanoparticles [80].

In order to increase aquaculture production rates, manufacturers are trying to use nanoparticles to remove obstacles relevant to growth, waterborne food, culturing of species, their health, reproduction and water treatment [81]. It was observed that nanomaterials, especially gold, silver and silver oxide, serve as novel antimicrobial agents against fungal, bacterial and parasitic fish diseases [82].

Nanotechnology has a wide range of application in aquaculture, including obtaining taste and color; prevention of decomposition; enhancing bioavailability of functional compounds; controlled release of food materials; protection of food products against microbial contamination; enhancing shelf-life and stability of sensitive ingredients; and carrier vehicles of nutraceuticals, nutrients, food additives, enzymes and food antimicrobials. This technology is used to improve chemical, physical and nutritional quality of fish feed and their respective ingredients for aquaculture species, antifouling in aquaculture and fishing nets, new packaging material for seafood products and antibacterial substances for aquaculture tanks. It was reported by the Russian Academy of Sciences that sturgeon and young carp showed faster growth rates (24% and 30%, respectively) when they were fed nanomaterials of iron [83].

11.3.7.1   Applications in Vaccines for Aquaculture

Vaccines have been used for many years as a defensive mechanism to cope with infections of aquaculture. The most effective and reliable methods of vaccines administration is either by injection or oral administration. The former, a traditional adjuvant system, has many adverse side effects, as vaccine preparation needs many water or oil formulations. Such formulations along with administration generally leads to high mortality rate of fish [84]. To cope with such issues, scientists has recently introduced a relatively safer and effective nanodelivery system as an alternative method for delivery of vaccines in fish. Up until now, different encapsulation techniques have been introduced and tried. Alginate particles were reported as one of the preliminary candidates for delivery of vaccines in aquatic animals [85]. Alginate is a copolymer α-L-guluronic acid (G) and β-D-mannuronic acid (M) that is present as a polysaccharide in some bacteria and in different species of brown algae. It has been known for its physical and mechanical stability, as well as muco-adhesive characteristics, allowing easy oral administration due its contact with walls of epithelial cells [86, 87]. Alginate particles are generally produced by emulsification, which is one of the fastest ways for nanoparticle production [88]. The spray method and orifice-ionic gelation methods are generally used to a lesser extent [89]. It was reported from different researchers that alginate is used as a survival and weight enhancer of fish [90, 91] as well as an antigen adjuvant [92, 93]. Moreover, alginate administration has also shown enhanced immune stimulant response of turbot (*Scophthalmus maximus* L.) against *Vibrio anguillarum* [94] as well as enhanced defense of the brown-marbled grouper (*Epinephelus fuscoguttatus*) and carp (*Cyprinus carpio* L.) [95, 96, 97]. Chitosan (CS), generally obtained from the exoskeleton of insects and crustaceans, is regarded as the second most abundant natural biopolymer. Due to its unique biological properties, like biodegradable, nontoxic, biocompatible and bioadhesive, chitosan-based formulations are generally used as bionanosensors, edible coatings and drug carrier vehicles, as well as in different medical disciplines such as surgical procedures and dentistry [98, 99]. Meshikini and colleagues highlighted beneficial effects of chitosan in fisheries by describing that a chitosan-supplemented diet increased resistance of rainbow trout against environmental pressure and immunological variables, therefore resulting in decreased counts of eosinophils and neutrophils and increased count of lymphocytes [100]. On the other hand, chitosan has also been used as a vehicle for different kinds of vaccines and DNA in fish using different routes of administration like injection and oral routes. For example, major capsid protein (MCP) of lymphocystis disease virus (LCDV) in Japanese flounder

(*Paralichythys olivaceus*) [101], *Vibrio anguillarum* in Asian sea bass (*Lates cal-carifer*) [102], dietary RNA in *Labeo rohita* [103], *Vibrio parahaemolyticus* in *Acanthopagrus schlegelii* [104] and *Philasterides dicentrarchi* in *Scophthalmus maximus* [105] has been successfully delivered and encapsulated in chitosan-based systems. Similarly another biodegradable polymer known as poly D, L-lactic-co-glycolic acid (PLGA), produced from two different monomers (lactic and glycolic acid), has been broadly used for delivery and encapsulation of different compounds in fish. In a study, PLGA was used for delivery of *Aeromonas hydrophilus* in rohu, and showed significant antibody response and immune-stimulatory response in this fish as compared to the control group [106]. Another study of Japanese flounder showed similar results, where PLGA-encapsulated DNA vaccine showed enhanced inducing effects on immunological parameters against lymphocytes [107]. Another formulation of "liposomes" have also been extensively used in various research fields in fish farming. Liposomes-loaded *Aeromonas salmonicida* antigen showed skin ulcers and improved survival rate (83%) in carp (*Cyprinus carpio*) as compared to the control group [108]. Moreover, liposomes-encapsulated *Aeromonas hydrophilus* showed significant enhanced antibody counts in serum and boosted the immune system of common carp (*Cyprinus carpio*) as demonstrated by improved protection against live *Aeromonas hydrophilla* [109].

11.3.7.2 Applications in Water Treatment

Water treatment is one of the most important pillars required for sustainable aquaculture. In recent times, water pollution is regarded as one of the most important health hazards throughout the world, which is continuously increasing due to disposal of waste materials from agriculture, industries and cities, as well as the misuse of antibiotics and other synthetic compounds in fisheries. Water pollution not only affects human health directly but also indirectly affects aquatic animals leading to different foodborne diseases in humans upon consumption. In addition to this, the fish industry also faces some economic losses due to microorganisms and heavy metals in waterbodies affecting growth and leading to death of fish. In aquaculture, nanoparticles have crucial applications in deterioration of water pollution and providing safe and favorable habitat for fish breeding. In this viewpoint, scientists endorse photo catalysis and adsorption as the most affordable and efficient approaches for water treatment.

In a study, magnetic konjac glucomannan (KGM) aerogels were used to sanitize water from arsenite using pH-dependent characteristics [110]. However, in the last few years, graphene nanosheets (GNs) and graphene oxide (GO) have garnered tremendous attention worldwide for their significant role in decontamination of various water pollutants [111, 112, 113]. The coating of hybrid GO-TiO_2 for various energy and environmental applications like adsorption, organic dyes and evacuating heavy metal ions from wastewater has been particularly studied [114, 115]. Being biologically and chemically stable, nontoxic, cost-effective and an efficient photocatalyst makes TiO_2 a potential wastewater treatment tool. Many studies dealing with photocatalytic activities of TiO_2 for inactivation of pathogenic microorganisms like bacteria, viruses and fungi have been reported [116, 117, 118]. Scientists have been concerned about the issue of removal of contaminants from water. Water treatment is mandatory, because high concentrations of contaminants can lead to life-threatening effects on human health due to accumulation of these contaminants in tissues of aquatic animals, especially in fish, which come at the end of aquatic food chain, and consumption of fish is highly recommended to treat cancer and cardiovascular diseases (CVDs) [119]. Due to residing and feeding in an aquatic environment, fish are generally most exposed and susceptible to the detrimental influence of these pollutants [120, 121]. There are some studies showing accumulation of higher levels of some heavy metals like Cd, Pb and Hg in the tissues of aquatic animals due to anthropogenic activities or natural processes (e.g., volcanic activities) [122]. Furthermore, toxicity F^{-1} was noticed to be responsible for habitat destruction of the freshwater snail *Physella acuta* [123], and the malfunctioning of gastric function, immune system and enzyme actions of fish under study [124]. Wu et al. removed F^- and Hg from aqueous solution using three-dimensional reduced grapheme oxide (3D RGO) hydrogel prepared by the hydrothermal process. The results of this study showed the significant role of hydrogel for F^{-1} and Hg^{+2} that reached to 31.3 and 185 mg g^{-1}, respectively. This system was regarded as one of the most suitable environmental pollution management system [125]. Nanocrystalline cellulose (NCC) was synthesized by Azdabakht and his colleagues for removal of nitrate from aqueous solution. They concluded that bagasse-based NCC could be an efficient method for removal of nitrate from both wastewater and water reservoirs, obtaining a peak level of nitrate removal up to 25% at pH 6 [126]. In another study, magnetic iron-aluminum oxide/grapheme oxide (IAO/GO) nanoparticle-based selective adsorbent was used for water treatment from F^-. The absorbent was particularly characterized to have high stability in an acid-base environment, enhanced selective adsorption capability for F^{-1} and superparamagnetism features. Therefore, IAO/GO-based adsorbents were suggested as a promising applicant for removal of F^{-1} from natural water reservoirs [112]. Another investigation used TiO_2-SiO_2 and TiO_2 nanocomposite for F^{-1} removal from aqueous solution. The adsorbents so coated showed F^{-1} adsorption reaching a level of 94.3 mg g^{-1} with TiO_2 [127]. Liu et al. reported nanonet treatment as one of the best methods after testing a number of nanosystems, with the ability to improve the survival rate of fish up to 100%. In addition to improved pH and water quality, they also noticed a significant decrease in water nitrite and nitrate levels [128].

11.4   CONCLUSIONS

Nanomaterials have extensive applications in various fields of science and technology. A synthesis approach for nanomaterials can lead to optimized production of materials. Nanoparticles have promising applications in disease prevention, treatment and control, the most recent examples being anti-COVID-19 PPE, diagnostic assays and vaccines employing different nanomaterials. The increasing number of commercialized nanobased products affirms the support of industry toward nanoscience. However, studies on nanotoxicology for nontarget organisms and systems is still warranted. Industry readiness to support the development of nanobased products could be enhanced by evaluating the potential risks associated with their use. Analysis of safety concerns of nanoparticles from the One Health perspective can further encourage research and progress in nanoscience.

REFERENCES

1. J. Jeevanandam, K. Pal, M.K. Danquah, Virus-like nanoparticles as a novel delivery tool in gene therapy. *Biochime* 157 (2019) 38–47.
2. J.N. Tiwari, R.N. Tiwari, K.S. Kim, Progress in materials science three-dimensional nanostructured materials for advanced electrochemical energy devices. *Progress in Material Science* 57 (2012) 724–803. https://doi.org/10.1016/j.pmatsci.2011.08.00357
3. K. Pal, S. Sajjadifar, M. Abd Elkodous, Y.A. Alli, F. Gomes, J. Jeevanandam, S. Thomas, A. Sigov, Soft, self-assembly liquid crystalline nanocomposite for superior switching. *Electronic Materials Letters* 15(1) 84–101 (2019). https://doi.org/10.1007/s13391-018-0098-y
4. L. Tian, L. Li, A review on the strengthening of nanostructured materials. *International Journal of Current Engineering and Technology* 8 (2018) 236–249. https://doi.org/10.14741/ijcet/v.8.2.7
5. I. Ijaz, E. Gilani, A. Nazir, A. Bukhari, Detail review on chemical, physical and green synthesis, classification, characterizations and applications of nanoparticles. *Green Chemistry Letters and Reviews* 13 (2020) 223–245. https://doi.org/10.1080/17518253.2020.1802517
6. J. Jeevanandam, A. Barhoum, Y.S. Chan, A. Dufresne, M.K. Danquah, Review on nanoparticles and nanostructured materials : History, sources, toxicity and regulations. *Beilstein Journal of Nanotechnology* 9 (2018) 1050–1074. https://doi.org/10.3762/bjnano.9.98
7. S. Gazibegovic, D. Car, H. Zhang, S.C. Balk, J.A. Logan, M.W.A. de Moor, M.C. Cassidy, R. Schmits, D. Xu, G.Z. Wang, P. Krogstrup, R.L.M.O.H. Veld, K. Zuo, Y. Vos, J. Shen, D. Bouman, B.S. Hojaei, D. Pennachio, J.S. Lee, P.J. van Veldhoven, S. Koelling, M.A. Verheijen, L.P. Kouwenhoven, C.J. Palmstrom, E.P.A.M. Bakkers, Epitaxy of advanced nanowire quantum devices. *Nature* 548(7668) 434–438 (2017). https://doi.org/10.1038/nature23468
8. R.E. Hummel, *Understanding Materials Science History, Properties, Applications* (2nd ed.). 2004. New York, NY: Springer-Verlag New York, LLC. ISBN 978-0-387-26691-6.
9. D. Schaming, H. Remita, Nanotechnology: From the ancient time to nowadays. *Foundations of Chemistry* 17(3) 187–205 (2015). https://doi.org/10.1007/s10698-015-9235-y
10. G. Artioli, I. Angelini, A. Polla, Crystals and phase transitions in protohistoric glass materials. *Phase Transitions* 81(2–3) 233–252 (2008).
11. M.M. Dewidar, H.C. Yoon, J.K. Lim, Mechanical properties of metals for biomedical applications using powder metallurgy process: A review. *Metals and Materials International* 12(3) (2006) 193. https://doi.org/10.1007/BF03027531
12. A.K. Tiwari, S. Nasreen, M. Shahbaz, S. Hammoudeh, Time-frequency causality and connectedness between international prices of energy, food, industry, agriculture and metals. *Energy Economics* 85 (2020) 104529. https://doi.org/10.1016/j.eneco.2019.104529
13. G. Yongqiang, K. Ruan, X. Shi, X. Yang, J. Gu, Factors affecting thermal conductivities of the polymers and polymer composites: A review. *Composites Science and Technology* (2020) 108134. https://doi.org/10.1016/j.compscitech.2020.108134

14. T.E. Graedel, L. Erdmann, Will metal scarcity impede routine industrial use? *MRS Bulletin* 37(4) 325–331 (2012). https://doi.org/10.1557/mrs.2012.34

15. H. Essabir, S.N. Oui, M. Bensalah, R. Bouhfid, A.K. Qaiss, 6 - Shape memory based on composites and nanocomposites materials: From synthesis to application. In *Polymer Nanocomposite-Based Smart Materials*, Woodhead Publishing Series in Composites Science and Engineering, 2020, 103–120.

16. H. Hahn, Gas phase synthesis of nanocrystalline materials. *Nanostructured Materials* 9 (1997) 3–12. https://doi.org/10.1016/S0965-9773(97)00013-5

17. J.M. Lackner, W. Waldhauser, Inorganic PVD and CVD coatings in medicine—A review of protein and cell adhesion on coated surfaces. *Journal of Adhesion Science and Technology* 24(5) 925–961 (2010). https://doi.org/10.1163/016942409X12598231568023

18. H. Zhu, X. Wang, Y. Li, Z. Wang, F. Yang, X. Yang, Microwave synthesis of fluorescent carbon nanoparticles with electrochemiluminescence properties. *Chemical Communications* 34 5118–5120 (2009). https://doi.org/10.1039/B907612C

19. I. Khan, K. Saeed, I. Khan, Nanoparticles: Properties, applications and toxicities. *Arabian Journal of Chemistry* 12 (7) 908–931 (2017) https://doi.org/10.1016/j.arabjc.2017.05.011

20. F.K. Liu, P.W. Huang, T.C. Chu, F.H. Ko, Gold seed-assisted synthesis of silver nanomaterials under microwave heating. *Materials Letters* 59(8–9) 940–944. https://doi.org/10.1016/j.matlet.2004.10.070

21. S. Dagher, Y. Haik, A.I. Ayesh, N. Tit, Synthesis and optical properties of colloidal CuO nanoparticles. *Journal of Luminescence* 151 (2014) 149–154. https://doi.org/10.1016/j.jlumin.2014.02.015

22. B. Hu, S. Wang, K. Wang, M. Zhang, S. Yu, Microwave-assisted rapid facile "Green" synthesis of uniform silver nanoparticles: Self-assembly into multilayered films and their optical properties. *The Journal of Physical Chemistry C* 112 (2008) 11169–11174. https://doi.org/10.1021/jp801267j

23. M. Valodkar, S. Modi, A. Pal, S. Thakore, Synthesis and anti-bacterial activity of Cu, Ag and Cu–Ag alloy nanoparticles: A green approach. *Materials Research Bulletin* 46 384–389 (2011). https://doi.org/10.1016/j.materresbull.2010.12.001

24. H. Cheng, J. Ma, Z. Zhao, L. Qi, Hydrothermal preparation of uniform nanosize rutile and anatase particles. *Chemistry of Materials* 7(4) 663–671 (1995). https://doi.org/10.1021/cm00052a010

25. P. Majerič, R. Rudolf, Advances in ultrasonic spray pyrolysis processing of noble metal nanoparticles-review. *Materials (Basel, Switzerland)* 13(16) 3485 (2020). https://doi.org/10.3390/ma13163485

26. N.M. Shaalan, D. Hamad, A.Y. Abdel-Latief, M.A. Abdel-Rahim, Preparation of quantum size of tin oxide: Structural and physical characterization. *Progress in Natural Science: Materials International* 26 145–151 (2016). https://doi.org/10.1016/j.pnsc.2016.03.002

27. D. Perednis, O. Wilhelm, S.E. Pratsinis, L.J. Gauckler, Morphology and deposition of thin yttria-stabilized zirconia films using spray pyrolysis. *Thin Solid Films* 474 (2005) 84–95. https://doi.org/10.1016/j.tsf.2004.08.014

28. S.A.M. Ealia, M.P. Saravanakumar, A review on the classification, characterisation, synthesis of nanoparticles and their application. *IOP Conference Series: Materials Science and Engineering* 263 (2017) 032019. https://doi.org/10.1088/1757-899X/263/3/032019

29. R. Taziwa, L. Ntozakhe, E. Meyer, Structural, morphological and Raman scattering studies of Carbon doped ZnO nanoparticles fabricated by PSP technique. *Journal of Nanoscience and Nanotechnology Research* 1 (2017) 1–8, iMedPub journals.

30. T. Satyanarayana, S.S. Reddy, A review on chemical and physical synthesis methods of nanomaterials. *International Journal for Research in Applied Science & Engineering Technology* 6 (1) 2885–2889 (2018). https://doi.org/10.22214/ijraset.2018.1396

31. E.K. Goharshadi, S. Hashem, R. Mehrkhah, P. Nancarrow, Sonochemical synthesis and measurement of optical properties of zinc sulfide quantum dots. *Chemical Engineering Journal* 209 (2012) 113–117. https://doi.org/10.1016/j.cej.2012.07.131

32. K. Vijayarangamuthu, S. Rath, Nanoparticle size, oxidation state and sensing response of tin oxide nanopowders using Raman spectroscopy. *Journal of Alloys and Compounds* 610 (2014) 706–712. https://doi.org/10.1016/j.jallcom.2014.04.187

33. I.A. Rahman, V. Padavettan, Synthesis of Silica nanoparticles by sol-gel: Size-dependent properties, surface modification and applications in silica-polymer nano-composites–a review. *Journal of Nanomaterials* (2012) 132424. (2012) https://doi.org/10.1155/2012/132424

34. G. Elango, S.M. Kumaran, S.S. Kumar, S. Muthuraja, S.M. Roopan, Green synthesis of SnO2 nanoparticles and its photocatalytic activity of phenolsulfonphthalein dye. *Spectrochimica Acta Part A: Molecular and Biomolecular Spectroscopy* 145 (2015) 176–180. https://doi.org/10.1016/j.saa.2015.03.033

35. Z.A. Kalaki, R. Safaeijavan, M.M. Ortakand, Biosynthesis of silver nanoparticles, using *Mentha longifolia* (L.) Hudson leaf extract and study its antibacterial activity. *Journal of Paramedical Sciences* 8 (2017) 24–30. https://doi.org/10.22037/JPS.V8I2.14432

36. S. Gunalan, R. Sivaraj, V. Rajendran, Green synthesized ZnO nanoparticles against bacterial and fungal pathogens. *Progress in Natural Science: Materials International* 22 (2013) 693–700. https://doi.org/10.1016/j.pnsc.2012.11.015

37. K.-S. Lee, M.A. El-Sayed, Gold and silver nanoparticles in sensing and imaging: Sensitivity of plasmon response to size, shape, and metal composition. *The Journal of Physical Chemistry B* 110 (2006) 19220–19225. https://doi.org/10.1021/jp062536y

38. T. Zaheer, M. Imran, K. Pal, M.S. Sajid, R.Z. Abbas, A.I. Aqib, M.A. Hanif, S.R. Khan, M.K. Khan, Z.D. Sindhu, S. Rahman, Synthesis, characterization and acaricidal activity of green-mediated ZnO nanoparticles against *Hyalomma* ticks. *Journal of Molecular Structure* 1227 (2021) 129652. https://doi.org/10.1016/j.molstruc.2020.129652

39. A. Yasuhara, K. Kubo, S. Yanagimoto, T. Sannomiya, Thermodynamic tuning of Au–Ag–Cu nanoparticles with phase separation and ordered phase formation. *The Journal of Physical Chemistry C* 124 (2020) 15481–15488. https://doi.org/10.1021/acs.jpcc.0c02834

40. W. Zhang, J. Li, J. Zhang, J. Sheng, T. He, M. Tian, Y. Zhao, C. Xie, L. Mai, S. Mu, Top-down strategy to synthesize mesoporous dual carbon armored MnO nanoparticles for lithium-ion battery anodes. *ACS Applied Materials & Interfaces* 9 (2017) 12680–12686. https://doi.org/10.1021/acsami.6b16576

41. Y. Ishida, R.D. Corpuz, T. Yonezawa, Matrix sputtering method: A novel physical approach for photoluminescent noble metal nanoclusters. *Accounts of Chemical Research* 50 (2017) 2986–2995. https://doi.org/10.1021/acs.accounts.7b00470

42. M.W. Chung, I.Y. Cha, M.G. Ha, Y. Na, J. Hwang, H.C. Ham, H.-J. Kim, D. Henkensmeier, S.J. Yoo, J.Y. Kim, S.Y. Lee, H.S. Park, J.H. Jang, Enhanced CO 2 reduction activity of polyethylene glycol-modified Au nanoparticles prepared via liquid medium sputtering. *Applied Catalysis B: Environmental* 237 (2018) 673–680. https://doi.org/10.1016/j.apcatb.2018.06.022

43. M. Salavati-Niasari, N. Mir, F. Davar, A novel precursor in preparation and characterization of nickel oxide nanoparticles via thermal decomposition approach. *Journal of Alloys and Compounds* 493 (2010) 163–168. https://doi.org/10.1016/j.jallcom.2009.11.153

44. C. Sungjun, W. Jung, G.Y. Jung, K.S. Eom, High-performance boron-doped silicon micron-rod anode fabricated using a mass-producible lithography method for a lithium ion battery. *Journal of Power Sources* 454 (2020) 227931. https://doi.org/10.1016/j .jpowsour.2020.227931

45. C.-J. Chang, C.-F. Wang, J.-K. Chen, C.-C. Hsieh, P.-A. Chen, Fast formation of hydrophilic and reactive polymer micropatterns by photocatalytic lithography method. *Applied Surface Science* 286 (2013) 280–286. https://doi.org/10.1016/j.apsusc.2013.09.071

46. M. Karrina, S.A.M. Tofail, Nanoparticles in biomedical applications. *Advances in Physics: X* 1 (2017) 54–88. https://doi.org/10.1080/23746149.2016.1254570

47. R. Feynman, There's plenty of room at the bottom. *Science* 254 (1991) 1300–1301.

48. H.-R. Lim, H.S. Kim, R. Qazi, Y.-T. Kwon, J.-W. Jeong, W.-H. Yeo, Wearable Flexible hybrid electronics: Advanced soft materials, sensor integrations, and applications of wearable flexible hybrid electronics in healthcare, energy, and environment. *Advanced Materials* 32 (2020) 1901924. https://doi.org/10.1002/adma.202070116

49. K.Gajanan, S.N.Tijare, Application of nanomaterials, materials today proceedings, 5(1, Part 1) (2018) 1093–1096.

50. S. Fatikow, M. Bartenwerfer, O.C. Haenssler. Towards Robot-Based Manipulation, Characterization and Automation at the Nanoscale, 2019 International Conference on Industrial Engineering, Applications and Manufacturing (ICIEAM), Sochi, Russia, 2019, pp. 1–5. https://doi.org/10.1109/ICIEAM.2019.8742974.

51. M. Li, L. Liu, N. Xi and Y. Wang, Applications of micro/nano automation technology in detecting cancer cells for personalized medicine. *IEEE Transactions on Nanotechnology* 16(2) (2017) 217–229. https://doi.org/10.1109/TNANO.2017.2654320.

52. V. Kant, Nanotechnology and HFE: Critically engaging human capital in small-scale robotics research. *Cognition, Technology & Work* 19 (2017) 419–444. https://doi.org/10 .1007/s10111-017-0414-6

53. A. Iqbal, *Sparking Innovations with Biomimetics: An Integration of Bioscience, Robotics and Nanotechnology.* India: Applied Science Innovations Pvt. Ltd.

54. S. Aggarwal, D. Gupta, S. Saini. A Literature Survey on Robotics in Healthcare, 2019 4th International Conference on Information Systems and Computer Networks (ISCON), Mathura, India, 2019, pp. 55–58. https://doi.org/10.1109/ISCON47742.2019 .9036253.

55. A.T. Azar, A. Madian, H. Ibrahim, M.A. Taha, N.A. Mohammad, Z. Fathy, B.M.A. Al Naga. Control Systems Design of Bio-Robotics and Bio-mechatronics with Advanced Applications, 2020, Pages 329–394.

56. C.L. Albert, F. Rubio, S. Zeng, H. Liao, Applied mathematics for engineering problems in biomechanics and robotics. *Mathematical Problems in Engineering* 2019 (2019) 2578916. https://doi.org/10.1155/2019/2578916

57. S. Bayda, M. Adeel, T. Tuccinardi, M. Cordani, F. Rizzolio, The history of nanoscience and nanotechnology: From chemical–physical applications to nanomedicine. *Molecules* 25 (2020) 112.

58. S.I. Asiya, K. Pal, S. Kralj, G.S. El-Sayyad, F.G. de Souza, T. Narayanan, Sustainable preparation of gold nanoparticles via green chemistry approach for biogenic applications. *Materials Today Chemistry* 17 (2020) 100327. https://doi.org/10.1016/j.mtchem .2020.100327

59. O. Salata Applications of nanoparticles in biology and medicine. *Journal of Nanobiotechnology* 2 (2004) 3.

60. W. Sun, X. Zhang, H.-R. Jia, Y.-X. Zhu, Y. Guo, G. Gao, Y.-H. Li, F.-G. Wu, Water-dispersible candle soot–derived carbon nano-onion clusters for imaging-guided photothermal cancer therapy. *Small* 15 (2019): 1804575. https://doi.org/10.1002/smll .201804575

61. J. Pan, P. Hu, Y. Guo, J. Hao, D. Ni, Y. Xu, Q. Bao, H. Yao, C. Wei, Q. Wu, J. Shi, Combined magnetic hyperthermia and immune therapy for primary and metastatic tumor treatments. *ACS Nano* 14 (2020) 1033–1044. https://doi.org/10.1021/acsnano .9b08550

62. V.C. Karagkiozaki, S.D. Logothetidis, S.N. Kassavetis, G.D. Giannoglou, Nanomedicine for the reduction of the thrombogenicity of stent coatings. *International Journal of Nanomedicine* 5 (2010) 239–248. https://doi.org/10.2147/ijn.s7596

63. F. Riccardo, K.-Y. Chu, A.Q. Shen, Detection of antibodies against SARS-CoV-2 spike protein by gold nanospikes in an opto-microfluidic chip. *Biosensors and Bioelectronics* 169 (2020) 112578. https://doi.org/10.1016/j.bios.2020.112578

64. J. Shao, H. Xie, H. Huang, Z. Li, Z. Sun, Y. Xu, Q. Xiao, X.-F. Yu, Y. Zhao, H. Zhang, H. Wang, P.K. Chu, Biodegradable black phosphorus-based nanospheres for in vivo photothermal cancer therapy. *Nature Communications* 7 (2016) 12967. https://doi.org /10.1038/ncomms12967

65. O. Fellahi, R.K. Sarma, M.R. Das, R. Saikia, L. Marcon, Y. Coffinier, T. Hadjersi, M. Maamache, R. Boukherroub, The antimicrobial effect of silicon nanowires decorated with silver and copper nanoparticles. *Nanotechnology* 24 (2013) 495101. https://doi.org /10.1088/0957-4484/24/49/495101

66. R. Chen, L. Daishun, L. Zhao, S. Wang, Y. Liu, R. Bai, S. Baik, Y. Zhao, C. Chen, T. Hyeon, Parallel comparative studies on mouse toxicity of oxide nanoparticle-and gadolinium-based T1 MRI contrast agents. *ACS Nano* 9 (2015) 12425–12435. https:// doi.org/10.1021/acsnano.5b05783

67. A. Bolhassani, Shabnam Javanzad, Tayebeh Saleh, Mehrdad Hashemi, M.R.A. Deghi, S.M. Sadat, Polymeric nanoparticles, human vaccine and immunotherapeutics, Feb 1; 10(2) (2014) 321–332.

68. W. Zhang, J. Xu, P. Li, X. Gao, W. Zhang, H. Wang, B. Tang, Treatment of hyper-phosphatemia based on specific interaction between phosphorus and active center Zr(IV) of nano-MOFs. *Chemical Science* 9 (2018) 7483–7487. https://doi.org/10.1039/ C8SC02638F

69. A.M. Li, Ecological determinants of health: Food and environment on human health. *Environmental Science and Pollution Research* 24 (2017) 9002–9015. https://doi.org /10.1007/s11356-015-5707-9

70. D. Markovic, I. Branovic, R. Popovic, Smart Grid and nanotechnologies: A solution for clean and sustainable energy. *Energy and Emission Control Technologies* 3 (2015) 1–13 https://doi.org/10.2147/EECT.S48124

71. R.K. Ibrahim, M. Hayyan, M.A. AlSaadi, A. Hayyan, S. Ibrahim, Environmental application of nanotechnology: Air, soil, and water. *Environmental Science and Pollution Research* 23 (2016) 13754–13788. https://doi.org/10.1007/s11356-016-6457-z

72. X. Qu, P.J.J. Alvarez, Q. Li, Applications of nanotechnology in water and wastewater treatment. *Water Research* 47 (2013) 3931–3946. https://doi.org/10.1016/j.watres.2012 .09.058

73. I.D.L. Cavalcanti, M.C.B.L. Nogueira, Pharmaceutical nanotechnology: Which products are been designed against COVID-19? *Journal of Nanoparticle Research* 22 (2020) 276. https://doi.org/10.1007/s11051-020-05010-6

74. C.C. Coelho, L. Grenho, P.S. Gomes, P.A. Quadros, M.H. Fernandes, Nano-hydroxyapatite in oral care cosmetics: Characterization and cytotoxicity assessment. *Scientific Reports* 9 (2019) 11050. https://doi.org/10.1038/s41598-019-47491-z

75. Z.A.A. Aziz, H. Mohd-Nasir, A. Ahmad, S.H.M. Setapar, W.L. Peng, S.C. Chuo, A. Khatoon, K. Umar, A.A. Yaqoob, M. Ibrahim, M. Nasir, Role of nanotechnology for design and development of cosmeceutical: Application in makeup and skin care. *Frontiers in Chemistry* 7 (2019) 739. https://doi.org/10.3389/fchem.2019.00739

76. A.C. Santos, D. Rodrigues, J.A.D Sequeira, I. Pereira, A. Simões, D. Costa, D. Peixoto, G. Costa, F. Veiga, Nanotechnological breakthroughs in the development of topical phytocompounds-based formulations. *International Journal of Pharmaceutics* 572 (2019) 118787. https://doi.org/10.1016/j.ijpharm.2019.118787.

77. Nanotechnology in food, Center for Food Safety (2017). Retrieved from: www.centerforfoodsafety.org.http://salsa3.salsalabs.com/o/1881/p/salsa/web/common/public/content?content_item_KEY=14112%20#showJoin

78. A. Gramza-Michałowska, D. Kmiecik, J. Kobus-Cisowska, A. Żywica, K. Dziedzic, A. Brzozowska, Phytonutrients in oat (*Avena sativa L.*) Drink : Effect of plant extract on antiradical capacity, nutritional value and sensory characteristics. *Polish Journal of Food and Nutrition Sciences* 68 (2018) 63–71. https://doi.org/10.1515/pjfns-2017 -0009

79. Nanotechnology Products Database. COVID-19. (2020). https://product.statnano.com/.

80. K. Khosravi-Katuli, E. Prato, G. Lofrano, M. Guida, G. Vale, G. Libralato, Effects of nanoparticles in species of aquaculture interest. *Environmental Science and Pollution Research* 24(21) (2017) 17326–17346.

81. J.C.M. Márquez, A.H. Partida, M. del Carmen, M. Dosta, J.C. Mejía, J.A.B. Martínez, Silver nanoparticles applications (AgNPS) in aquaculture. *International Journal of Fisheries and Aquatic Studies* 6(2) (2018) 05–11.

82. F. Ahmed, F.M. Soliman, M.A. Adly, H.A. Soliman, M. El-Matbouli, M. Saleh, Recent progress in biomedical applications of chitosan and its nanocomposites in aquaculture: A review. *Research in Veterinary Science* 126 (2019) 68–82.

83. E. Can, V. Kizak, M. Kayim, S.S. Can, B. Kutlu, M. Ates, ... N. Demirtas, Nanotechnological applications in aquaculture-seafood industries and adverse effects of nanoparticles on environment. *Journal of Materials Science and Engineering* 5(5) (2011) 605–609.

84. J. Ji, D. Torrealba, A. Ruyra, N. Roher, Nanodelivery systems as new tools for immunostimulant or vaccine administration: Targeting the fish immune system. *Biology* 4 (2015) 664–696.

85. P. Joosten, E. Tiemersma, A. Threels, C. Dhieux-Caumartin, J. Rombout, Oral vaccination of fish against Vibrio anguillarum using alginate microparticles. *Immunological Aspects of Oral Vaccination in Fish* 56 (1997) 471–485.

86. W.R. Gombotz, Protein release from alginate matrices. *Advanced Drug Delivery Reviews* 31 (1998) 267–285.

87. A. Sosnik, Alginate particles as platform for drug delivery by the oral route: State-of-the-art. *ISRN Pharmaceutics* 2014 (2014) 1–17.

88. C.P. Reis, R.J. Neufeld, F. Veiga, Preparation of drug-loaded polymeric nanoparticles. In: L.P. Balogh (ed.) *Nanomedicine in Cancer*, 2017, pp. 197–240. Singapore: Pan Stanford Publishing Pte Ltd.

89. D. Bi, Oral vaccination against lactococcosis in rainbow trout (*Oncorhynchus mykiss*) using sodium alginate and poly (lactide-co-glycolide) carrier. Kafkas Universitesi Veteriner Fakultesi Dergisi, 16 (2010).

90. K. Fujiki, H. Matsuyama, T. Yano, Protective effect of sodium alginates against bacterial infection in common carp, *Cyprinus carpio L. Journal of Fish Diseases* 17 (1994) 349–355.

91. S.-T. Chiu, R.-T. Tsai, J.-P. Hsu, C.-H. Liu, W. Cheng, Dietary sodium alginate administration to enhance the non-specific immune responses, and disease resistance of the juvenile grouper *Epinephelus fuscoguttatus. Aquaculture* 277 (2008) 66–72.

92. M. Tafaghodi, S.A. Sajadi Tabasi, M. Payan, Alginate microsphere as a delivery system and adjuvant for autoclaved *Leishmania major* and *Quillaja saponin*: Preparation and characterization. *Iranian Journal of Pharmaceutical Sciences* 3 (2007) 61–68.

93. O. Borges, M. Silva, A. de Sousa, G. Borchard, H.E. Junginger, A. Cordeiro-da-Silva, Alginate coated chitosan nanoparticles are an effective subcutaneous adjuvant for hepatitis B surface antigen. *International Immunopharmacology* 8 (2008) 1773–1780.

94. J. Skjermo, Ø. Bergh, High-M alginate immunostimulation of Atlantic halibut (*Hippoglossus hippoglossus* L.) larvae using Artemia for delivery, increases resistance against vibriosis. *Aquaculture* 238 (2004) 107–113.

95. A.-C. Cheng, Y.-Y. Chen, J.-C. Chen, Dietary administration of sodium alginate and j-carrageenan enhances the innate immune response of brown-marbled grouper *Epinephelus fuscoguttatus* and its resistance against *Vibrio alginolyticus*. *Veterinary Immunology and Immunopathology* 121 (2008) 206–215.

96. H.B.T. Huttenhuis, A.S.P. Ribeiro, T.J. Bowden, C. Van Bavel, A.J. Taverne-Thiele, J.H.W.M. Rombout, The effect of oral immuno-stimulation in juvenile carp (*Cyprinus carpio* L.). *Fish and Shellfish Immunology* 21 (2006) 261–271.

97. S.-P. Yeh, C.-A. Chang, C.-Y. Chang, C.-H. Liu, W. Cheng, Dietary sodium alginate administration affects fingerling growth and resistance to *Streptococcus* sp. and iridovirus, and juvenile non-specific immune responses of the orange-spotted grouper, *Epinephelus coioides*. *Fish & Shellfish Immunology* 25 (2008) 19–27.

98. P.K. Dutta, J. Dutta, V. Tripathi, Chitin and chitosan: Chemistry, properties and applications. *Journal of Scientific and Industrial Research* 63 (2004) 20–31.

99. I.A. Sogias, A.C. Williams, V.V. Khutoryanskiy, Why is chitosan mucoadhesive? *Biomacromolecules* 9 (2008) 1837–1842.

100. S. Meshkini, A.-A. Tafy, A. Tukmechi, F. Farhang-Pajuh, Effects of chitosan on hematological parameters and stress resistance in rainbow trout (*Oncorhynchus mykiss*). *Veterinary Research Forum* 3(1) (2012) 49–54.

101. J. Tian, J. Yu, X. Sun, Chitosan microspheres as candidate plasmid vaccine carrier for oral immunisation of Japanese flounder (*Paralichthys olivaceus*). *Veterinary Immunology and Immunopathology* 126 (2008) 220–229.

102. S.R. Kumar, V.I. Ahmed, V. Parameswaran, R. Sudhakaran, V.S. Babu, A.S. Hameed, Potential use of chitosan nanoparticles for oral delivery of DNA vaccine in Asian sea bass (*Lates calcarifer*) to protect from *Vibrio (Listonella) anguillarum*. *Fish and Shellfish Immunology* 25 (2008) 47–56.

103. S. Ferosekhan, S. Gupta, A. Singh, M. Rather, R. Kumari, D. Kothari et al., RNA-loaded chitosan nanoparticles for enhanced growth, immunostimulation and disease resistance in fish. *Current Nanoscience* 10 (2014) 453–464.

104. L. Li, S.L. Lin, L. Deng, Z.G. Liu, Potential use of chitosan nanoparticles for oral delivery of DNA vaccine in black seabream *Acanthopagrus schlegelii* Bleeker to protect from Vibrio parahaemolyticus. *Journal of Fish Diseases* 36 (2013) 987–995.

105. L. Leon-Rodrıguez, A. Luzardo-Alvarez, J. Blanco-Mendez, J. Lamas, J. Leiro, Biodegradable microparticles covalently linked to surface antigens of the scuticociliate parasite *P. dicentrarchi* promote innate immune responses in vitro. *Fish and Shellfish Immunology* 34 (2013) 236–243.

106. T. Behera, P.K. Nanda, C. Mohanty, D. Mohapatra, P. Swain, B.K. Das et al., Parenteral immunization of fish, *Labeo rohita* with Poly D, L-lactide-co-glycolic acid (PLGA) encapsulated antigen microparticles promotes innate and adaptive immune responses. *Fish and Shellfish Immunology* 28 (2010) 320–325.

107. J. Tian, J. Yu, Poly (lactic-co-glycolic acid) nanoparticles as candidate DNA vaccine carrier for oral immunization of Japanese flounder (*Paralichthys olivaceus*) against lymphocystis disease virus. *Fish & Shellfish Immunology* 30 (2011) 109–117.

108. T. Irie, S. Watarai, T. Iwaski, H. Kodama, Protection against experimental *Aeromonas salmonicida* infection in carp by oral immunisation with bacterial antigen entrapped liposomes. *Fish and Shellfish Immunology* 18 (2005) 235–242.

109. S. Yasumoto, T. Yoshimura, T. Miyazaki, Oral immunization of common carp with a liposome vaccine containing *Aeromonas hydrophila* antigen. *Fish Pathology* 41 (2006) 45–49.

110. S. Ye, W. Jin, Q. Huang, Y. Hu, B.R. Shah, Y. Li, et al. Development of Mag-FMBO in clay-reinforced KGM aerogels for arsenite removal. *International Journal of Biological Macromolecules* 87 (2016) 77–84.

111. E. Motamedi, M. Talebi Atouei, M.Z. Kassaee, Comparison of nitrate removal from water via graphene oxide coated Fe, Ni and Co nanoparticles. *Materials Research Bulletin* 54 (2014) 34–40.

112. L. Liu, Z. Cui, Q. Ma, W. Cui, X. Zhang, One-step synthesis of magnetic iron–aluminum oxide/graphene oxide nanoparticles as a selective adsorbent for fluoride removal from aqueous solution. *RSC Advances* 6 (2016) 10783–10791.

113. L. Kuang, Y. Liu, D. Fu, Y. Zhao, FeOOH-graphene oxide nanocomposites for fluoride removal from water: Acetate mediated nano FeOOH growth and adsorption mechanism. *Journal of Colloid and Interface Science* 490 (2017) 259–269.

114. C. Hu, T. Lu, F. Chen, R. Zhang, A brief review of graphene–metal oxide composites synthesis and applications in photocatalysis. *Journal of the Chinese Advanced Materials Society* 1 (2013) 21–39.

115. R. Atchudan, T.N. Edison Jebakumar Immanuel, S. Perumal, D. Karthikeyan, Y.R. Lee, Effective photocatalytic degradation of anthropogenic dyes using graphene oxide grafting titanium dioxide nanoparticles under UV-light irradiation. *Journal of Photochemistry and Photobiology A: Chemistry* 333 (2017) 92–104.

116. C. Hu, Y. Lan, J. Qu, X. Hu, A. Wang, Ag/AgBr/TiO2 visible light photocatalyst for destruction of azodyes and bacteria. *The Journal of Physical Chemistry B* 110 (2006) 4066–4072.

117. J. Park, E. Kettleson, W.-J. An, Y. Tang, P. Biswas, Inactivation of *E. coli* in water using photocatalytic, nanostructured films synthesized by aerosol routes. *Catalysts* 3 (2013) 247–260.

118. K. Ouyang, K. Dai, S.L. Walker, Q. Huang, X. Yin, P. Cai, Efficient photocatalytic disinfection of *Escherichia coli* O157: H7 using C 70-TiO 2 hybrid under visible light irradiation. *Scientific Reports* 6 (2016) 25702.

119. I. Sioen, S.D. Henauw, F. Verdonck, N.V. Thuyne, J.V. Camp, Development of a nutrient database and distributions for use in a probabilistic risk-benefit analysis of human seafood consumption. *Journal of Food Compositions and Analysis* 20 (2007) 662– 670.

120. S. Mahboob, H. Al-Balawi, F. Al-Misned, S. Al-Quraishy, Z. Ahmad, Tissue metal distribution and risk assessment for important fish species from Saudi Arabia. *Bulletin of Environment Contamination and Toxicology* 92 (2014): 61–66.

121. Y. Saleh, M.-A.S. Marie, Assessment of metal contamination in water, sediment, and tissues of *Arius thalassinus* fish from the Red Sea coast of Yemen and the potential human risk assessment. *Environmental Science and Pollution Research* 22 (2014) 5481–5490.

122. G. Dugo, L. La Pera, A. Bruzzese, T.M. Pellicano, V.L. Turco, Concentration of Cd (II), Cu (II), Pb (II), Se (IV) and Zn (II) in cultured sea bass (*Dicentrarchus labrax*) tissues from Tyrrhenian Sea and Sicilian Sea by derivative stripping potentiometry. *Food Control* 17 (2006) 146–152.

123. J.A. Camargo, A. Alonso, Ecotoxicological assessment of the impact of fluoride (F) and turbidity on the freshwater snail *Physella acuta* in a polluted river receiving an industrial effluent. *Environmental Science and Pollution Research* 24 (2017) 15667–15677.

124. P. Manna, M. Sinha, P.C. Sil, A 43 kD protein isolated from the herb *Cajanus indicus* L. attenuates sodium fluoride-induced hepatic and renal disorders in vivo. *Journal of Biochemistry and Molecular Biology* 40 (2007) 382–395.

125. S. Wu, L. Kong, J. Liu, Removal of mercury and fluoride from aqueous solutions by three-dimensional reduced-graphene oxide aerogel. *Research on Chemical Intermediates* 42 (2016) 4513–4530.

126. P. Azadbakht, H. Pourzamani, S.J. Petroudy, B. Bina, Removal of nitrate from aqueous solution using nanocrystalline cellulose. *International Journal of Environmental Health Engineering* 5 (2016) 17.

127. Y. Zeng, Y. Xue, S. Liang, J. Zhang, Removal of fluoride from aqueous solution by TiO2 and TiO2–SiO2 nanocomposite. *Chemical Speciation & Bioavailability* 29 (2017) 25–32.

128. A. Liu, Y. Cao, M. Dai, Z.W. Liao, Nanomaterial application in carp aquiculture experiment. *Fish Modernization* 2 (2008) 24–27.

Index

A

Antimicrobial, 19, 27, 29, 52–55, 120–122, 124, 129, 131, 133, 165–167, 174, 188, 189, 193–195, 210, 231, 232, 236, 291, 294
Antimicrobial resistance, 120–122
Antiviral drugs, 207
Applications, 2–5, 7, 9, 11–13, 19–21, 24, 28, 30, 38, 52, 53, 58–60, 71, 102, 104, 106, 107, 122, 124, 129, 142, 145, 162, 176, 177, 187, 188, 192, 195, 197, 209, 213, 230–232, 234, 239, 241, 257, 259, 290–292, 294, 296, 297

B

Bacteria, 19, 21, 25, 27–29, 32, 38, 54, 55, 57–60, 62, 71, 103, 105, 121–124, 127, 129–134, 136, 137, 139, 145, 165, 166, 169, 170, 175, 177, 190–192, 194–197, 206, 295, 297
Biomedical applications, 19, 213, 234, 262, 289

C

Carbon, 8, 10, 69, 95, 96, 102, 103, 107, 165, 166, 168, 169, 172, 173, 210, 215, 293
Challenges, 120, 164, 213
COVID-19, 214

D

Drug delivery, 13, 18–20, 26, 38, 52, 54, 103, 106, 123, 138, 165, 166, 174, 177, 192, 194, 195, 215, 217, 218, 233, 239, 241, 292
Dynamic light scattering, 68

E

Emulsion-evaporation, 138
Excitation, 142

F

Fungi, 17

G

Green synthesis, 24, 27, 57, 71, 168

H

Health, 145
Hybridization, 166

I

In vitro, 32, 134, 166, 169, 177, 189, 192, 195, 196, 211, 217, 233–237, 242, 264, 273
In vivo, 124, 134, 139, 166, 190, 191, 195, 208, 215, 217, 218, 236, 237, 242, 264, 273

L

Lipid membranes, 165
Lithium-sulfur batteries, 166
Long tubular, 165

M

Metabolites, 52, 62, 210
Metal, 5, 6, 71, 121–124, 126–132, 137, 139, 144, 162, 164, 238, 257, 259, 263, 266, 268–272
Metallic, 18, 19, 21, 23–26, 30, 35, 38, 53, 55–60, 64, 100, 102, 122, 134, 140, 142, 145, 164–167, 169–171, 173, 175–177, 193, 207–211, 213, 215, 217–219, 229, 242, 258, 260, 265–267, 285, 294
Metallic nanoparticles, 27–29, 31–34, 52, 61, 63, 65, 229
Microbes, 21, 52, 62, 120–123, 131, 165, 167, 169, 174, 188, 190, 191, 193, 195, 196
Microorganisms, 18

N

Nanomaterials, 1–14, 17–38, 51–71, 83–107, 119–146, 161–178, 187–197, 205–220, 229–242, 257–273, 283–298
Nanoparticle, 5, 127, 128, 162, 170, 171, 174, 187, 210, 211, 239–241, 268–273, 293
Nanosensors, 209
Nanotherapeutic, 208
Nanotubes, 3, 4, 8, 11, 95, 102, 165, 166, 168, 172, 173, 175, 178, 208, 288
Nanovaccines, 137–140

Natural resources, 71, 188
Non-metal, 164, 171, 178

O

Oxygen species, 166

P

Pathogens, 25, 27–29, 54, 122, 123, 137, 165, 170,
        172, 173, 206, 209, 213, 231, 233, 292
Photocatalysis, 188, 195, 260
Plant extracts, 19, 25, 26, 28, 54, 57, 58, 60, 71,
        188, 237

For Product Safety Concerns and Information please contact our EU
representative GPSR@taylorandfrancis.com
Taylor & Francis Verlag GmbH, Kaufingerstraße 24, 80331 München, Germany